PLASTIC POLLUTION IN THE GLOBAL OCEAN

Trends in Aquatic Systems

Series Editor: M. Debora Iglesias-Rodriguez
(University of California, Santa Barbara, USA)

The book series Trends in Aquatic Systems explores themes relevant to the state of the oceans with a focus on climate change, human perturbation of the marine environment and food security. The Series comprises already published and emerging themes including impacts of climate change on marine systems and plastic pollution. Additional themes: marine food insecurity, endocrine disruptors in marine systems, trends in coral ecosystems, and the state of marine aquaculture are expected to add to the current series.

Published:

Vol. 1 *Plastic Pollution in the Global Ocean*
edited by Alice A. Horton (National Oceanography Center, UK)

Trends in Aquatic Systems

Volume 1

PLASTIC POLLUTION IN THE GLOBAL OCEAN

Editor

Alice A. Horton

National Oceanography Center, UK

World Scientific

NEW JERSEY • LONDON • SINGAPORE • BEIJING • SHANGHAI • HONG KONG • TAIPEI • CHENNAI • TOKYO

Published by

World Scientific Publishing Co. Pte. Ltd.
5 Toh Tuck Link, Singapore 596224
USA office: 27 Warren Street, Suite 401-402, Hackensack, NJ 07601
UK office: 57 Shelton Street, Covent Garden, London WC2H 9HE

Library of Congress Cataloging-in-Publication Data
Names: Horton, Alice A., 1988– editor.
Title: Plastic pollution in the global ocean / Editor, Alice A. Horton, National Oceanography Center, UK.
Description: New Jersey : World Scientific, [2023] | Series: Trends in aquatic systems ; vol. 1 | Includes bibliographical references and index.
Identifiers: LCCN 2022022552 | ISBN 9789811259104 (hardcover) | ISBN 9789811259111 (ebook) | ISBN 9789811259128 (ebook other)
Subjects: LCSH: Marine pollution. | Plastics--Environmental aspects. | Marine resources conservation. | Plastic marine debris.
Classification: LCC GC1085 .P53 2023 | DDC 363.739/4--dc23/eng/20220805
LC record available at https://lccn.loc.gov/2022022552

British Library Cataloguing-in-Publication Data
A catalogue record for this book is available from the British Library.

For any available supplementary material, please visit
https://www.worldscientific.com/worldscibooks/10.1142/12912#t=suppl

Desk Editors: Nimal Koliyat/Adam Binnie/Shi Ying Koe

Typeset by Stallion Press
Email: enquiries@stallionpress.com

https://doi.org/10.1142/9789811259111_fmatter

About the Editor

Alice A. Horton is a Principal Investigator of Anthropogenic Contaminants at the National Oceanography Centre, UK. She completed her PhD at Leiden University in the Netherlands in 2019. Her expertise lies in the fate and ecological effects of microplastics in aquatic environments. Her research focusses on how the connections between land, lakes, rivers and the ocean lead to the transport and accumulation of microplastics within the environment, their associations with organic chemicals, and how these processes influence the interactions and effects of microplastics on aquatic organisms.

Dr Horton has published numerous papers and book chapters in this field, and has worked as a guest editor for the journal *Environment International* (Elsevier). She has chaired sessions and presented at a wide range of international conferences including MICRO, Society of Environmental Toxicology and Chemistry (SETAC), European Geosciences Union (EGU), World Water Week, and the Arctic Circle Assembly.

https://doi.org/10.1142/9789811259111_fmatter

Contents

https://doi.org/10.1142/9789811259111_0001

Introduction: Plastic Pollution in the Global Ocean

Alice A. Horton

National Oceanography Centre, European Way, Southampton SO14 3ZH, UK

alice.horton@noc.ac.uk

Plastic pollution is a growing environmental problem that is attracting increasing interest across society, from academics to the general public. A significant factor in the wide public interest in plastics is its visibility: present throughout urban and rural environments, washing up on beaches and even visible from space (Biermann *et al.*, 2020; Topouzelis *et al.*, 2019). With growing plastic production and usage, plastic waste within the environment will continue to increase. This increased input, along with its persistence, leads to accumulation and increasing ecosystem exposure, with as-yet-unknown consequences.

Public awareness of plastic pollution lies primarily in the portrayal that is presented by the media. This usually entails charismatic megafauna such as whales, dolphins and turtles entangled in plastic fishing lines or carrier bags, or washed up dead on beaches with stomachs found to be full of plastic. Indeed, this is part of the plastic pollution story, but certainly not the entirety. Recently, news stories have more frequently been presenting broader aspects of plastic pollution, including the abundance of microplastics (plastic particles 1 μm–5 mm in size) and possible ecological hazards posed by these, both to ecosystems and to human health. While useful for people to be aware of the wide aspects of environmental issues and ongoing areas of research, dramatic stories are often prioritised to garner attention, often despite incomplete evidence or only preliminary data. This presentation of provisional or uncertain information as hard facts can lead to scaremongering, leading people to feel overwhelmed, scared or powerless given the scale of the issue, thus leading to inaction, apathy or "environmental fatigue". While many media outlets will in fact endeavour to present an unbiased and accessible picture, the best way to present accessible and factually correct scientific information to the public is directly via communications from scientists themselves.

With respect to scientific research on plastics, this is a highly nuanced and complex field. In addition to the aforementioned topics of megafauna ingestion and plastic hazard, research is targeting a huge variety of aspects of the issue, including material sources, chemistry and degradation, plastic transport and fate, trophic transfer and ecosystem-scale effects, human health, and novel polymer alternatives, to name only a few. The field of plastic pollution research is moving incredibly quickly, with multiple new publications and data published every week. However, while the scientific literature is rapidly growing, these publications are often targeted at those with very specific expertise, and it can be difficult even for researchers themselves to determine which are the most pertinent, and of the highest quality, without sifting through large volumes of information. There is, therefore, a need for an accessible, up to date volume which draws together the key state-of-the-art knowledge and research in a comprehensive yet concise way.

This book brings together a collection of chapters written by world-leading experts in environmental plastic pollution inputs, fate, effects and solutions. This book provides a clear overview of the current scientific understanding, future implications and key considerations for the management and mitigation of plastic waste within the world's oceans. This is delivered through synthesis and presentation of current and relevant information in a logical and manageable structure. This book will, therefore, act as a key reference text which can be used by undergraduates to experienced academics wishing to learn more or update themselves on the subject, in addition to policymakers, regulators and industrial professionals who have a stake in better understanding, addressing and tackling the issue of plastic waste within the environment.

1. Current Understanding of Plastic Pollution

The amount of plastics being produced annually is exponentially increasing, from 1.7 million tonnes (Mt) in 1950 to 368 Mt in 2019

(Figure 1). However, waste management systems have not kept pace with this rapid development and production of plastic materials, and consequently, large amounts of plastics now enter the environment annually, as a result of both intentional and unintentional disposal. It is estimated that of the 6.3 billion metric tonnes of plastics produced globally up until 2015, 79% of this remains as waste within landfills or in the natural environment (with the remainder having been recycled or incinerated). By 2050, this could account for 12 Bt of waste plastics in the environment (Geyer *et al.*, 2017). Estimates of the volumes of plastic input to the environment vary depending on the factors used to predict losses (including population size, waste generation, waste mismanagement, regional income and more), but estimates currently range between 5 and 23 Mt of plastics entering aquatic systems (including rivers and lakes) per year (Borrelle *et al.*, 2020; Jambeck *et al.*, 2015).

As these estimates show, in line with human activities, it is recognised that majority of marine plastics come from land, and indeed studying land and rivers is imperative to understanding the bigger picture of plastic pollution. Nonetheless, the ocean is recognised as the final sink for a large proportion of land- and marine-derived plastics (such as fishing gear and shipping debris), whereby once plastics enter the oceans, they are unlikely

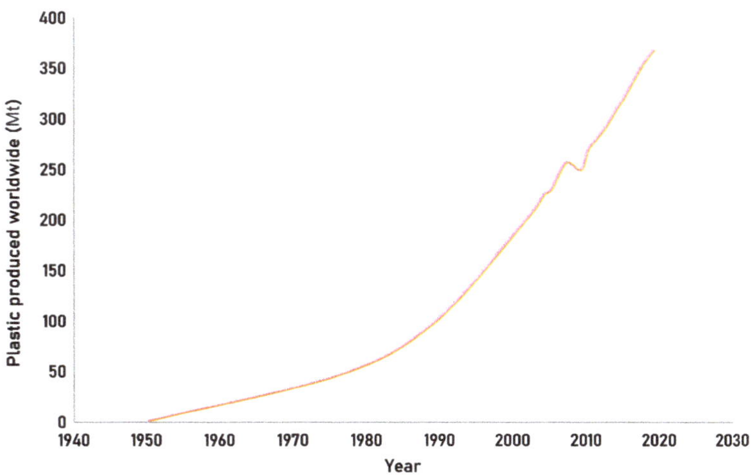

Figure 1. Global plastic production by year since commercial plastic production began in 1950.

Source: Adapted from PlasticsEurope (2012, 2015, 2016, 2018, 2020, 2021).

to be returned to land, except for the small proportion that washes up on beaches (note that the predicted amount of plastics beaching is variable depending on different models used, and whether beaches are actually included in the models — exclusion from models does not necessarily mean that no beaching occurs (Lebreton *et al.*, 2019)).

Plastics on land will be subject to environmental processes that accelerate their degradation, such as exposure to UV radiation, proportionally higher temperatures achieved on land, biodegradation via microorganisms and mechanical degradation (for example, on road surfaces or via agricultural activities). However, plastics in the oceans, especially sub-surface, will not be subject to such dramatic variations in conditions and are thus likely to degrade at a far slower rate (Andrady, 2015; Barnes *et al.*, 2009) (Table 1). While microorganisms may be more likely to colonise plastics within an aquatic environment than on land and can secrete enzymes which facilitate biodegradation, the actual decomposition resulting from microorganism associations is negligible in the marine environment, due to limited oxygen and low temperatures (Andrady, 2015). Especially where plastics reach the ocean floor, they will also be exposed to little or no sunlight plus low temperatures, and thus degradation will be considerably restricted. This can lead the deep sea floor to become a sink for microplastics (Woodall *et al.*, 2014), leading to accumulation hotspots (Courtene-Jones *et al.*, 2020; Kane *et al.*, 2020).

In 2015, a publication by van Sebille *et al.* (2015) stated that due to the scarcity of ocean plastic data, 99% of the plastics believed to have been input to the ocean were unaccounted for, i.e., only 1% had been detected and quantified. Since then, research has developed significantly, however we still do not have a full understanding of the fate and final sinks of the majority of marine plastics. Recent research has shown that there are in fact greater volumes of plastic by mass in the oceans (in a study focussed on the Atlantic Ocean) than was previously believed to have been input. Of only three common polymers (polyethylene, polypropylene and polystyrene) and even if only considering microplastic-sized items, it was calculated that over five times as much plastic was present within the Atlantic than predictions would suggest (Pabortsava & Lampitt, 2020). This may be due to the combined factors of analysing far smaller particles than the majority of studies to date (>25 μm compared to >100 or >300 μm) and also analysing sub-surface waters. In many previous studies, only surface plastics were considered within ocean plastic budgets,

Table 1. Processes affecting plastic degradation in different environments.

	Environment			
Driver of degradation	**Land**	**Beach**	**Surface water**	**Deep water or sediment**
UV exposure (sunlight)	Yes	Yes	Yes	No
High temperatures	High	High	Moderate	Low
Oxygen levels	High	High	High/Moderate	Low
Biofouling[a]	No	No	Yes	Yes

Note: [a]May contribute to biodegradation through presence of microorganisms or may shield particle from UV light.
Source: Modified from Andrady (2015).

based on an assumption that the majority of plastics will float. We now know this not to be the case: in addition to large amounts of non-buoyant plastics within the ocean (i.e., plastics with a greater density than seawater, for example, PET and PVC; Table 2), many supposedly buoyant polymers can be found within the water column and deep sea (Bergmann *et al.*, 2017; Pabortsava & Lampitt, 2020; Zhang *et al.*, 2020), as a result of biofouling, aggregation or incorporation into organic matter, such as faecal pellets and marine snow, all of which can enhance an item's density (Cole *et al.*, 2016; Porter *et al.*, 2018). Publications such as these highlight the rapid development of our understanding in this area over the last few years but also highlight the crucial need to continue to develop our understanding and research in this field, given how significantly our understanding can change with new data and synthesis.

2. Key Themes for Understanding Plastics in the Oceans

2.1. *Sources of plastics to the environment*

Plastics are a diverse and heterogeneous material group. They are synthetically produced using a carbon-hydrogen (C–H) backbone, with different functional groups. Each permutation of chemical structure, formed as a result of the composition and position of atoms, thus forms a different polymer type. To the polymers are added colourants and additive chemicals to give each plastic its desired characteristics for the intended application, from packing to textiles, construction materials to paints, and many more in between.

As a result of the growing global demand for plastics due to their cheap, convenient and resilient nature, sources of plastics to the environment are also widespread, as a result of careless discard, mismanagement of waste streams, or intentional dumping. The key sources vary regionally, however the largest contributors also tend to correlate with the polymers and items most commonly produced (Andrady, 2011; PlasticsEurope, 2020) and those products designed for a short lifetime and rapid replacement (regardless of how long the material is capable of continuing to do its intended job). The most common polymer types produced in 2019 were polypropylene (PP), low-density polyethylene (LDPE) and high-density polyethylene (HDPE) (Table 2) (PlasticsEurope, 2020), which correlates strongly with polymer types found in environmental studies (Jones *et al.*, 2020; Naidu *et al.*, 2021; Zhang *et al.*, 2017). Examples of the most common items found within the environment, or fragments thereof, include pre-production pellets, single-use packaging, vehicle tyres, paints and coatings, fishing gear and textile fibres (Boucher & Friot, 2017; Hann *et al.*, 2018).

Confounding our understanding of the abundance, fate and effects of plastics in the environment is the fact that plastics are commonly

Table 2. Common polymer types produced in Europe in 2020.

Polymer name	Abbreviation	Polymer density (g/cm^3)	Proportional demand (%)
Polypropylene	PP	0.89–0.91	19.7
Low-density polyethylene	LDPE	0.89–0.94	17.4
High-density polyethylene	HDPE	0.94–0.97	12.9
Polyvinyl chloride	PVC	1.3–1.58	9.6
Polyethylene terephthalate	PET	1.29–1.4	8.4
Polyurethane	PUR	1.17–1.28	7.8
Polystyrene	PS	1.04–1.08 (expanded PS = 0.015–0.03)	6.1
Other	—	—	18.1

Note: Density of seawater ranges from 1.02–1.03 g/cm^3.
Source: Demand data from PlasticsEurope (2021); Density data from Nuelle *et al.* (2014).

considered as a single pollutant type ("plastic"), regardless of variations in polymer type, size or shape. In early plastics research, this very broad characterisation was sufficient — studies were simply interested to report, for example, plastic particles ("microplastics") within the environment where these had not been previously observed (Carpenter & Smith, 1972; Thompson *et al*., 2004). However, it is now clearly known that these varied characteristics can significantly influence the behaviour, fate and ecological effects of plastics and are, therefore, crucial to take into account when trying to understand plastic abundance and distribution within the oceans (Bansal *et al*., 2021; Skalska *et al*., 2020; Walsh *et al*., 2021). With our greatly improved knowledge, these days, this generalisation is, therefore, no longer appropriate. Much more constructive is distinguishing by polymer type and shape (e.g., "polystyrene microplastic beads"). This is becoming widely recognised by researchers in this field, who are more frequently measuring and reporting the specific and varied characteristics of particles both found in the field and those used in experimental laboratory studies.

2.2. *Regional differences*

It is fundamental to our understanding of plastics in the environment to acknowledge that plastics, while globally ubiquitous in their presence, are extremely heterogeneously distributed, both in terms of abundance and type. This depends on a combination of factors including local sources and environmental processes affecting the transport of plastic (wind, rainfall, water currents, natural or artificial structures and anthropogenic activities). For example, in the River Ganges, a very large proportion of large (macro)plastic debris has been identified to be derived from fishing gear (Nelms *et al*., 2021), translating, along with laundered clothing, to an abundance of microplastic fibres also observed in this river (Napper *et al*., 2021). In the Great Pacific Garbage Patch, 46% of debris is derived from fishing nets, likely transported from Asia (Lebreton *et al*., 2018). These observations are likely due to the importance of artisanal fishing in countries whose income depends on abundant natural resources derived from rivers and coastal areas, and thus the high level of fishing activity compared to countries who do not have such abundant resources or rely on them less. For example, the River Mekong supports the largest inland fishery in the world, while the Ganges–Brahmaputra–Meghna Delta fisheries (India and Bangladesh) contribute 4–5% GDP (Lauria *et al.*, 2018). In other regions, other sources of plastic dominate. For example, in the River Thames, UK, wet wipes and sanitary products (made from synthetic textiles) derived from untreated sewage are the dominant form of macroplastic litter, in addition to single-use items (Bernardini *et al.*, 2020; McCoy *et al.*, 2020; Morritt *et al.*, 2014). Also in Europe, the dominant sources of plastic (if only considering microplastics) are tyre wear particles, road marking paints, and pre-production pellets (Hann *et al.*, 2018). This regional variability is very important to be aware of, especially when developing models of plastic abundance and sources using field data, as data from one study may not be directly representative of the wider area or areas being modelled.

2.3. *Factors influencing plastic fate*

The diverse nature of plastics detailed earlier has a significant influence on their degradation, fate and interactions within the environment. The principal consideration is the density of the polymer, which will determine whether an item sinks or floats in seawater or freshwater. If pristine, a polymer that is denser than 1 g/cm^3 will sink in freshwater, while a polymer denser than 1.02–1.03 g/cm^3 will sink in seawater. However, environmental studies have shown that polymers which would be expected to be buoyant are commonly found within the water column and sediment, far below the water surface (Kanhai *et al.*, 2019; Pabortsava & Lampitt, 2020). This is because their density, and thus their fate, is also significantly influenced by biological interactions, for example, the development of biological communities (biofilms) on the surface of plastics, ingestion by organisms, or aggregation with organic matter, such as marine snow or faecal pellets (Cole *et al.*, 2016; Porter *et al.*, 2018). The behaviour of plastics will also be influenced by their initial shape and size, which will influence physical interactions of plastics with the surrounding environment, for example, whether they are transported easily by currents and winds. Furthermore, their size will influence their surface to volume ratio, and thus their surface area, determining the area available for biological and chemical interactions (Chen *et al.*, 2021; Deng *et al.*, 2021; Endo & Koelmans, 2019).

While plastic is persistent, degradation does occur, leading plastics to age, crack and break down into smaller pieces. In this way, large items will eventually form microplastics and nanoplastics, which have different toxicological effects compared to larger, pristine items (Lambert *et al.*, 2017). For example, smaller particles are often observed to be more toxic than larger particles due to their enhanced bioavailability and ability to translocate within tissues and cells (especially in the case of nanoplastics) (An *et al.*, 2021; Yang *et al.*, 2020), although this size effect is not consistent across studies (Wu *et al.*, 2021). Ageing also leads to the association of microplastics with pathogens, organic chemicals or metals not present on pristine particles, and thus toxicity may increase by association (Castro-Castellon *et al.*, 2021; Wang *et al.*, 2020). Nonetheless, ageing can reduce the inherent toxicity of plastics, as incorporated chemicals are leached out and thus plastics become more inert over time (Pflugmacher *et al.*, 2021). Exposure to leached chemicals can also cause toxicity in itself, even where the physical plastic item is no longer in the vicinity (Luo *et al.*, 2020; Zimmermann *et al.*, 2020). This highlights that the characteristics of the particles, in addition to the amount of time they have been in the environment, can significantly influence the hazard that plastics pose to organisms and ecosystems.

2.4. *Ecological and toxicological effects*

Complicating our understanding of the hazard posed by plastics is the fact that under some experimental conditions and with some species, plastics (especially small-scale particles such as microplastics and nanoplastics) cause rapid and extreme health effects or mortality, while under other conditions and to other species, no effect is observed. Experimental ecotoxicological studies are commonly carried out using microplastics and nanoplastics (as opposed to macroplastics) due to the practical and ethical limitations of working with large items and organisms. However, it is not often observed that microplastics and nanoplastics cause outright mortality to organisms, unless at very high (unrealistic) concentrations. Instead, chronic exposure can lead to the manifestation of sub-lethal effects, such as hormonal changes, oxidative stress, and reduced growth and reproduction (Piccardo *et al.*, 2021; Wang *et al.*, 2019; Yu *et al.*, 2020). This means that the acute ecotoxicological studies that are commonly carried out are unlikely to provide a good representation of the chronic effects of extended low-level exposure that will be characteristic of the real environment. Nonetheless, they can give an indication of toxicity thresholds, above which harm will occur for different species.

One way to assess the effects of microplastics across a range of species is to use a species sensitivity distribution (SSD) curve. This uses toxicity data for a given endpoint (for example, mortality or growth) to give an indication of likely community-level effects across multiple species (Posthuma *et al.*, 2001). By starting with multiple data points of microplastic toxicity for one species, a species-specific probability function can be used to assess the likelihood of microplastics being toxic (or harmful) to this species (Adam *et al.*, 2019). While it is possible to use a single toxicity measurement for one species, it is not useful to use only data from one study, as many factors other than simply the microplastics concentration (such as temperature and food availability) can also influence the toxicity. Calculating these species-specific probabilities across a range of different species then enables an SSD curve to be plotted, to compare species sensitivities and predict the cumulative proportion of species that will be affected at a given concentration of microplastics (Adam *et al.*, 2019; Burns & Boxall, 2018). This also enables the calculation of a predicted no effect concentration (PNEC), i.e., a "safe" environmental concentration at which, theoretically, no ecological harm occurs. It should be noted that the more data available, the more likely it is that this value can be predicted accurately, although some uncertainty is unavoidable (Posthuma *et al.*, 2001). These data can (and should) be further separated by polymer type, particle size and particle shape where possible, however this relies on sufficient available data across multiple species for these descriptors. SSDs can only work effectively where multiple data points are available across multiple species (Besseling *et al.*, 2019). For certain species, for example, model organisms such as *Daphnia magna* which are commonly tested upon, data are often available across a range of stressors including different types of microplastics, but such calculations become difficult or impossible where limited data exist.

Where data are available, further calculations can be undertaken to compare the PNEC resulting from the SSD to measured environmental concentrations (MECs) within different locations, to predict location-specific risks to ecosystems. Current predictions suggest that global risks resulting from microplastics are low (overall in the global surface ocean 0.17% of organisms inhabiting the surface layer are at risk), but thresholds may exceed in particularly contaminated regions, such as the Mediterranean Sea and the Yellow Sea. This risk is likely to increase, given predicted increases in contamination over coming decades (Everaert *et al.*, 2020).

However, it is worth bearing in mind that due to limited data, this particular analysis did not take into account particle characteristics in influencing their toxicity. It is for this reason that continued ecotoxicity testing, especially across varied species (not simply model organisms), is crucial for a broader understanding of long-term ecosystem-level effects of microplastics.

3. Structure of the Book

This book provides a wide overview and greater depth of some of these key issues and considerations related to plastics within the oceans. The structure is intended to logically flow through the fundamental topics, starting from an overview of the history and usage of plastics and the likely future of these materials moving through the state of the knowledge in different specific research areas and key oceanographic regions, with considerations on waste management, novel materials and potential solutions to the problem of plastic pollution. This format will give the reader a sound overview of the current state of the science in this highly topical and fast-moving field.

A brief overview of each chapter is given as follows.

Chapter 1 gives the backstory of plastics, how and why they were developed, and how they became viewed as a "wonder material" leading to huge consumer demand and the rapid growth in plastic production in recent decades. The chapter also outlines some of the challenges related to the handling of these materials as waste, potential future options for reducing our huge reliance on plastics, and thus angles for solutions to the huge volumes of plastic currently entering the ocean and wider environment.

Chapter 2 considers the amounts, sources and pathways of plastic waste into the ocean (including air, land and freshwater), especially considering the importance of rivers as part of the "plastic cycle". It is essential to be aware that not all plastics that enter rivers will reach the ocean, given that rivers can act as significant sinks for plastics. This is crucial to account for in global plastic transport models, many of which currently assume that all plastics derived from land and entering rivers will ultimately reach the ocean. This chapter thus better helps the reader to understand how location, environment and annual/seasonal conditions can influence plastic transport and retention in the environment, and what ultimately reaches the sea.

Chapter 3 elucidates some of the key factors that influence the distribution and fate of plastics once they reach the oceans. There are a multitude of factors influencing plastics' transport, from physical, biological and chemical interactions of the particles with other materials and biota to environmental conditions, such as waves, wind and currents. All of these will influence where plastics end up, how they will degrade and ultimately their accumulation and longevity in the oceans.

Multiple reports have now highlighted the perils of large plastics within the marine environment, posing a hazard to marine megafauna, such as mammals and reptiles. These plastics may be in the form of large litter lost from land or discarded from ships, or abandoned or discarded fishing gear (also known as ghost gear). These plastics have the potential to last for hundreds of years, capturing, entangling and injuring countless organisms during their time at sea. **Chapter 4** details some of the current knowledge and considerations for megafauna interactions with plastics, and future research recommendations.

Chapter 5 introduces microplastics: what they are, where they come from, how they are formed, and the factors that lead them to further fragment and degrade within the marine environment. Given that all plastics in the environment will eventually degrade into microplastics (and eventually, nanoplastics), microplastics form a fundamental part of the plastic issue and must be equally studied and understood if we are to understand the long-term implications, mitigation strategies and solutions for plastic pollution.

Following the introduction to microplastics, **Chapter 6** covers the various interactions of marine invertebrates, including zooplankton, with microplastics. Detail is provided on studies which have measured the abundance of microplastics in different invertebrate organisms globally, as well as some of the observed effects microplastics can have across different species. This chapter provides context for lower trophic organism exposure to plastics and also how these organisms might themselves influence the distribution of plastics within the environment.

Building on the previous chapter, **Chapter 7** provides a detailed overview of microplastic and nanoplastic toxicity across different organisms, and how toxicity is influenced by specific particle characteristics including polymer type, shape and size. Local environmental factors will affect concentration and thus ecosystem exposure, therefore microplastic effects on ecosystems are highly unlikely to be consistent worldwide, instead varying at a regional or local scale. Consideration is also given to the possible long-term ecological implications of the increasing volumes of plastic waste within the environment.

The Arctic has a low human population compared to most of the rest of the world but is nonetheless incredibly ecologically and economically important, including for fisheries, shipping and tourism. The Arctic is also one of the most rapidly changing global regions, as a result of a changing climate leading to air and sea surface warming, and melting sea ice. Despite comparatively few local sources, the nature of ocean currents, converging in the Arctic, means that once plastics reach the Arctic, they are unlikely to leave, leading the Arctic to become a sink for plastics of all sizes. **Chapter 8** details the implications that this contamination may have and the importance of studying the Arctic for plastic pollution.

Considered by many to be a pristine environment or "the last untouched ocean", Antarctica and the surrounding Southern Ocean are nonetheless subject to anthropogenic influence and contamination in the form of plastic (and other) waste. While the only continent without a permanent human population, and still comparatively clean compared to other regions, contamination results from tourism, research operations and long-range transport from far-afield regions. **Chapter 9** considers how the specific activities within, and characteristics of, this region influence the presence and abundance of plastics, and ways in which this might be managed in coming years, to ensure this important region remains as unpolluted as possible.

Plastics are known to act as substrates for a range of organisms which use these items as novel habitats. This extends not only from large organisms such as invertebrates and plants but also to microbes which build communities on the surface of macroplastics and microplastics. It has been observed that different communities form on plastics compared to on natural substrates, and that the colonisation and subsequent transport of lightweight plastics can lead to the transport of species to locations in which they are not native, known as "the vector effect". **Chapter 10** covers some of the issues surrounding this association of microbial communities with plastics.

Biodegradable plastics have recently been in the spotlight as a suggested solution to the plastic pollution problem, whereby these materials could replace conventional plastics. However, there are common public misconceptions that these materials are less hazardous than plastics and will naturally degrade if thrown anywhere within the environment. **Chapter 11** considers the pros and cons of these novel materials as plastic alternatives, from the perspective of the chemistry of the materials, their characteristics, and the ecological, social and economic and waste management considerations that must be taken into account when considering replacing plastics with these alternative materials.

No volume on plastic pollution would be justified or compete without expert thoughts and suggestions on how the plastic problem can be tackled from different angles. **Chapter 12** provides details and considerations of policies, industry action, novel technologies and innovative solutions required to reduce unnecessary plastic use and loss, all with the aim of preventing plastics from entering the environment.

The issue of plastics is complex and impossible to fully cover in detail in one book alone. Multiple different material types, production techniques and uses mean that plastics, while incredibly useful, pose long-term challenges for our use and disposal of them. No one solution alone will be capable of solving the problem of plastics in the ocean, and indeed many of the related issues will be far more costly to tackle (both economically and environmentally) than may be feasible or beneficial. Instead, a suite of interlinked approaches must be taken by various stakeholders to ensure plastics do not pose long-term risks to human and ecosystem health. This is especially crucial in the oceans given the fundamental importance of oceanic ecosystems for maintaining life on earth, through carbon sequestration, oxygen production and food supplies. Many marine ecosystems have yet to be fully explored or even discovered, and thus through anthropogenic activities, we risk damaging or totally changing these ecosystems before we are even aware of their existence. This book, therefore, aims to give the reader a broad overview of the topic of plastic pollution from the perspective of different experts, from which point the reader will be better informed to make decisions, pursue future research or simply discuss the topic with peers.

References

Adam, V., Yang, T., & Nowack, B. (2019). Toward an ecotoxicological risk assessment of microplastics: Comparison of available hazard and exposure data in freshwaters. *Environmental Toxicology and Chemistry*, *38*(2), 436–447.

An, D., Na, J., Song, J., & Jung, J. (2021). Size-dependent chronic toxicity of fragmented polyethylene microplastics to *Daphnia magna*. *Chemosphere*, *271*, 129591.

Andrady, A. L. (2011). Microplastics in the marine environment. *Marine Pollution Bulletin*, *62*(8), 1596–1605.
Andrady, A. L. (2015). In Bergmann, M., Gutow, L., & Klages, M. (Eds.), *Marine Anthropogenic Litter*, pp. 57–72. Cham: Springer International Publishing.
Bansal, M., Santhiya, D., & Sharma, J. G. (2021). Behavioural mechanisms of microplastic pollutants in marine ecosystem: Challenges and remediation measurements. *Water, Air, & Soil Pollution*, *232*(9), 372.
Barnes, D. K. A., Galgani, F., Thompson, R. C., & Barlaz, M. (2009). Accumulation and fragmentation of plastic debris in global environments. *Philosophical Transactions of the Royal Society B: Biological Sciences*, *364*(1526), 1985–1998.
Bergmann, M., Wirzberger, V., Krumpen, T., Lorenz, C., Primpke, S., Tekman, M. B., & Gerdts, G. (2017). High quantities of microplastic in Arctic deep-sea sediments from the HAUSGARTEN observatory. *Environmental Science & Technology*, *51*(19), 11000–11010.
Bernardini, G., McConville, A., & Castillo, A. C. (2020). Macro-plastic pollution in the tidal Thames: An analysis of composition and trends for the optimization of data collection. *Marine Policy*, *119*, 104064.
Besseling, E., Redondo-Hasselerharm, P., Foekema, E. M., & Koelmans, A. A. (2019). Quantifying ecological risks of aquatic micro- and nanoplastic. *Critical Reviews in Environmental Science and Technology*, *49*(1), 32–80.
Biermann, L., Clewley, D., Martinez-Vicente, V., & Topouzelis, K. (2020). Finding plastic patches in coastal waters using optical satellite data. *Scientific Reports*, *10*(1), 1–10.
Borrelle, S. B., Ringma, J., Law, K. L., Monnahan, C. C., Lebreton, L., McGivern, A., Murphy, E., Jambeck, J., Leonard, G. H., & Hilleary, M. A. (2020). Predicted growth in plastic waste exceeds efforts to mitigate plastic pollution. *Science*, *369*(6510), 1515–1518.
Boucher, J., & Friot, D. (2017). *Primary Microplastics in the Oceans: A Global Evaluation of Sources*. Gland, Switzerland: IUCN. pp. 1–43.
Burns, E. E., & Boxall, A. B. A. (2018). Microplastics in the aquatic environment: Evidence for or against adverse impacts and major knowledge gaps. *Environmental Toxicology and Chemistry*, *37*(11), 2776–2796.
Carpenter, E. J., & Smith, K. (1972). Plastics on the Sargasso Sea surface. *Science*, *175*(4027), 1240–1241.
Castro-Castellon, A. T., Horton, A. A., Hughes, J. M. R., Rampley, C., Jeffers, E. S., Bussi, G., & Whitehead, P. (2021). Ecotoxicity of microplastics to freshwater biota: Considering exposure and hazard across trophic levels. *Science of the Total Environment*, *816*, 151638.
Cole, M., Lindeque, P. K., Fileman, E., Clark, J., Lewis, C., Halsband, C., & Galloway, T. S. (2016). Microplastics alter the properties and sinking rates of

zooplankton faecal pellets. *Environmental Science & Technology, 50*(6), 3239–3246.

Courtene-Jones, W., Quinn, B., Ewins, C., Gary, S. F., & Narayanaswamy, B. E. (2020). Microplastic accumulation in deep-sea sediments from the Rockall trough. *Marine Pollution Bulletin, 154*, 111092.

Deng, H., Fu, Q., Li, D., Zhang, Y., He, J., Feng, D., Zhao, Y., Du, G., Yu, H., & Ge, C. (2021). Microplastic-associated biofilm in an intensive mariculture pond: Temporal dynamics of microbial communities, extracellular polymeric substances and impacts on microplastics properties. *Journal of Cleaner Production, 319*, 128774.

Endo, S., & Koelmans, A. A. (2019). In Takada, H., & Karapanagioti, H. K. (Eds.), *Hazardous Chemicals Associated with Plastics in the Marine Environment*, pp. 185–204. Cham: Springer International Publishing.

Everaert, G., De Rijcke, M., Lonneville, B., Janssen, C., Backhaus, T., Mees, J., van Sebille, E., Koelmans, A., Catarino, A. I., & Vandegehuchte, M. B. (2020). Risks of floating microplastic in the global ocean. *Environmental Pollution, 267*, 115499.

Geyer, R., Jambeck, J. R., & Law, K. L. (2017). Production, use, and fate of all plastics ever made. *Science Advances, 3*(7), e1700782.

Hann, S., Sherrington, C., Jamieson, O., Hickman, M., Peter, K., Bapasola, A., & Cole, G. (2018). Investigating options for reducing releases in the aquatic environment of microplastics emitted by (but not intentionally added in) products. pp. 1–321. *Eunomia* London/Bristol.

Jambeck, J., Geyer, R., Wilcox, C., Siegler, T. R., Perryman, M., Andrady, A. L., Narayan, R., & Law, K. L. (2015). Plastic waste inputs from land into the ocean. *Science, 347*(6223), 768–771.

Jones, J. I., Vdovchenko, A., Cooling, D., Murphy, J. F., Arnold, A., Pretty, J. L., Spencer, K. L., Markus, A. A., Vethaak, A. D., & Resmini, M. (2020). Systematic analysis of the relative abundance of polymers occurring as microplastics in freshwaters and estuaries. *International Journal of Environmental Research and Public Health, 17*(24), 9304.

Kane, I. A., Clare, M. A., Miramontes, E., Wogelius, R., Rothwell, J. J., Garreau, P., & Pohl, F. (2020). Seafloor microplastic hotspots controlled by deep-sea circulation. *Science, 368*(6495), 1140–1145.

Kanhai, L. D. K., Johansson, C., Frias, J., Gardfeldt, K., Thompson, R. C., & O'Connor, I. (2019). Deep sea sediments of the Arctic central basin: A potential sink for microplastics. *Deep Sea Research Part I: Oceanographic Research Papers, 145*, 137–142.

Lambert, S., Scherer, C., & Wagner, M. (2017). Ecotoxicity testing of microplastics: Considering the heterogeneity of physicochemical properties. *Integrated Environmental Assessment and Management, 13*(3), 470–475.

Lauria, V., Das, I., Hazra, S., Cazcarro, I., Arto, I., Kay, S., Ofori-Danson, P., Ahmed, M., Hossain, M. A. R., Barange, M., & Fernandes, J. A. (2018). Importance of fisheries for food security across three climate change vulnerable deltas. *Science of the Total Environment, 640–641*, 1566–1577.

Lebreton, L., Egger, M., & Slat, B. (2019). A global mass budget for positively buoyant macroplastic debris in the ocean. *Scientific Reports, 9*(1), 12922.

Lebreton, L., Slat, B., Ferrari, F., Sainte-Rose, B., Aitken, J., Marthouse, R., Hajbane, S., Cunsolo, S., Schwarz, A., Levivier, A., Noble, K., Debeljak, P., Maral, H., Schoeneich-Argent, R., Brambini, R., & Reisser, J. (2018). Evidence that the Great Pacific Garbage Patch is rapidly accumulating plastic. *Scientific Reports, 8*(1), 1–15.

Luo, H., Li, Y., Zhao, Y., Xiang, Y., He, D., & Pan, X. (2020). Effects of accelerated aging on characteristics, leaching, and toxicity of commercial lead chromate pigmented microplastics. *Environmental Pollution, 257*, 113475.

McCoy, K. A., Hodgson, D. J., Clark, P. F., & Morritt, D. (2020). The effects of wet wipe pollution on the Asian clam, *Corbicula fluminea* (Mollusca: Bivalvia) in the River Thames, London. *Environmental Pollution, 264*, 114577.

Morritt, D., Stefanoudis, P. V., Pearce, D., Crimmen, O. A., & Clark, P. F. (2014). Plastic in the Thames: a river runs through it. *Marine Pollution Bulletin, 78*(1–2), 196–200.

Naidu, S. A., Mawii, L., Rao, V. R., Anitha, G., Mishra, P., Narayanaswamy, B. E., Kumar, V. A., Murthy, M. V. R., & Gupta, G. V. M. (2021). Characterization of plastic debris from surface waters of the eastern Arabian Sea-Indian Ocean. *Marine Pollution Bulletin, 169*, 112468.

Napper, I. E., Baroth, A., Barrett, A. C., Bhola, S., Chowdhury, G. W., Davies, B. F. R., Duncan, E. M., Kumar, S., Nelms, S. E., Hasan Niloy, M. N., Nishat, B., Maddalene, T., Thompson, R. C., & Koldewey, H. (2021). The abundance and characteristics of microplastics in surface water in the transboundary Ganges River. *Environmental Pollution, 274*, 116348.

Nelms, S. E., Duncan, E. M., Patel, S., Badola, R., Bhola, S., Chakma, S., Chowdhury, G. W., Godley, B. J., Haque, A. B., & Johnson, J. A. (2021). Riverine plastic pollution from fisheries: Insights from the Ganges River system. *Science of the Total Environment, 756*, 143305.

Nuelle, M. T., Dekiff, J. H., Remy, D., & Fries, E. (2014). A new analytical approach for monitoring microplastics in marine sediments. *Environmental Pollution, 184*, 161–169.

Pabortsava, K. & Lampitt, R. S. (2020). High concentrations of plastic hidden beneath the surface of the Atlantic Ocean. *Nature Communications, 11*(1), 1–11.

Pflugmacher, S., Tallinen, S., Kim, Y. J., Kim, S., & Esterhuizen, M. (2021). Ageing affects microplastic toxicity over time: Effects of aged polycarbonate on germination, growth, and oxidative stress of Lepidium sativum. *Science of the Total Environment*, *790*, 148166.

Piccardo, M., Bertoli, M., Pastorino, P., Barceló, D., Provenza, F., Lesa, D., Anselmi, S., Elia, A. C., Prearo, M., Pizzul, E., & Renzi, M. (2021). Lethal and sublethal responses of *Hydropsyche pellucidula* (Insecta, Trichoptera) to commercial polypropylene microplastics after different preconditioning treatments. *Toxics*, *9*(10), 256.

PlasticsEurope. (2012). *Plastics — The Facts 2012. An Analysis of European Latest Plastics Production, Demand and Waste Data.* Brussels: Plastics Europe, Association of Plastic Manufacturers.

PlasticsEurope. (2012). Plastics — The Facts 2012. *An Analysis of European Latest Plastics Production, Demand and Waste Data.* pp. 1–36. Brussels: Plastics Europe, Association of Plastic Manufacturers.

PlasticsEurope. (2016). Plastics — The Facts 2016. *An Analysis of European Latest Plastics Production, Demand and Waste Data.* pp. 1–35. Brussels: Plastics Europe, Association of Plastic Manufacturers.

PlasticsEurope. (2018). Plastics — The Facts 2018. *An Analysis of European Latest Plastics Production, Demand and Waste Data.* pp. 1–58. Brussels: Plastics Europe, Association of Plastic Manufacturers.

PlasticsEurope. (2020). Plastics — The Facts 2020. *An Analysis of European Latest Plastics Production, Demand and Waste Data.* pp 1–64 Brussels: Plastics Europe, Association of Plastic Manufacturers.

PlasticsEurope. (2021). *Plastics — The Facts 2021. An Analysis of European Latest Plastics Production, Demand and Waste Data.* pp. 1–33. Brussels: Plastics Europe, Association of Plastic Manufacturers.

Porter, A., Lyons, B. P., Galloway, T. S., & Lewis, C. (2018). Role of marine snows in microplastic fate and bioavailability. *Environmental Science & Technology*, *52*(12), 7111–7119.

Posthuma, L., Suter II, G. W., & Traas, T. P. (2001). *Species Sensitivity Distributions in Ecotoxicology*. pp. 1–616. CRC Press, Boca Raton, London, New York.

Skalska, K., Ockelford, A., Ebdon, J. E., & Cundy, A. B. (2020). Riverine microplastics: Behaviour, spatio-temporal variability, and recommendations for standardised sampling and monitoring. *Journal of Water Process Engineering*, *38*, 101600.

Thompson, R. C., Olsen, Y., Mitchell, R. P., Davis, A., Rowland, S. J., John, A. W. G., McGonigle, D., & Russell, A. E. (2004). Lost at sea: Where is all the plastic? *Science*, *304*, 838.

Topouzelis, K., Papakonstantinou, A., & Garaba, S. P. (2019). Detection of floating plastics from satellite and unmanned aerial systems (Plastic Litter Project 2018). *International Journal of Applied Earth Observation and Geoinformation*, *79*, 175–183.

van Sebille, E., Wilcox, C., Lebreton, L., Maximenko, N., Hardesty, B. D., van Franeker, J. A., Eriksen, M., Siegel, D., Galgani, F., & Law, K. L. (2015). A global inventory of small floating plastic debris. *Environmental Research Letters*, *10*(12), 124006.

Walsh, A. N., Reddy, C. M., Niles, S. F., McKenna, A. M., Hansel, C. M., & Ward, C. P. (2021). Plastic formulation is an emerging control of its photochemical fate in the ocean. *Environmental Science & Technology*, *55*(18), 12383–12392.

Wang, J., Li, Y., Lu, L., Zheng, M., Zhang, X., Tian, H., Wang, W., & Ru, S. (2019). Polystyrene microplastics cause tissue damages, sex-specific reproductive disruption and transgenerational effects in marine medaka (*Oryzias melastigma*). *Environmental Pollution*, *254*, 113024.

Wang, Q., Wangjin, X., Zhang, Y., Wang, N., Wang, Y., Meng, G., & Chen, Y. (2020). The toxicity of virgin and UV-aged PVC microplastics on the growth of freshwater algae *Chlamydomonas reinhardtii*. *Science of the Total Environment*, *749*, 141603.

Woodall, L. C., Sanchez-Vidal, A., Canals, M., Paterson, G. L., Coppock, R., Sleight, V., Calafat, A., Rogers, A. D., Narayanaswamy, B. E., & Thompson, R. C. (2014). The deep sea is a major sink for microplastic debris. *Royal Society Open Science*, *1*(4), 140317.

Wu, D., Wang, T., Wang, J., Jiang, L., Yin, Y., & Guo, H. (2021). Size-dependent toxic effects of polystyrene microplastic exposure on Microcystis aeruginosa growth and microcystin production. *Science of the Total Environment*, *761*, 143265.

Yang, H., Xiong, H., Mi, K., Xue, W., Wei, W., & Zhang, Y. (2020). Toxicity comparison of nano-sized and micron-sized microplastics to Goldfish Carassius auratus Larvae. *Journal of Hazardous Materials*, *388*, 122058.

Yu, S.-P. Y., Cole, M. C., & Chan, B. K. (2020). Chapter 7 Review: Effects of microplastic on zooplankton survival and sublethal responses. In *Oceanography and Marine Biology: An Annual Review*, Vol. 58, pp. 351–394. Boca Raton, London, New York: CRC Press.

Zhang, D., Liu, X., Huang, W., Li, J., Wang, C., Zhang, D., & Zhang, C. (2020). Microplastic pollution in deep-sea sediments and organisms of the Western Pacific Ocean. *Environmental Pollution*, *259*, 113948.

Zhang, K., Xiong, X., Hu, H. J., Wu, C. X., Bi, Y. H., Wu, Y. H., Zhou, B. S., Lam, P. K. S., & Liu, J. T. (2017). Occurrence and characteristics of microplastic pollution in Xiangxi Bay of Three Gorges Reservoir, China. *Environmental Science & Technology*, *51*(7), 3794–3801.

Zimmermann, L., Göttlich, S., Oehlmann, J., Wagner, M., & Völker, C. (2020). What are the drivers of microplastic toxicity? Comparing the toxicity of plastic chemicals and particles to *Daphnia magna*. *Environmental Pollution*, *267*, 115392.

https://doi.org/10.1142/9789811259111_0002

Chapter 1

The History and Future of Plastics

Rachael Rothman and Anthony J. Ryan*

The Grantham Centre for Sustainable Futures,
The University of Sheffield,
Sheffield S10 2TN, UK

**a.ryan@sheffield.ac.uk*

1. The History of Plastics

1.1. *The dawn of plastics*

It takes two elephant tusks to make a set of billiard balls. The super-consumers of the late 19th century (the English gentry and the American nouveau riche) were driving elephants towards extinction and so it was the price of ivory that drove the R&D of the first thermoplastics. In 1863, Michael Phelan, "the father of American billiards", took out an advert in the name of the firm Phelan and Collender offering $10,000 (roughly $200,000 today) for the invention of a substitute for ivory.

Previously, an Englishman, Alexander Parkes, had attempted to commercialise mouldable cellulose nitrate, under the name of Parkesine, by mixing it with vegetable oil, but his business failed. His colleague Daniel Spill went on to market the same material as Xylonite and was eventually successful with a company that became BX Plastics (Rasmussen, 2021).

An American, John Wesley Hyatt, had picked up these innovations, but his camphor containing plastic, celluloid, wasn't the perfect substitute for billiard balls. Nitrocellulose lacquer, the soluble form of gun cotton which is an explosive made from cellulose trinitrate, had been around for a couple of decades already and so every impact of celluloid billiard balls led to a small explosion. Something sounding like a gun going off could prove fatal in American saloons to this day. So whilst it's true that no one ever won the prize, it did lead to the development of two industries that shaped the 20th century: plastics and photography/cinema (Seymour & Kauffman, 1992); see the following.

Hyatt and Spill's companies developed a wide range of nitrocellulose-based plastic products with a range of properties that made plastics attractive and, eventually, indispensable to modern life. They could be moulded, be hard or flexible as required and be coloured, were waterproof and very durable, and, most importantly, were amazingly cheap. In an incredible irony, an 1878 sales pamphlet for celluloid boasted: "As petroleum came to the relief of the whale, so has celluloid given the elephant, the tortoise and the coral insect a respite in their native haunts; and it will no longer be necessary to ransack the earth in pursuit of substances which are constantly growing scarcer" (Freinkel, 2011).

Ten years later, George Eastman and Hannibal Goodwin applied for patents for flexible films based on plasticised cellulose nitrate. Initially used for still photography stock, with a coating of gelatin emulsion containing light-sensitive silver halide, it revolutionised the process of taking pictures and eventually led to the movie industry (del Amo, 2000). Even though it presented an enormous fire risk, celluloid film was used as movie film right up to the 1950s when it was eventually replaced by cellulose acetate "safety" film and thereafter PET (polyethylene terephthalate).

At first, we embraced plastic with a clean conscience and an eagerness for material goods that knew no bounds. 150 years of materials science and engineering has led us to a cornucopia of plastic materials embedded in our daily lives. Grafted on this is more than 60 years of behavioural training to irresponsibly dispose of cheap plastics. This is famously exemplified by the 1955 edition of *LIFE* magazine that waxed lyrical about the benefits of throwaway living (*Life Magazine*, 1955). Central to the article is an image (too expensive to reproduce herein) of a happy family being showered by various single-use, disposable plastic items, which has the caption: "The objects flying through the air in this picture would take 40 hours to clean — except no housewife need bother. They are all meant to be thrown away after use".

It is not only synthetic plastics that have impacts on our environment, with natural rubber also having detrimental effects, especially in the context of ocean pollution. Natural rubber, a polyisoprene latex from the tree *hevea brasiliensis*, effectively started the polymer industry (Loadman, 2005). First, through the production of waterproof fabrics for the military, invented by Charles Macintosh using a naphtha solution of natural rubber to glue two pieces of cotton together, it gave us the eponymous Mac — a term now commonly used for any waterproof coat. But the rubber industry really expanded with the independent discovery of sulfur crosslinking by Charles Goodyear and Thomas Hancock in the 1840s (Morton, 1999). This meant that John Boyd Dunlop's invention of the pneumatic tyre was already there for the development of motorised vehicles, and an abundant source of microplastic (Parker-Jurd *et al.*, 2021) that was mature 100 years before Richard Thompson and colleagues (Thompson *et al.*, 2004) named them.

1.2. *The evolution of fossil plastic — Late 1800s and early 1900s*

All of the plastics discussed thus far are natural polymers or at least come from renewable polymer feedstocks. The natural fibres that were used in clothing in this period are also polymers; silk and wool are polypeptides (proteins), and cotton and linen are polysaccharides (cellulose). Cellulose provides the mechanical structure to plant cell walls and woods, and a combination of stiff cellulose microfibres in an amorphous polysaccharide matrix can be considered the original fibre reinforced composite.

The first truly synthetic plastic was invented by a commercially driven chemist who had cut his teeth in the photographic industry, Leo Baekeland (Crespy *et al.*, 2008). The idea that the natural resins and fibres were big molecules was gaining some traction, and the reactions between phenols and formaldehydes had been explored in search for synthetic dyes but all they made were a dark coloured mess. Baekeland approached the field systematically, carefully controlling and examining the effects of temperature and the types and proportions of materials used, to make a synthetic replacement for shellac (a varnish made from a resin secreted by the lac beetle). By introducing fillers such as wood flour and asbestos to the syrupy resin, and controlling the pressure and temperature during moulding, he produced a hard plastic which he called Bakelite.

Compression moulded artefacts were manufactured as well as sheets and rods that could be machined. Baekeland's process patent for making insoluble products of phenol and formaldehyde was filed in July 1907 and granted in 1909 (Mercelis, 2012). Perhaps Baekeland's greatest achievement though was transforming the word plastic from an adjective into a noun; its original meaning being formable or shapeable derived from the Latin word *plasticus*. Now, this new plastic material Bakelite was aggressively marketed as "the material of a thousand uses" and its electrical insulating properties and hardness found many applications in electrical (telephones, radios and switches) and other domestic goods.

In the 1920s, Herman Staudinger used natural polymers (cellulose, silk and most importantly natural rubber) to demonstrate that they consisted of long-chain molecules (Frey & Johann, 2019). He was awarded the Nobel Prize for this in 1953 (NobelPrize.org, 1953), but his landmark paper in 1920 *Über Polymerisation* (Staudinger, 1920) was greeted with derision by academic (organic) chemists. The prevailing view was that these materials were colloids, held together by non-covalent bonds and not high molecular weight molecules made by linking together a large number of small molecules. Whilst Staudinger didn't invent polystyrene or poly(ethylene oxide), he did demonstrate their reactions, which he called "polymerisation", whereby individual repeating units were joined together by covalent bonds. He predicted the development of synthetic fibres and many other applications for plastics from a sound theoretical and experimental basis.

1.3. *Classification of polymers & their properties*

Polymer molecules that make plastics can be classified into three main groups: thermoplastics, thermosets and elastomers (or rubbers) (Cowie & Arrighi, 2007). This type of division is based on the architecture of the polymer and the way in which it is processed (Wilkinson & Ryan, 1999). The most important distinction between polymers is whether they are cross-linked or not. A cross-linked polymer has molecules that span the whole volume of the artefact, and the molecular weight can be considered to be infinite.

A schematic structure for a cross-linked polymer is shown in Figure 1 where the black lines represent the polymer backbone, the circles the cross-links, and the arrows the structure repeating out to the edges of the sample. This type of structure is often referred to as a polymer network because the covalent bonds that exist between the chains mean that the polymer network can ever flow. If it has a high cross-link density and is glassy, then it's called a thermoset; if it has a low cross-link density and is elastic, it's known as a rubber or an elastomer. The molecules referred to as linear and branched in Figure 1 have finite molecular weights and, given sufficient mobility along the chain and between chains, can become liquid-like and flow. These polymers are comprised of thermoplastics and can be moulded into an amazing plethora of shapes.

The temperature at which a linear or branched polymer becomes liquid-like depends on its chemical structure and hence mobility. Amorphous polymer chains undergo a glass to rubber to liquid transition depending on their chemical structure and architecture. Think of them behaving like a hard toffee that's brittle when you take it out of the bag but becomes soft and pliable as it warms up in your mouth. When a polymer like polystyrene is heated, the range of molecular motion increases until eventually

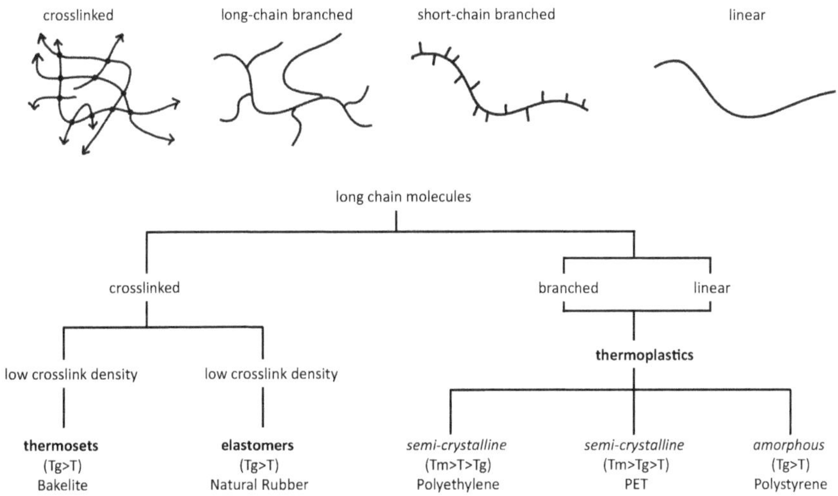

Figure 1. Classification of polymers according to their molecular architecture and the relationship between the service temperature (T), the melting point (T_m) and the glass transition temperature (T_g). The operational labels thermoplastics, thermosets and elastomers are highlighted in bold, and examples of specific plastics are given for each of the final classes.

co-operative molecular motions are allowed and the material becomes soft and deformable; eventually, the polymer molecules can become so fluid that the plastic can be moulded.

Some polymers can crystallise and it depends on how the atoms are organised along the polymer chain, whether they are symmetrical or not (Young & Lovell, 2011). Molecular segments of chains line up in order to pack into an atomic lattice forming crystalline domains separated by regions that are not crystalline but amorphous. Candle wax is essentially low in molecular weight polyethylene. At low temperatures, it is quite hard, but as soon as you go above the melting point, it becomes a liquid and can be poured, and if you put it in a mould, it will take on the new shape. And that is essentially how thermoplastics work, but with molecules that are many orders of magnitude bigger. Whether the amorphous part is glassy or rubbery defines the properties and the polymer's service temperature. However, the nature of synthetic polymers means they are much tougher and stronger than candle wax.

1.4. *Plastics between the wars*

The DuPont company set out on a fundamental research programme built on Staudinger's work and, in 1927, hired as its leader the talented chemist from Harvard, Wallace Hulme Carothers (Hermes, 1996). Not only did Carothers invent the first commercial thermoplastic, nylon, but he also developed the theoretical tools to predict the length of the polymer based on the composition of the monomers it is made from. Their early work on polyesters, from the polycondensation reaction of acids with alcohols, didn't lead to many interesting polymers. To get the high molecular weights required to make fibres, the reactions had to be driven to very high conversions, and even then, the melting points were too low and they were susceptible to water degradation. Following pressure from management toward targeted research to produce something useful, they developed a completely different polymer, polychloroprene, from the chlorinated analogue of isoprene, and this first synthetic rubber is still produced today as neoprene (Smith, 1985). They moved on to work on polyamides, reacting diamines with diacids, and these produced high melting point polymers that could be readily made into fibres. Adipic acid was reacted with hexamethylene diamine to make the polyamide nylon 66. In 1935, a whole team of scientists and engineers put to work to commercialise it, and by 1938, it was used as the bristles on a toothbrush, followed in 1940 by the production of women's stockings, "nylons", which sold 64 million pairs in their first year. Nylons were in short supply though as production was shifted to parachute production for the Second World War (Gaines, 2002).

The search for synthetic fibres, to replace silk, was also underway in Germany at IG Farben. To get around the DuPont patents, Paul Schlack developed the ring-opening polymerisation of caprolactam to make the polyamide nylon 6 (Matthies & Seydl, 1986), and Otto Bayer developed polyurethanes (Seymour *et al.*, 1989) from the polyaddition of isocyanates and alcohols.

It is interesting to note that this old polymer, nylon 6, shows the way how plastics might be chemically upcycled (Korley *et al.*, 2021). You can replace lots of engineering plastics with nylon 6, and, in fact, material selection is something that's often driven by marketing or by optimisation of a design for reasons other than its reuse and repurposing, e.g., lightweighting. Nylon 6 can return back into its monomer, caprolactam, by heating under vacuum in the presence of a catalyst (Pilati & Toselli, 2002). It liberates the monomer and purifies it at the same time and is the basis of EcoNyl, a business set up to recycle carpets and deliver chemically recycled nylon fibres back into the market.

Polystyrene, PS, was first manufactured by IG Farben in 1931 despite the polymerisation of styrene having been known since 1839 when German pharmacist Eduard Simon reported its conversion into a solid later named metastyrol. Staudinger used it as an example of polymerisation in his *Über Polymerisation* paper, but developments were slow. It was the cheap and plentiful polystyrene that gave plastic its derogatory meaning — cheap, nasty and disposable. By 1937, Dow Chemical Company had an efficient way to make high-purity styrene monomers, and by 1938, polystyrene was being produced commercially. It quickly became one of the most important modern plastics owing to the low cost of producing large volumes of styrene monomers and subsequently polymers, the development of injection moulding technology, and the optical and physical properties of the material (Scheirs & Priddy, 2003).

Other linear vinyl polymers were also discovered early, poly(vinyl chloride) (PVC) in 1872 and poly(methyl acrylate) (PMA) in 1880, but were not produced commercially until much later, for example, poly(vinyl acetate) in 1920, poly(methyl acrylate) PMMA in 1927 and poly(vinyl chloride) in 1930. Many of the other conventional linear polymers of today formed by chain polymerisation were developed from the 1930s to the 1950s with something of a hiatus during the Second World War (Mosley, 2017).

PMMA was commercialised in the late 1930s under the trade names Perspex and Plexiglas. It was invented in the early 1930s by Rowland Hill and John Crawford at Imperial Chemical Industries (ICI) in the UK who registered it with the trademark Perspex. Contemporaneously, Otto Rohm, of Rohm and Haas AG in Germany, attempted to produce safety glass by polymerizing the monomer methyl methacrylate between two layers of glass. The polymer separated from the glass as a clear plastic sheet, which Röhm gave the trademarked name Plexiglas in 1933. In the United States, E.I. duPont de Nemours & Company (now DuPont Company) subsequently introduced its own PMMA product under the trademark Lucite.

PVC was first synthesised inadvertently in 1872 by Eugen Baumann when a white solid was reported to appear inside a flask of vinyl chloride that had been left in the dark. In the early 20th century, the German chemical company Griesheim-Elektron attempted to use PVC in commercial products, but difficulties in processing the rigid, brittle polymer thwarted their efforts. It was Waldo Semon at the BF Goodrich company who developed a method to plasticise PVC by blending it with various additives, notably phthalate esters (Titow, 1984). This reduces the glass transition of PVC, making it flexible and easy to process with a wide range of applications. It is the plasticisers that led to the widespread commercial use of PVC that subsequently alerted medics and ecologists to the potential damage that plastics can cause as a result of plasticisers leaching into the environment (Billings *et al.*, 2021).

The polymers that are produced in the biggest volumes, the polyolefins (thermoplastics formed by polymerisation of polymers with a carbon–carbon double bond, e.g., polyethylene), were relatively late arrivals. The first industrially practical polyethylene synthesis was accidentally discovered in 1933 at ICI (Nowlin, 2014). That high molecular weight alkane formed tough waxy solids had been known for a while, but their synthesis was expensive and unwieldy. Just like DuPont, ICI ran a fundamental science research programme. Eric Fawcett and Reginald Gibson were studying the effects of high pressure on mixtures of gases including ethylene. The loss of pressure in the equipment was traced to the production of a white, waxy material, but because the reaction had been initiated by oxygen contamination, it was difficult to reproduce. It wasn't until 1935 that Michael Perrin, another ICI chemist, developed the high-pressure industrial synthesis of polyethylene with production starting in 1939. Commercialisation was suspended on the outbreak of World War II and it became a state secret. It turned out to be a fantastic high-frequency electrical insulator and was used in the production of the coaxial cables in radar sets. In common with many other polymers, there was lots of R&D effort during the war, and by 1944, DuPont and the Bakelite Corporation began large-scale commercial production under licence from ICI.

1.5. *The 1950s and beyond — The plastic age*

It was the chain straighteners Karl Ziegler and Julio Natta (McMillan, 1979) who really got the polyolefins business going in the 1950s. They discovered catalysts that could control the stereochemistry of polymerisation. This meant that semi-crystalline polypropylene could come into being and now polyethylene had a real partner. The technologies developed meant that there was better and better control over both the size and architecture of polyolefins and so they came to be the most produced plastics ever. It was the development of metallocene, soluble catalysts, in the 90s that led to a complete overhaul of the range of polyolefin materials, and many and varied copolymers with all sorts of molecular sizes and shapes could be produced.

The combination of chemistry, physics and engineering isn't the reason we have so much plastic pollution in the sea. It's the combination of cheap, durable and most importantly malleable material that has led us to where we are today. The very meaning of plastic is that it can be reshaped; this meant that designers could design a whole host of previously unimaginable products and that marketeers could sell them to a population that had its consumerism unlocked by the plethora of cheap goods available post-war.

More than 80% of the current plastic production is of four types: polyolefins, polystyrene (PS), PVC and PET (Bucknall, 2020). Whilst Carothers and his team made polyesters at DuPont in the USA from 1928, the major polyester discovery was made at the Calico Printers Association in Manchester by James Dickson and John Whinfield who were searching for new fibre-forming polymers. In 1941, they patented poly(ethylene terephthalate), PET, made from the reaction of ethylene glycol with terephthalic acid (Scheirs & Long, 2004). Patent rights were granted to ICI in the UK and to DuPont in the US by the late 1940s, and they produced PET fibres under their trademarks Terylene and Dacron. The UK had a well-established business in natural fibres. Cotton in Lancashire and wool in Yorkshire were world-leading industries so it's not surprising that the development in nylon and polyester occurred there. Companies that built their businesses on natural fibres moved into these new synthetic fibres, the (ICI) crimplene suit was a badge of honour, and the hard-wearing nylon shirt underneath meant less laundry.

PET film stock enabled the widespread use of magnetic tape, revolutionised music recording and the scene was set for "The man in the white suit", a 1951 satirical movie (Mackendrick, 1951). It had a really big star of the time (Alec Guinness) and was absolutely in tune with technological advances that were driving economic and social innovations. It tells the story of a scientist who invents an everlasting fibre; the company he works for is delighted until it realises that it will put them out of business. If you make a suit that lasts forever, why would anyone want to buy another suit? To the minds of the pre-consumerist '50s, this would cause the business to fail. So, they make him an offer. Lock your invention away and we'll make you rich, and so, they recruit his girlfriend to persuade him to do so. But he won't do it and he runs away in his white suit, which, slowly but surely, starts to fall apart, leaving him standing in his underwear, the object of ridicule, as he realises the fatal flaw in what he'd made. The prescience of this film is not lost on us today because there are many companies full of scientific researchers trying to realise the dream of marketeers for plastics that will environmentally decay into benign products that leave no trace behind, not realising that they can cause more actual damage, in terms of greenhouse gas (GHG) emissions, than the materials they are intended to replace.

As important as synthetic fibres are, it is plastic moulding that took polyethylene, polystyrene and PVC out to the masses. The cornucopia of plastic products was absolutely blinding as seen in Figure 1. Initially, they were found to be of poor quality because the designers hadn't come to terms with how to use these materials. But, as knowledge developed, designs got better, materials got better, processes got better and plastic invaded every aspect of modern life.

Plastics are ubiquitous and have transformed our modern world. With numerous useful and often unparalleled properties, plastics are a highly adaptable class of materials exploited for an ever-widening range of applications. The annual global plastic production has continually increased, reaching ~350 Mt in 2019 (Geyer *et al.*, 2017). It is no exaggeration to say that plastics permeate almost everything we see and use on a daily basis, from the clothes we wear to products we use (plastic bottles) and reuse (credit cards), to the less visible binders of our infrastructure (water pipes and electrical cable insulation). The exponential increase in plastic usage since the 1950s stems from their deliberate increasing exploitation, resulting in a fiftyfold increase in their use due to their low cost, ease of production and transport and the full cost not being borne by producers or users.

There are all sorts of applications of plastics that enable things that would have been inconceivable to our grandparents. Non-stick pans rely on controlling the surface energy of a material; Teflon (polytetrafluoroethylene) is one of those amazing materials (Ebnesajjad, 2016). Not only will it stop your breakfast egg from sticking to the pan but, if you live somewhere that's wet, the same Teflon material can make Gore-Tex, a microporous fabric that allows water vapour to escape but not water liquid to enter a garment, thus making it ideal for waterproof clothing.

The ubiquity and diversity of plastics are best demonstrated by their use in sport and leisure applications as an example (Jenkins, 2003). The range of materials inside a set of modern running shoes is unbelievable, from the energy-absorbing poly(ethylene-vinylacetate) EVA foam in the midsole to the high-grip rubber out-sole beneath it, and the PU memory foam insole on top. There will be a PU synthetic leather on the PET of the outer and nylon in the laces, and the clothes you go running in, from Lycra shorts to high wicking and breathable nylon and polyester T-shirts. The shape of the fibres is even manipulated in order that they hide dirt and promote evaporation at one and the same time. It's not only outside your body that you will find plastics. If you have white fillings, they are glass-filled acrylic. Hip replacements are made from polyethylene, heart valves from polyurethane and silicone implants are the same material that's used to make non-stick muffin tins.

The examples above are all about structural materials, but transistors, LEDs and solar cells are also made from functional plastic. Your mobile phone wouldn't work without plastic; it's the insulator that enables all of the solid-state electronics to work. The battery relies on a membrane that's made from a very special type of plastic that allows the conduction of ions but not the penetration of metals. The liquid crystal display has lots of molecules trapped in cages made of plastic, switched by plastic transistors, and the case that holds it all together is often plastic made to look like metal. If you have a state-of-the-art OLED television, or a foldable phone, then the light-emitting diodes that make up the screen are even plastic themselves.

It is, however, packaging that consumes the greatest proportion of plastic (Figure 2), but it is also the application with the shortest use phase. It's often seen as the most egregious use of plastic, but it contributes enormously to preventing food waste. People always ask why cucumbers are wrapped in plastic and the answer is quite simple: they last much longer when wrapped in plastic (Dhall *et al.*, 2012) and you can put them in and out of the fridge for a period of a week or more and they will still be crisp and tasty every time you cut off a slice for your sandwich. In fact, some food products only exist because they were developed in symbiosis with their packaging materials. Forty years ago, crisps were wrapped in cellophane and had a small blue bag inside containing salt. Marketing said it was so you could sprinkle salt on your crisps before you eat them, but in reality, that little blue bag was there to absorb the moisture that came through the cellophane bag and would have made the crisps go soggy.

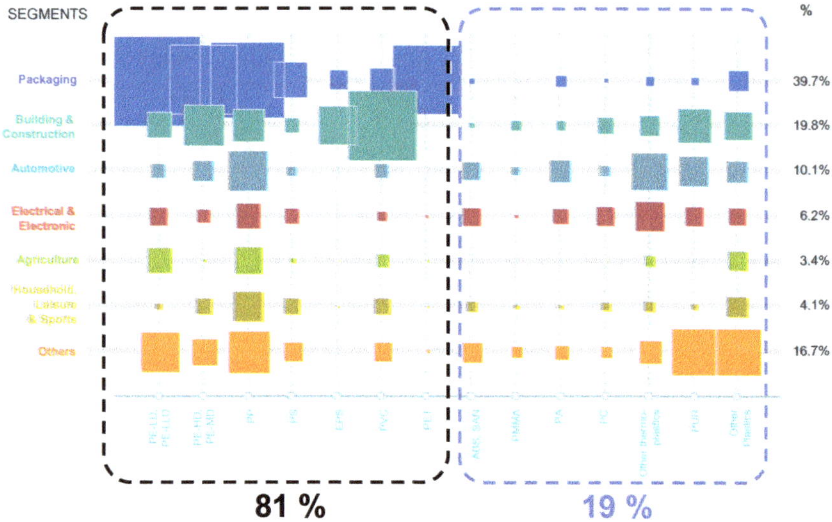

Figure 2. The consumption of plastics as a function of plastic type and usage. Note that 81% of plastic is in four classes, polyolefins (low-density polyethylene, PE-LD; linear-low-density polyethylene, PE-LLD; high-density polyethylene, PE-HD; medium-density polyethylene, PE-MD; and polypropylene, PP), styrenics (PS and expanded polystyrene, EPS), PVC and PET, and nearly 40% is in the fastest turnover (shortest use-phase) sector of packaging.

The crisps were fried in saturated fat because saturated fat doesn't go rancid so quickly, but nowadays we have foods containing polyunsaturated fats that effectively last forever. They rely on metallised PET packets that don't allow oxygen or water ingress and the product itself sits in a nitrogen atmosphere, maintaining its quality and edibility for over a year, even if the product had little or no nutritional qualities in the first place.

One of the drivers for packaging development has been lightweighting, meaning the development of polymer films that give the maximum protection for the minimum weight. To enable this, you need to combine many different plastics in the same thin film. Something like PET or a polyolefin will give mechanical strength, an oxygen barrier polymer such as polyvinyl alcohol (PVOH) and surface treatment to enable printing. All these different functional layers are held together by interfacial layers of an adhesive polymer. There can even be active layers to emit CO_2 or absorb O_2 (to preserve meat) or ethylene (to delay the ripening of fruit). This means that you can have a piece of steak that looks bright red even though it's been cut and packaged for over a week. The oxygen barrier properties of the packaging prevent oxidation of the iron in the haemoglobin, meaning the meat looks fresh and attractive (not brown) for the consumer; the consumer doesn't understand the high technology (Han *et al.*, 2018) involved in the plastic packaging and want to recycle it, but instead have to throw in the bin.

This ubiquity has also led to unprecedented ecological and environmental risks, as the low cost and lack of perceived value of plastic have led to a single-use throwaway culture with improper treatment at end-of-life (EoL). The historic and current mismanagement and unlegislated release/disposal of plastics impose negative impacts on our environment, health and well-being. The lack of circularity means that around 90% of the plastics ever produced have been incinerated, sent to landfills or just discarded. It is estimated that ~2.5% of these landfilled plastics (88–226 Mt) have been lost from the managed waste system into the marine environment. This creates a perpetually growing environmental and ecological disaster exemplified by the Mt of floating plastics forming the Great Pacific Garbage Patch (Lebreton *et al.*, 2018). Marine plastic has direct implications for marine flora and fauna as well as the health and well-being of society. Whilst the excellent material properties of plastics create value across global supply chains, delivering goods in better condition and at lower prices for end-users, plastic production and incineration emit 850 Mt of CO_2/yr[2] and are simply deferred fossil fuel combustion.

The largest fraction of plastics are produced, used once and then disposed of. The lifetime from production to end-of-life (EoL) varies considerably, from a few weeks (e.g., packaging) to many decades (e.g., construction). For managed plastics within the UK, legislative changes over the last decade have decreased the amount of EoL plastics sent to landfills to 30%, collection for recycling has increased to 32% and incineration for energy recovery is an increasing, yet unsustainable, pathway for the remaining 38%. But regional variations exist: Wales has the highest recycling rates in Europe (for municipal solid wastes including plastics) around ~63% accelerated by national policies and legislation. Recycling and recovery approaches for plastics are governed by extended producer responsibility (EPR), packaging recovery notes (PRN) and packaging export recovery note (PERN) schemes. The EPR scheme is in process of being reformed, and PRN/PERN notes do not promote best practices. The introduction of plastic waste import bans by countries, including China and Malaysia, highlights both the misuse of the system and the UK's lack of waste management infrastructure. Consequently, waste plastics' flow in the UK is in a dynamic state of flux, influenced by international policies and government consultations, and the COVID-19 pandemic is demonstrating the importance and dependence on single-use plastics.

2. The Challenges

2.1. *Why is recycling so difficult?*

Plastics are difficult to recycle from post-consumer waste because they need to be separated to be reprocessed — it's cheaper for the manufacturer to make new plastic packaging and get the consumer to throw the used plastic away — because those costs are borne by society in general and not the packaging producer or waste disposer, in particular.

The majority of plastic waste collected is put through open-loop mechanical recycling, where post-consumer plastics undergo mechanical processes to first separate the plastic into different types and recover the polymers with minimal changes to their chemical and physical nature. But it is not just the different types of polymers that need to be separated, it is different colours too. A lot of the recycled plastics from mechanical recycling are re-used in different products than those they were originally recovered from, in an open-loop system. Mechanical recycling also causes problems in terms of loss of the plastic's properties; they become degraded in terms of their molecular length and chemical composition. This often necessitates the addition of virgin material to achieve usable materials.

The product with the highest recycling rate, PET bottles, is typically recycled into fibre for the manufacture of polyester clothing. PET has the advantage over other common polymers of being able to be chemically upgraded after separation and cleaning to give materials that are close to virgin properties. Of course, closed-loop recycling, where the recycled plastic is used for its original purpose, i.e., PET bottle recyclate re-used to produce new PET bottles, is the preferred option for a circular economy, but to date surprisingly only little closed-loop recycling is achieved (less than 10%) (Lonca *et al.*, 2020).

PE and PP recycling represent different challenges, and the recycling of different kinds of PE and PP types varies considerably. Recycling of mixed PP waste, in general, and even separate recycling of individual PP waste packaging types, will not technically facilitate recycling into new packaging products, whereas efficient separation of PE bottles can allow closed-loop processes. The homogeneity of PE and PP recyclates is crucial for reprocessing. Such homogeneity may be achieved through additional plastic sorting and regulatory harmonisation of product design, accounting for polymer properties and recyclability. But degradation of PP during recycling is another important limitation for PP recycling, necessary to address in the future (Eriksen *et al.*, 2019).

Different kinds of plastics have to be separated for recycling because they contain long-chain molecules (polymers) that don't mix at the molecular level. That polyethylene is incompatible with polystyrene might not be surprising, but even graduate chemists fail to comprehend that polyethylene (PE) and polypropylene (PP) don't mix even though they both have the same empirical formula of $n(CH_2)$ (Lodge & Hiemenz, 2021). That's why separating plastics at the recycling centre is so important and choices made on our behalf mean that some plastics can never be recycled.

The properties of polymer blends mean that they need to have compatibiliser molecules to stitch the interfaces of the phases together — uncompatibilised blends have very poor properties (Creton, 2017) — and the holy grail for post-consumer waste is to make compatibilisers that can stitch everything together from a mixed recyclate. Until now, this has proven to be impossible, but a recent paper has shown one way that this might be possible (Eagan *et al.*, 2017).

2.2. *The role of the Persuasion Industries* (McKevitt, 2018)

Why did we move away from glass bottles with deposits? We were sold an ephemeral lightweight, recyclable concept from an environmental perspective. Certainly, the amount of packaging that is transported has gone down — 25 g of PET versus 1 kg of glass for a litre of sparkling water — but the reason it happened was because it was cheaper for the supplier and helped the companies' bottom line, not any real desire to benefit the environment. Costs were reduced because it is less expensive to make a new PET bottle than to collect, clean and refill an existing bottle, and the disposal and/or recycling of the bottles was externalised. In fact, it has led to the degradation of previous safe municipal (tap) water systems in some countries (Hawkins *et al.*, 2015).

Taking a holistic, whole system approach to life cycle assessment (LCA) is really important when considering packaging systems; both the environmental impacts of the packaging itself and the impact (positive or negative) of the packaging on the environmental impacts of the contents need to be considered. This is particularly pertinent for food, yet the role of packaging systems to reduce food waste is rarely modelled in LCA studies. This means that packaging formats with a lower environmental impact but that cause high food waste may appear to be a better alternative than a packaging system with a higher environmental impact that reduces food waste. However, food generally has a much higher environmental impact than the packaging system and the function of "avoiding food waste" is a critical packaging issue (Wikström *et al.*, 2014).

Our recent life cycle assessment, LCA, on milk delivery shows that glass bottles only need to be used around 15 times to have a lower carbon footprint than their single-use HDPE counterparts. With an average number of uses (280 in a five-year lifetime), four pints of milk in glass bottles have a carbon footprint of about 60% of that of a four-pint HDPE carton. However, the milk in glass would cost the consumer around 15% more than the same volume in plastic, putting the onus on the customer to decide if they are prepared to pay more to be more sustainable. Milk can benefit from local, distributed supply chains, therefore the heavier packaging has little impact on transport emissions as the travel distance is short. Glass bottles are not suitable for all applications, for example, many cafes use over 40 litres of milk a day. We recently published work comparing reusable steel milk churns (that can be connected directly to a delivery system like beer barrels in a pub) (Błażejewski *et al.*, 2021).

The churns have a carbon footprint of around 30% of that of single-use plastic, assuming a 50% recycling rate of the HDPE bottles. The water footprint of the reusable churns is less than half of the HDPE bottles despite the need for cleaning water. Glass bottles and churns are not new — the milk churn came into being in the mid-1800s — so, in essence, to reduce the carbon footprint of milk, we need to go backwards to get to the future.

Many recent developments have moved us in the wrong direction: sports drink cans have three different (incompatible) polymers: PET bottle, shrink-wrapped PE film and PP snap-top lid. All three components can be individually recycled but they are rarely separated other than by shredding. The only function of black plastic trays is to amplify the colour of a product and its outcome is to prevent recycling. But in many cases, the functionality of the packaging, for example, specific moisture and gas barriers, or printing and sealing techniques, has real benefits in terms of reducing food waste, but they also mean that the packaging can't be recycled because they are multilayer films that are virtually impossible and prohibitively expensive to separate.

Fruits and vegetables are egregious examples of excess plastic because they already come in a protective skin. Bananas are a great example of packaging design, individuals can be snapped off from a bigger pack, they have oriented properties so that the skin splits lengthways to expose the product, and they are truly biodegradable. The packaging serves no practical purpose yet less than 20% of supermarket fresh fruits and vegetables are offered "loose".

A tale of two cucumbers provides a counterfactual example, the behaviour of unwrapped versus shrink-wrapped (Manjunatha & Anurag, 2014). Leave them in the fridge. What happens? For a couple of days, both will look OK, but the unwrapped loses water faster than the shrink-wrapped cucumber and after four days will become wrinkled and flaccid. Two days later, the unwrapped cucumber will show obvious signs of (microbial) decay; a day later, a mushy mess is leaking fluid, whilst the wrapped cucumber is still crisp and will remain so for another week. Rolling the wrapping down as it is used means very little waste. In this case (and others like soft fruits), the packaging can prevent spoiling and wastage. A recent study found that the environmental benefit of food waste reduction due to plastic wrapping the cucumbers was found to be 4.9 times higher than the negative environmental impact due to the packaging itself (Shrivastava *et al.*, 2021). Put another way, each cucumber that has to be thrown away has the equivalent environmental impact of 93 plastic cucumber wraps.

The shelf life of pre-prepared, peeled fruits is normally much less than its unprepared counterpart (Barrios *et al.*, 2014). Orange segments (PE tray, modified atmosphere and PET sealing film) last about four days, whereas a whole orange can last for six months. Compare the environmental lifetime of orange peel and polyethylene — for the convenience of not peeling an orange! Oranges may seem like a strange candidate for peeling, but in reality, it's no different from bagged salads, pomegranate seeds or baby carrots. Consumers are waking up to the egregious nature of these packaged products, e.g., the cauliflower steak that was pulled because of consumer reaction (*BBC News*, 2018).

But it's not that simple a question though. One person's lazy and profligate behaviour is another person's lifesaver; pre-packed fruits and vegetables enable disabled people to access fresh food otherwise prevented by reduced mobility/dexterity, and we must always consider social as well as environmental sustainability when making decisions.

3. The Future of Plastics

3.1. *What can we do about it?*

So what can we do to reduce single-use plastics? Change the way we live, be mindful and make evidenced-based decisions. Let's use a typical academic's lunch — a falafel wrap, a banana and a bottle of water — and the choices made. Social media often advocates reusing aluminium foil rather than cling film for the wrap. But foil has to be re-used nearly 200 times to have a lower carbon footprint than cling film (5 g of Al versus 0.2 g of cling film, at six times more embedded energy and nine times more GHG per gram). Compare this to a reusable plastic product — a thick PE film with a velcro closure made from 14 g of the same material as the cling film — this only needs to be used 70 times to get ahead (on GHG emissions) of using new cling film every time. Moreover, there is no daily cling film, or weekly Al foil, going to landfills. Obviously, the banana needs nothing. But the water? Reuse a PET bottle (Greenwood *et al.*, 2021). The trouble is it's not as convenient!

3.2. *What are the alternatives?*

There are bio-based and biodegradable plastics available, but these are often "greenwash". At present, the least environmentally impactful plastic is made from oil, and the PET bottle market provides the worst examples of greenwash. Fully bio-based PET is never better than petrochemical PET because geology (time, heat and pressure) has already done nearly half the processing from biomass to plastics. LCAs show that irrespective of the source of the T (terephthalic acid) — corn, non-food crops or agricultural waste — the energy inputs and GHG emissions are always higher (Chen *et al.*, 2016). Until we stop burning fossil fuels for energy, that will remain to be the case, and the circular economy will have to wait for renewable energy to take over.

When bioplastics degrade, the predominant products are carbon dioxide and water. There will be a small amount of carbon left in biomass in the compost, but the quantity and environmental impacts of this are varied depending on the composting environment and are hard to quantify. If the benefit to the compost/soil is small or non-existent, then composting is akin to low-temperature incineration. Considerable progress needs to be made on understanding the life cycle impacts of biodegradable plastics to be able to accurately forecast their future use (Bhagwat *et al.*, 2020).

If biodegradable plastics end up in the plastic recycling streams (used for HDPE, PET, etc.), they cause contamination and decrease the value and use of the recyclate, sometimes to the point that it is no longer usable. In the short-medium term, biodegradable plastics make most sense for food packaging; once food waste collection is mainstream, the packaging can be disposed of with the food waste and they can be composted together (Nandakumar *et al.*, 2021). This avoids contamination of other waste streams, but it will be really important to ensure that users understand and engage with the system and that packaging is disposed of in the appropriate way.

Many biodegradable plastics are not degradable in water, so if they end up in the ocean, they will remain there for a long time. If they become litter on land, they will degrade but often slowly, and there is nothing to stop them from harming wildlife in the interim before full degradation occurs (Lamberti *et al.*, 2020). The key here is for plastics to not become litter in the first place — we need to give this incredibly versatile and durable material value so that people keep hold of it. In an ideal world, plastic will be reused again, and again, and again. Only when it is not possible to be reused any more should it be recycled or sent to waste. This is the best way to decrease the carbon footprint of the products. It requires a change of mindset of both producers and citizens (not consumers, our linguistics colleagues explained that consumers have a right to consume, whereas citizens have a responsibility). Producers of plastic products must design for reuse and for end-of-life, not just for use, as is most often done currently. Customers must accept a new style of living, and prioritise the environment over convenience.

Bioplastics could also be a route to carbon sequestration (Ryan & Rothman, 2021). Photosynthesis remains the most efficient Negative Emissions Technology available to us and by a considerable margin. Where Mother Nature lets us down, however, is at the end of the cycle when plants die and decompose, and the carbon dioxide they temporarily stored, and even more potent greenhouse gases, like methane, are released back into the atmosphere. We could make durable (i.e., non-degradable) bioplastics from photosynthetic biomass thereby taking CO_2 out of the atmosphere. If that bioplastic is sufficiently durable, is used again and again, then buried and not allowed to decompose, we would sequester carbon. In fact, if we converted all the current 300 Mt of annual plastic production to non-degradable bioplastics, using 100% renewable energy and agricultural waste as the feedstock, we would be able to remove billion tonnes (Bt) of CO_2 from the atmosphere every year.

That's where science and technology can come in. At present, with crops for humans and livestock, we use maybe 20% of a plant, possibly even less, the rest winds up as agricultural waste. To prevent this agricultural waste from continuing to contribute to climate change, we can simply ferment it into ethanol, dehydrate the ethanol into ethylene, and convert the ethylene to polyethylene and via existing petrochemical technology all the other plastic feedstocks. We could produce as much plastic as we like, use it, and bury it, flipping the long-lasting nature of polymer waste on its head as a benefit, to permanently sequester carbon from the atmosphere.

The only reason we aren't doing this already is the process comes with a carbon cost, but it doesn't have to. All that needs to happen for this model to actually work, and contribute to the fight against climate change, is to defossilise the energy system and use renewable energy to power the process, and we're not that far away. Plastics in curated landfills would stop being an obstacle in the way of sustainability and will instead be considered a great tool for it.

4. Summary and Outlook

There is currently >7 Bt of plastic in the Earth System, rising to 40 by 2050. About 10 Mt of plastics reach the oceans every year; 12 Bt are projected to be in landfills by 2050. This book looks at what happens when they are released, as the flux of plastics to surface waters, coastlines and oceans is amplified, with significant environmental, social and economic costs.

As we've discussed, plastics are ubiquitous and have transformed our modern world, and we cannot just wish them away. With unique and useful properties, plastics are a highly adaptable class of materials exploited for an ever-widening range of applications. Moreover, they are often the best material for the job and have the lowest environmental impact provided that they are properly dealt with at the end-of-life. The problem is that it is both difficult and expensive to do so. Recycling plastics from post-consumer waste is essentially uneconomic because they need to be separated to be reprocessed — it is cheaper for the manufacturer to make new disposable plastic packaging because costs are borne by society, in general, and not the packaging producer or waste disposer. In the global north, plastic ends up in the environment through leakages from waste collection, recycling and disposal, driven by the growth in single-use plastic, with a significant amount of egregious litter and informal disposal. In the global south, there is often no formal waste collection, recycling or disposal and litter is worse. For plastics that have residual economic value, there are systems to capture and reuse or recycle. However, in most cases, post-consumer waste polymers have no intrinsic economic value so if there are no policies in place, and consequently no collection and recycling, they are destined to end up in the sea.

In some situations, single-use plastic may be the only option, e.g., dentures and sanitary products, and they should clearly be recovered. It is obvious, however, that we should make the best use of both our fossil and renewable resources, so the move towards a zero-waste, circular economy is essential. This move should not be driven by ideology though; it must be based on evidence. We should develop new business models for reuse and recycling based on the idea that "you can buy the product but rent the packaging" rather than legitimise a throw-away culture with degradable polymers. Moreover, we need to work out how to deliver this at a global scale developing policies and practices in the context of local cultures. A new plastic system that prioritises reuse, and then recycling, of durable materials requires a step change in behaviour for nearly everyone.

Current understanding of the pathways and fate of plastic in the environment is poorly constrained, with only fragmentary evidence restricted to parts of the cycle. The problem is now part of public consciousness and is on the political agenda nationally and internationally, so natural and environmental scientists must work together to understand the mechanisms of plastic degradation and project the ecological consequences of these processes over decades — centuries. These scientists must also work with sociologists, economists, political scientists and psychologists to understand human behaviour and promote change at individual, societal and cultural levels. This change can only be enabled by working directly with stakeholders — nationally and globally — from across the supply chain. Circular transformations require co-developed solutions with plastic producers, processors and recyclers, cross-sector businesses, environmental groups, waste management companies, NGOs, third sector, policymakers and other citizens and communities. Only by adopting a system-based, multidisciplinary approach can we implement effective and lasting solutions to plastic in the environment and minimise the impact from next-generation plastics.

References

Barrios, S., De Aceredo, A., Chao, G., De Armas, V., Ares, G., Martín, A., Soubes, M., & Lema, P. (2014). Passive modified atmosphere packaging extends shelf life of enzymatically and vacuum-peeled ready-to-eat valencia orange segments. *Journal of Food Quality*, *37*(2), 135–147. https://doi.org/10.1111/JFQ.12074.

BBC News. (2018). Marks & Spencer to stop selling "cauliflower steak". Retrieved from https://www.bbc.co.uk/news/uk-42639479.

Bhagwat, G., Gray, K., Wilson, S. P., Muniyasamy, S., Vincent, S. G. T., Bush, R., & Palanisami, T. (2020). Benchmarking bioplastics: A natural step towards a sustainable future. *Journal of Polymers and the Environment*, *28*(12), 3055–3075. https://doi.org/10.1007/S10924-020-01830-8.

Billings, A., Jones, K. C., Pereira, M. G., & Spurgeon, D. J. (2021). Plasticisers in the terrestrial environment: Sources, occurrence and fate. *Environmental Chemistry*, *18*(3), 111–130. https://doi.org/10.1071/EN21033.

Błażejewski, T., Walker, S. R., Muazu, R. I., & Rothman, R. H. (2021). Reimagining the milk supply chain: Reusable vessels for bulk delivery. *Sustainable Production and Consumption*, *27*, 1030–1046. https://doi.org/10.1016/J.SPC.2021.02.030.

Bucknall, D. G. (2020). Plastics as a materials system in a circular economy. *Philosophical Transactions of the Royal Society A*, 378(2176), 20190268.

Chen, L., Pelton, R. E. O., & Smith, T. M. (2016). Comparative life cycle assessment of fossil and bio-based polyethylene terephthalate (PET) bottles.

Journal of Cleaner Production, *137*, 667–676. https://doi.org/10.1016/J.JCLEPRO.2016.07.094.

Cowie, J. M. G. & Arrighi, V. (Valeria). (2007). *Polymers Chemistry and Physics of Modern Materials*, 3rd edn. Boca Raton: CRC Press.

Crespy, D., Bozonnet, M., & Meier, M. (2008). 100 years of bakelite, the material of a 1000 uses. *Angewandte Chemie International Edition*, *47*(18), 3322–3328. https://doi.org/10.1002/ANIE.200704281.

Creton, C. (2017). Molecular stitches for enhanced recycling of packaging. *Science (New York, N.Y.)*, *355*(6327), 797–798. https://doi.org/10.1126/SCIENCE.AAM5803.

del Amo, A. (2000). A brief history of plasticized cellulose nitrate or celuloid. *Journal of Film Preservation*, 60/61, 41–45.

Dhall, R. K., Sharma, S. R., & Mahajan, B. V. C. (2012). Effect of shrink wrap packaging for maintaining quality of cucumber during storage. *Journal of Food Science and Technology*, *49*(4), 495. https://doi.org/10.1007/S13197-011-0284-5.

Eagan, J. M., Xu, J., Di Girolamo, R., Thurber, C. M., Macosko, C. W., LaPointe, A. M., Bates, F. S., & Coates, G. W. (2017). Combining polyethylene and polypropylene: Enhanced performance with PE/ *i* PP multiblock polymers. *Science*, *355*(6327), 814–816. https://doi.org/10.1126/science.aah5744.

Ebnesajjad, S. (2016). *Expanded PTFE Applications Handbook: Technology, Manufacturing and Applications*. Amsterdam: Elsevier Inc.

Eriksen, M. K., Christiansen, J. D., Daugaard, A. E., & Astrup, T. F. (2019). Closing the loop for PET, PE and PP waste from households: Influence of material properties and product design for plastic recycling. *Waste Management*, *96*, 75–85. https://doi.org/10.1016/J.WASMAN.2019.07.005.

Freinkel, S. (2011). A brief history of plastic's conquest of the world. *Scientific American*. Retrieved from https://www.scientificamerican.com/article/a-brief-history-of-plastic-world-conquest/06/05/2022.

Frey, H. & Johann, T. (2019). Celebrating 100 years of "polymer science": Hermann Staudinger's 1920 manifesto. *Polymer Chemistry*, *11*(1), 8–14. https://doi.org/10.1039/C9PY90161B.

Gaines, A. (2002). *Wallace Carothers and the Story of DuPont Nylon*. Pembroke Park: Mitchell Lane.

Geyer, R., Jambeck, J. R., & Law, K. L. (2017). Production, use, and fate of all plastics ever made. *Science Advances*, *3*(7), e1700782. https://doi.org/10.1126/sciadv.1700782.

Greenwood, S. C., Walker, S., Baird, H. M., Parsons, R., Mehl, S., Webb, T. L., Slark, A. T., Ryan, A. J., & Rothman, R. H. (2021). Many happy returns: Combining insights from the environmental and behavioural sciences to understand what is required to make reusable packaging mainstream.

Sustainable Production and Consumption, *27*, 1688–1702. https://doi.org/10.1016/J.SPC.2021.03.022.

Han, J. W., Ruiz-Garcia, L., Qian, J. P., & Yang, X. T. (2018). Food packaging: A comprehensive review and future trends. *Comprehensive Reviews in Food Science and Food Safety*, *17*(4), 860–877. https://doi.org/10.1111/1541-4337.12343.

Hawkins, G., Potter, E., & Race, K. (Eds.). (2015). *Plastic Water. The Social and Material Life of Bottled Water. Plastic Water*. Cambridge: MIT Press.

Hermes, M. E. (1996). *Enough for One Lifetime: Wallace Carothers, Inventor of Nylon*. American Chemical Society and the Chemical Heritage Foundation.

Jenkins, M. (2003). *Materials in Sports Equipment: Volume 1. Materials in Sports Equipment*, 1st edn. Amsterdam: Elsevier Inc. https://doi.org/10.1201/9781439824108.

Korley, L. S. T. J., Epps, T. H., Helms, B. A., & Ryan, A. J. (2021). Toward polymer upcycling-adding value and tackling circularity. *Science (New York, N.Y.)*, *373*(6550), 66–69. https://doi.org/10.1126/SCIENCE.ABG4503.

Lamberti, F. M., Román-Ramírez, L. A., & Wood, J. (2020). Recycling of bioplastics: Routes and benefits. *Journal of Polymers and the Environment*, *28*(10), 2551–2571. https://doi.org/10.1007/S10924-020-01795-8.

Lebreton, L., Slat, B., Ferrari, F., Sainte-Rose, B., Aitken, J., Marthouse, R., Hajbane,S., Cunsolo, S., Schwarz, A., Levivier, A., Noble, K., Debeljak, P., Maral, H., Schoeneich-Argent, R., Brambini, R., & Reisser, J. (2018). Evidence that the Great Pacific Garbage Patch is rapidly accumulating plastic. *Scientific Reports*, *8*(1), 1–15. https://doi.org/10.1038/s41598-018-22939-w.

Life Magazine. (1955). Throwaway living: Disposable items cut down household chores. *Life* (August).

Loadman, J. (2005). *Tears of the Tree: The Story of Rubber*. Oxford: OUP Oxford.

Lodge, T. & Hiemenz, P. C. (2021). *Polymer Chemistry*. Boca Raton: CRC Press.

Lonca, G., Lesage, P., Majeau-Bettez, G., Bernard, S., & Margni, M. (2020). Assessing scaling effects of circular economy strategies: A case study on plastic bottle closed-loop recycling in the USA PET market. *Resources, Conservation and Recycling*, *162*, 105013. https://doi.org/10.1016/J.RESCONREC.2020.105013.

Mackendrick, A. (1951). *The Man in the White Suit*. London: Ealing Studios.

Manjunatha, M. & Anurag, R. K. (2014). Effect of modified atmosphere packaging and storage conditions on quality characteristics of cucumber. *Journal of Food Science and Technology*, *51*(11), 3470. https://doi.org/10.1007/S13197-012-0840-7.

Matthies, P. & Seydl, W. F. (1986). History and development of nylon 6. In *High Performance Polymers: Their Origin and Development*, pp. 39–53. Dordrecht: Springer. https://doi.org/10.1007/978-94-011-7073-4_4.

McKevitt, S. (2018). *The Persuasion Industries: The Making of Modern Britain.* Oxford: Oxford University Press. https://doi.org/10.1093/oso/9780198821700.001.0001.

McMillan, F. M. (1979). *The Chain Straighteners. The Chain Straighteners*. London, UK: Palgrave Macmillan. https://doi.org/10.1007/978-1-349-04430-6.

Mercelis, J. (2012). Leo Baekeland's transatlantic struggle for bakelite: Patenting inside and outside of America. *Technology and Culture*, *53*(2), 366–400. https://doi.org/10.1353/TECH.2012.0067.

Morton, M. (1999). *Rubber Technology*. Berlin: Springer Science & Business Media. https://doi.org/10.1007/978-94-017-2925-3.

Mosley, A. P. (2017). Chapter 16 — Acrylic plastics. In Gilbert, M. (Ed.), *Brydson's Plastics Materials*, 8th edn., pp. 441–456. Oxford: Butterworth-Heinemann.

Nandakumar, A., Chuah, J. A., & Sudesh, K. (2021). Bioplastics: A boon or bane? *Renewable and Sustainable Energy Reviews*, *147*, 111237. https://doi.org/10.1016/J.RSER.2021.111237.

NobelPrize.org. (1953). Hermann Staudinger — Biographical. Retrieved from https://www.nobelprize.org/prizes/chemistry/1953/staudinger/biographical/.

Nowlin, T. E. (2014). *Business and Technology of the Global Polyethylene Industry: An In-depth Look at the History, Technology, Catalysts, and Modern Commercial Manufacture of Polyethylene and Its Products.* Hoboken: John Wiley & Sons.

Parker-Jurd, F. N. F., Napper, I. E., Abbott, G. D., Hann, S., & Thompson, R. C. (2021). Quantifying the release of tyre wear particles to the marine environment via multiple pathways. *Marine Pollution Bulletin*, *172*, 112897. https://doi.org/10.1016/J.MARPOLBUL.2021.112897.

Pilati, F. & Toselli, M. (2002). Chapter 10. In La Mantia, F. P. (Ed.), *Handbook of Plastics Recycling*, pp. 297–336. Shrewsbury UK: Smithers Rapra Technology Ltd.

Rasmussen, S. C. (2021). From parkesine to celluloid: The birth of organic plastics. *Angewandte Chemie International Edition*, *60*(15), 8012–8016. https://doi.org/10.1002/anie.202015095.

Ryan, A. J., & Rothman, R. H. (2021). Engineering chemistry to meet COP26 targets. *Nature Reviews Chemistry*, *6*, 1–3.

Scheirs, J. & Long, T. E. (Eds.). (2004). *Modern Polyesters: Chemistry and Technology of Polyesters and Copolyesters*. Hoboken: John Wiley & Sons, Ltd. https://doi.org/10.1002/0470090685.

Scheirs, J. & Priddy, D. (Eds.). (2003). *Modern Styrenic Polymers: Polystyrenes and Styrenic Copolymers*, Vol. 6. Hoboken: John Wiley & Sons. Retrieved from https://www.wiley.com/en-gb/modern+Styrenic+Polymers%3A+Polystyrenes+and+Styrenic+Copolymers-p-9780471497523.

Seymour, R. B. & Kauffman, G. B. (1992). The rise and fall of celluloid. *Journal of Chemical Education*, *69*(4), 311–314. https://doi.org/10.1021/ED069P311.

Seymour, R. B., Mark, H. F., Pauling, L., Fisher, C. H., Stahl, G. A., Sperling, L. H., Marvel, C. S., & Carraher, C. E. (1989). Otto Bayer father of polyurethanes. In *Pioneers in Polymer Science*, pp. 213–219. Dordrecht: Springer. https://doi.org/10.1007/978-94-009-2407-9_22.

Shrivastava, C., Crenna, E., Schudel, S., Shoji, K., Onwude, D., Hischier, R., & Defraeye, T. (2021). To wrap or to not wrap cucumbers? *engrXiv*. https://doi.org/10.31224/OSF.IO/DYX9B.

Smith, J. K. (1985). The ten-year invention: Neoprene and DuPont research, 1930-1939. *Technology and Culture*, *26*(1), 34. https://doi.org/10.2307/3104528.

Staudinger, H. (1920). Über polymerisation. *Berichte Der Deutschen Chemischen Gesellschaft (A and B Series)*, *53*(6), 1073–1085. https://doi.org/10.1002/CBER.19200530627.

Thompson, R. C., Olson, Y., Mitchell, R. P., Davis, A., Rowland, S. J., John, A. W. G., McGonigle, D., & Russell, A. E. (2004). Lost at sea: Where is all the plastic? *Science*, *304*(5672), 838. https://doi.org/10.1126/SCIENCE.1094559/SUPPL_FILE/THOMPSON.SOM.PDF.

Titow, W. V. (1984). *PVC Technology*. https://doi.org/10.1007/978-94-009-5614-8.

Wikström, F., Williams, H., Verghese, K., & Clune, S. (2014). The influence of packaging attributes on consumer behaviour in food-packaging life cycle assessment studies — A neglected topic. *Journal of Cleaner Production*, *73*, 100–108. https://doi.org/10.1016/J.JCLEPRO.2013.10.042.

Wilkinson, A. N. & Ryan, A. J. (1999). *Polymer Processing and Structure Development*. Amsterdam: Kluwer Academic Publishers.

Young, R. J. & Lovell, P. A. (2011). *Introduction to Polymers*, 3rd edn. Boca Raton, London, New York: Taylor and Francis.

https://doi.org/10.1142/9789811259111_0003

Chapter 2

Overcoming Oceans of Ignorance: What You Should Know About Plastic Waste Before It Enters the Sea

Yvette Mellink*, Louise Schreyers*, Rahel Hauk*, Rose Pinto*, Khoa Thi*,†, Kryss Waldschläger* and Tim van Emmerik*,‡

**Hydrology and Quantitative Water Management Group, Wageningen University, Wageningen, The Netherlands*

†Faculty of Water Resources, Hanoi University of Natural Resources and Environment, Hanoi, Vietnam

‡tim.vanemmerik@wur.nl

1. Introduction

Ocean plastic pollution is an emerging threat to marine life and ecosystem health, in general. A rapidly growing amount of research focuses on exploring the transport pathways of plastic pollution in the world's oceans and assessing its associated negative impacts. For example, observation-based and modelling studies quantified concentrations of floating plastics in ocean gyres (Lebreton *et al.*, 2018) and revealed the fate of marine plastics (Olivelli *et al.*, 2020). Studies that aim to forecast or hindcast the trajectories of plastics in the ocean benefit from defining and characterizing plastic sources and defining exact entry points (Kaandorp *et al.*, 2020; Onink *et al.*, 2021). Effect and risk assessments rely on an accurate description of the type of plastic waste in various compartments of ecosystems.

In everyday practice, however, the complex dynamics of environmental plastics are often summarised in media-savvy facts. These include assumptions that 80% of marine plastics are land-based (Derraik, 2002), that 10 rivers cause 90–95% of total river plastic export into the ocean (Schmidt *et al.*, 2017), and that eight megatonnes of plastic are entering the ocean every year (Jambeck *et al.*, 2015). Most of these studies simplify plastics into a single material, ignoring the highly varying properties of the plastic items. Plastic pollution is, however, a non-trivial and complex challenge that spans across national borders, temporal and spatial scales, and environmental compartments. Plastic pollution comes in many different shapes, colours, polymers and sizes, which determine the interaction between plastics and the environment (van Emmerik, 2021; Schwarz *et al.*, 2019).

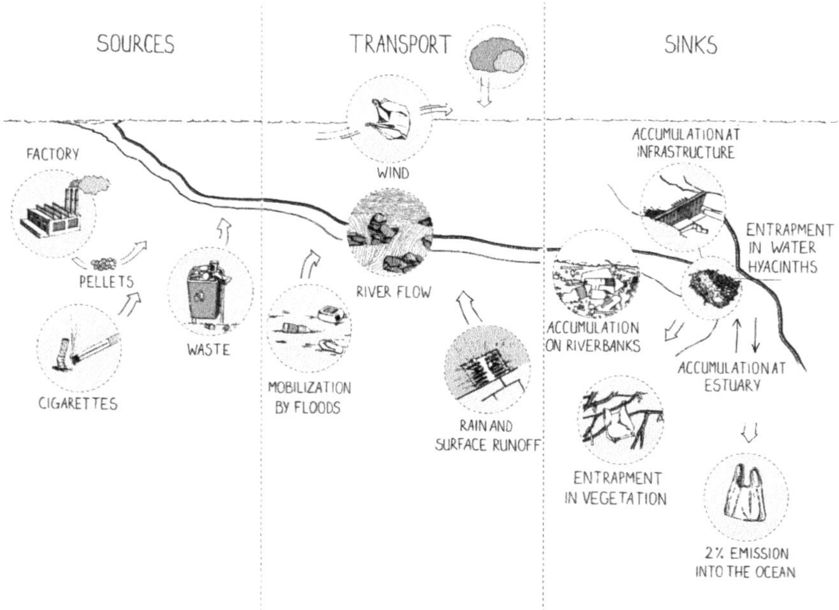

Figure 1. Schematic overview of the sources, transport processes and sinks of terrestrial and riverine plastics discussed in this chapter. Plastic waste enters rivers through, e.g., mismanaged plastic waste, leakage of pellets and cigarette butts. Transport from land to rivers and through rivers is partly driven by wind, rain-induced surface runoff and river flow dynamics. Sinks include riverbanks, vegetation, infrastructure, and, for some plastic waste, eventually the ocean.

Note: Artwork by Cher van den Eng.

In this chapter, we aim to provide a more nuanced view on what is behind the commonly used numbers and demonstrate that these can come with great uncertainties.

We demonstrate the complexity of land-based plastics by looking into (1) their main sources, (2) the transport processes from land to river to sea and (3) their sinks before they reach the ocean (Figure 1). The relationship between these three components largely depends on time and space, and we will illustrate this through several examples from around the world. Only through a comprehensive understanding of the sources, transport and fate of land-based plastics can we accurately estimate riverine plastic emissions into the oceans, which is crucial for the parameterisation of oceanic models on plastic pollution. Accordingly, we will particularly elaborate on the land-based plastics that do not make it into the ocean. At both local and global levels, it has been estimated that less than 2% of the total amount of plastic waste that leaks into the environment is eventually emitted into the ocean (Meijer *et al.*, 2021). But what happens to the rest? To answer this question, we will have a closer look at plastics from land-based sources before they enter the ocean and discuss the fate of the plastics that do not make it into the ocean.

2. Where Does Land-Based Plastic Waste Come from?

2.1. *How does plastic end up in the environment?*

Typically, it is believed that plastic waste mainly ends up in the environment due to the deliberate, improper disposal, i.e., littering, of single-use disposable plastic products (Hartley *et al.*, 2018; Soares *et al.*, 2021). Whereby, littering includes both the small quantities disposed of by consumers as well as the larger quantities dumped illegally. However, littering is not the only way for plastics to end up in the environment. A study by Lechthaler *et al.* (2020) provides an overview of the main entry paths for land-based macroplastics into the environment. These paths include, apart from littering, industry, waste management infrastructure and systems; natural extreme events and river transport account for large leakages of plastics into the environment as well. Plastic waste that comes from the industry is, for example, plastic industrial packaging materials from construction work and from industrial facilities but also includes (fragments of) plastic mulch and silage films used by the agricultural sector (Piehl *et al.*,

2018). The destructive forces that occur during natural extreme events, such as floods, tsunamis, hurricanes and storms damage infrastructure and can lead to an input of (fragments of) plastics from inventories or (unprotected) open landfills into the environment. Plastic items can also enter the environment via waste management infrastructure, for example, surface runoff in urban areas carries plastic waste into the sewage system or the disposal of sanitary napkins at sanitary facilities. Additionally, plastic waste can enter the environment in an accident during the transport of municipal solid waste to landfills or leakage from overfull landfills (Lechthaler *et al.*, 2020). A study by Ryberg *et al.* (2019) made a global estimation of the losses of plastics to the environment across the entire plastic value chain by using existing literature and databases coupled with improved and additional methodological modelling of the losses. They found that the main macroplastic loss source was mismanaged municipal solid waste and that of the 6.2 Mt of macroplastics globally lost to the environment in 2015, 4.1 Mt came from mismanaged municipal solid waste.

2.2. *Everywhere and overlooked: Plastic pellets and cigarette butts*

As discussed above, land-based plastics can enter the environment in many different ways: as lost material that would otherwise have had a purpose (losses during plastic production, lost cargo and incorrectly stored construction materials) and as waste that is intentionally (littering) or unintentionally (improper waste management and waste from municipal landfills) introduced into the environment (Lechthaler *et al.*, 2020). As there are already numerous review papers on those sources and pathways (Lechthaler *et al.*, 2020; Li *et al.*, 2016; Geyer *et al.*, 2017), we here highlight two regularly overlooked sources: plastic pellets and cigarette butts.

2.2.1. *Plastic pellets*

Plastic pellets, also called plastic nurdles, are easy to transport and are a common intermediate product in the plastic production chain. Pellets are small in size, typically 2–5 mm (Karlsson *et al.*, 2018), and can easily leak into the environment during each step of production, handling, transport, processing and recycling (PlasticsEurope 2018, pp. 12–13) (Figure 2).

Pellets frequently enter the environment during transport via cargo ships, where sometimes entire containers filled with pellets are lost and huge amounts of pellets are spilled at once. Plastic pellets contribute to global plastic pollution and are found in aquatic ecosystems around the globe (Mani *et al.*, 2015; Takada *et al.*, 2005; Fidra, 2021). For the UK, an annual pellet loss between 105 and 1,054 metric tonnes is estimated, which equates to a total of 5–53 billion pellets found (Cole & Sherrington, 2016). For comparison, Meijer *et al.* (2021) estimated an annual amount of mismanaged plastic waste of 29,914 metric tonnes for the UK. For the entire European Union, a loss of 16,888–167,431 metric tonnes of plastic pellets per year is estimated (Hann *et al.*, 2018). Two notable pellet spill incidents occurred in 2020: one whereby about 25 metric tonnes of pellets entered the Mississippi River in New Orleans (NOLA, 2020) and another whereby about 13.3 metric tonnes of pellets were lost in the North Sea (Kystverket, 2020).

2.2.2. *Cigarette butts*

Cigarette butts contain a filter that must reduce inhalation of harmful elements by smokers. They are regularly discarded after use and have become one of the most common waste items found in the environment (Novotny & Slaughter, 2014; Torkashvand *et al.*, 2020; Asensio-Montesinos *et al.*, 2019; Bonanomi *et al.*, 2020). Some assumptions indicate that 4.5 trillion cigarette butts are released into the environment each year, which corresponds to approximately 80% of the annually consumed cigarettes (Novotny & Slaughter, 2014) (Figure 1). Based on surveys, 55.7–75% of smokers dump their cigarette butts on the ground

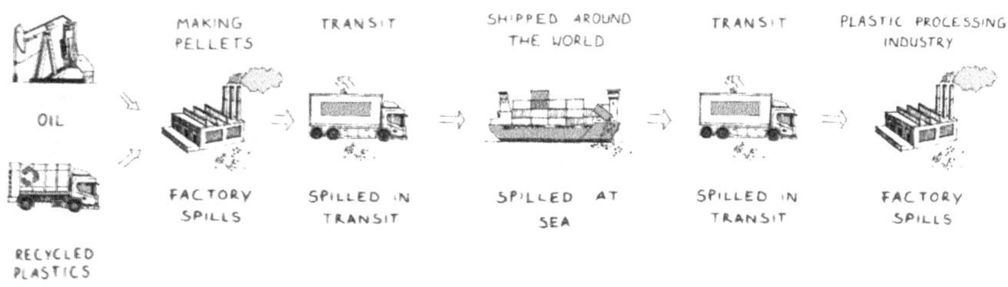

Figure 2. Plastic pellets are spilled into the environment during each step of production, handling, transport and processing.

Note: Artwork by Cher van den Eng.

or out of their car window (Rath *et al.*, 2012; Rahman *et al.*, 2020). This behaviour is highly influenced by the common misconception that cigarette butts do not cause any harm to the environment (Araújo & Costa, 2019; Rath *et al.*, 2012). However, this is not the case as most cigarette butts contain filters made out of the modified biopolymer-based plastic cellulose acetate (CA), which does not easily degrade in the environment due to the combined influences of plastic additives, the physical and chemical structure of the filters, and the occurring environmental conditions such as light irradiation and oxygen levels (Yadav & Hakkarainen, 2021). Degradation rates in the environment appear to be low, with weight losses between 20% and 30% after two years of ageing under laboratory and field conditions (Bonanomi *et al.*, 2015, 2020). Influenced by environmental forces, cigarette butts can disintegrate into microfibres with an average length of 20 mm (Torkashvand *et al.*, 2020). One filter consists of up to 15,000 individual fibres leading to a release of up to 0.3 million metric tonnes of microfibres only by smoked filters every year (Belzagui *et al.*, 2021). Due to the high difficulty of monitoring microfibres in the aquatic environment, the distribution and fate of those fibres are currently unknown (Woodall *et al.*, 2014; Suaria *et al.*, 2020). In addition to the avoidable input of plastics into the environment, these filters and fibres are heavily polluted with over 4,000 chemicals such as nicotine and heavy metals that are known to be carcinogenic and toxic. According to Slaughter *et al.* (2011), just one cigarette butt per litre of water leaches out enough toxins to kill half of the fresh or saltwater fish exposed to them.

2.3. *Ship away: Out of sight out of mind*

South and Southeast Asia are frequently designated as the largest plastic emitters due to poor waste management and low recycling rates (d'Ambrières, 2019). What is often overlooked is the contribution of high-income countries to global plastic emissions into the ocean as most estimates on global plastic emissions neglect solid waste exports from high-income countries to East and Asia-Pacific countries.

Previous studies estimated that China and Indonesia are the biggest contributors to marine plastic waste (Jambeck *et al.*, 2015) and that the United States only ranked 20th in global plastic emissions. However, other studies draw a different picture, with the US ranking as the world's first plastic producer generating 42 million metric tonnes of plastic waste in 2016 (Law *et al.*, 2020). The per capita contributions of the US and European Union members are estimated to be the highest globally, ultimately highlighting that the contributions of high-income countries have been substantially underestimated. One reason for this is that previous material flow analyses overlooked the export of (plastic) solid waste to low-income countries. Brooks *et al.* (2018) showed that high-income countries, especially North-American and European countries, have been the primary exporters of plastic waste since reporting began in 1988, contributing to 87% of all exports. East and Asia-Pacific countries, on the other hand, have been the main receivers of plastic waste, accounting for 75% of global plastic waste imports since 1988 (Brooks *et al.*, 2018). Within this region, China and Hong Kong combined imported as much as 72.4% of plastic waste although it should be noted that Hong Kong functions mainly as a port of entry for waste in China (Brooks *et al.*, 2018).

In addition to plastic exports, illegal dumping and poor waste management in high-income countries were not considered in previous estimates even though these practices are widespread (Law *et al.*, 2020). Jambeck *et al.* (2015), for instance, considered that 100% of the waste is properly managed in the US. Integrating field estimates, in order to account for waste mismanagement, illegal dumping practices and plastic exports in the US, have led to better assessments of the contribution of this country to global plastic emissions (Law *et al.*, 2020). Quantifying European countries' contribution to plastic emissions has also proven challenging due to complex trade flows. Countries with Europe's largest harbours, such as the Netherlands, Belgium, Germany and the UK, serve as plastic transit hubs within and outside Europe and thus register among the highest ocean plastic emissions per capita (Bishop *et al.*, 2020). This complexity in plastic export dynamics makes it ultimately difficult to trace back plastic emissions to specific countries of origin.

Future monitoring efforts should focus on developing methods to trace the movement of waste between nations. This is especially of interest now given that China has completely banned global waste import since 2017 and it remains unknown how global waste exports are affected by these policies (Brooks *et al.*, 2018).

3. Plastics on the Move

3.1. *Plastic transport processes: Gone with the wind or go with the flow?*

It is presumed that the transport pathways of macroplastic waste from source to sink are largely controlled by hydrometeorological variables, e.g., precipitation, water flow velocities, water discharges and wind velocities (Figure 1). Here, we present a short overview of our current state of knowledge on how and which (hydrometeorological) processes govern the transport of macroplastics through freshwater and terrestrial environments.

A great part of the journey of macroplastics runs through freshwater environments. Rivers have been assigned as the main transporters of plastics towards the oceans (Lebreton *et al.*, 2017; Schmidt *et al.*, 2017; van Emmerik & Schwarz, 2020). Hydrological variables such as river discharge, turbulence intensity, lateral (perpendicular to the river flow), longitudinal (parallel to the river flow) and vertical flow velocities largely control the transport and sinking of plastics within river channels. For example, several studies have demonstrated that a positive relationship exists between plastic transport emissions and river discharge (Schmidt *et al.*, 2017; van Emmerik *et al.*, 2019a; Wagner *et al.*, 2019), which is regularly accompanied by the exhibition of seasonality in river plastic emissions (Lebreton *et al.*, 2017; van Emmerik *et al.*, 2019b). A study by Haberstroh *et al.* (2021) on the influence of flow dynamics on plastic transport fluxes revealed that calm river flow conditions result in mainly advective, i.e., longitudinal, plastic fluxes, whereas high river discharges and high turbulence intensities enhance cross-sectional mixing, which results in a three to tenfold increase in lateral and vertical plastic fluxes.

The majority of marine plastics have a land-based source (Jambeck *et al.*, 2015), yet it is the mobilisation and transport of plastics through terrestrial environments that remain largely understudied. To date, the hydrometeorological variables that are hypothesised to control the transport of plastics in terrestrial environments are wind and precipitation-driven surface runoff. Despite the lack of a strong empirical basis, it is assumed that the (probability of) transport of plastic increases when the wind velocity and/or surface runoff flux increases (Lebreton *et al.*, 2017; Meijer *et al.*, 2021; Roebroek *et al.*, 2021b). For example, the plastic transport model by Meijer *et al.* (2021) assumed that the probability of plastic mobilisation by wind linearly increases with wind speed, whereby a wind speed of 32.7 m/s was presumed to be sufficient to mobilise 100% of the plastic items. A study by Roebroek *et al.* (2021b) revealed that macro litter abundances found on riverbanks more often significantly correlate with wind speed than with either precipitation or river discharge. In addition, they found that wind-driven and precipitation-driven transport mechanisms operate on very different timescales, whereby predicting litter item abundances from wind speed is optimal on the scale of days, whereas predicting item abundances from precipitation shows an optimal time lag of months (Roebroek *et al.*, 2021b). However, the most striking conclusion drawn by Roebroek *et al.* (2021b) is that a very substantial part of the variability in macro litter abundances found on riverbanks is actually a product of chance and human behaviour rather than being driven by hydrometeorological variables such as precipitation, wind speed or water level. Therefore, we have to remain cautious when assigning wind and surface runoffs as the transport agents of plastic litter through terrestrial environments.

What is often forgotten is that there are many non-hydrometeorological variables that have a major influence on the transport of plastics by hydrometeorological variables. For example, although this has mainly been demonstrated for microplastics (<5 mm), material properties like density, size, shape, surface smoothness and polymer composition have a great impact on the transport and distribution of plastics in aquatic environments (i.e., afloat at the surface, suspended in the water column or on the seafloor/riverbed) (Kowalski *et al.*, 2016; Chubarenko *et al.*, 2016; Kooi *et al.*, 2016; Besseling *et al.*, 2017; Khatmullina & Isachenko, 2017). In addition, numerous studies have shown that biofouling, i.e., algal attachment on the surface of plastics, has a major influence on the sinking rate and, therefore, on the transport of plastics (Fazey & Ryan, 2016; Kaiser *et al.*, 2017; Kooi *et al.*, 2017; Lobelle *et al.*, 2021). For the transport of plastics through terrestrial environments, terrain characteristics, such as slope and vegetation cover, determine the friction and, therefore, the ease with which plastics can move. For example, plastics are more likely to be transported over large distances along steep (smoothly) paved surfaces compared to flat more densely vegetated areas (Meijer *et al.*, 2021).

Humans play a key role in plastic transport as well, both by their behaviour, e.g., through clean-ups and relocating plastics, and by the structures they build. Examples of these structures are dams, sluices and pumping stations which obstruct the river flow (velocity) and thereby the transport of plastics. Consequently, they often develop into plastic accumulation zones (Honingh *et al.*, 2020). As these structures are not always easily recognisable on satellite imagery, it is important to conduct fieldwork and map these (unforeseen) obstacles in order to comprise a comprehensive view on the ability of a river to transport plastics towards the ocean.

To summarise, our knowledge on how hydrometeorological variables control the transport of plastic has greatly expanded over the last decade(s). However, the focus has mainly been on aquatic environments. Therefore, the next step is to explore how exactly hydrometeorological variables control the transport of plastics through terrestrial environments. More specifically, there is a great need to (empirically) determine the relationship between wind velocity and plastic transport velocity, and surface runoff flux and transport velocity. Are these relationships linear, exponential or asymptotic? Does, for example, the plastic transport velocity continue to increase with increasing wind speed or does the plastic transport velocity approach a maximum at some point? What is the influence of terrain friction caused by (the combination of) terrain slope and land use (vegetation density) on these relationships between wind velocity and plastic transport velocity, and surface runoff flux and transport velocity? Are these relationships different for different types of plastics, e.g., plastic bottles, bags, cups, food wrappers and foils? Answering these types of questions will reveal the transport mechanisms and patterns of plastics through terrestrial systems, which is crucial for the development of effective, targeted and close to the (land-based) source strategies that mitigate and prevent plastic pollution.

3.2. *Fast and furious: Plastic mobilisation by floods*

When plastic waste enters the river system, its transport is assumed to be mainly governed by hydrological processes such as flow velocity and water discharge. Observational studies have confirmed that high discharge conditions have the capacity to significantly increase plastic fluxes into the oceans. Increased stormwater runoff following heavy rainfall events was found to convey higher loads of plastic in urban rivers when compared with dry season (Moore *et al.*, 2011). Similarly, plastic transport in the downstream Seine (France) increased up to a factor of 10 during higher discharge conditions (van Emmerik *et al.*, 2019a). Given the demonstrated role of high discharge conditions in increased plastic emissions, extreme events such as floods are also hypothesised to be major drivers of plastic inputs into the world's oceans (Figure 1). However, no fundamental understanding of plastic (re)mobilizing mechanisms driven by floods has been reached so far and this field remains exploratory.

Several studies have highlighted the role that flood events can play in transporting and remobilizing plastics in different compartments of a river system. Hurley *et al.* (2018) found that microplastic concentrations in river sediments decreased by approximately 64% following a flood event in 2015/2016 (Hurley *et al.*, 2018). This suggests that microplastics previously deposited and buried in river beds can be remobilised during flood events and thus be released into the ocean. Release of plastic from the riverbed mainly occurs when the active layer of the riverbed is destabilised and the plastic particles get resuspended into the water column (Ockelford *et al.*, 2020).

Plastic waste temporarily stored in other compartments of the river system can also be remobilised during flood events. Riverbanks, floodplains and tidal flats have been identified as major plastic sinks and deposition areas, and could thus release large quantities of plastics into the river channel when the floodwater overtops them (van Emmerik *et al.*, 2020b; Liro *et al.*, 2020). Nevertheless, contrasting mobilisation and deposition mechanisms may also occur and ultimately result in solid waste aggradation into floodplains and riverbanks rather than remobilisation during flood events. Moreover, higher flow velocities in the river channel compared to the river banks could lead to the transfer of particles from faster flow (channel) to the lower flow areas (riverbanks) (Ciszewski & Grygar, 2016). The magnitude and frequency of such re-deposition mechanisms from rivers onto land remain to be explored.

In urban settings, flood and historical rainfall events are associated with increased quantities of plastic loads into rivers. Upon inundation, plastic waste accumulated in streets and alongside roads quickly becomes flushed away into urban stormwater drainage systems which drain into nearby rivers. Similarly, the overflowing of canals and ditches likely increases plastic loads into river channels. On the other hand, the blocking of urban hydraulic infrastructure by plastic waste also has the capacity to exacerbate flood risks (Honingh *et al.*, 2020). In addition, the damage caused by floods can in itself induce the production of new waste material from households, industries and other damaged infrastructure, including dams, reservoirs and artificial levees. For example, during the fluvial floods that devastated Kerala (India) in 2018, the sudden opening of the reservoirs' flood gates is considered to have led to the increased microplastic quantities observed on the coastal beaches (Kumar & Varghese, 2021).

Floods are often associated with higher plastic fluxes in rivers. The cumulative effect of debris remobilisation, entrainment of waste from flood-affected areas and the generation of new waste inputs prevails over localised plastic deposition mechanisms. In a Romanian river catchment, more than 80% of plastic dumped over a four-year period was observed to have flushed away during one period of flash floods (Mihai, 2018). This signifies the capacity of major floods to mobilise plastic waste across an entire river catchment. The global assessment to quantify the role of floods in plastic emissions showed that even low severity floods (with a 10-year return period) can substantially increase plastic mobilisation worldwide (tenfold increase), with large spatial variation among countries (Roebroek *et al.*, 2021a). These studies, by highlighting the major role that flood events play in plastic transport at different spatial scales, reaffirm the need for continuous field monitoring of plastic abundance in rivers and urban hydraulic networks.

In order to better understand and monitor the role of floods in ocean's plastic emissions, the influence of the flood characteristics, such as the return period (likelihood of a particular magnitude flood to occur) and the type of flooding (fluvial, coastal, flash floods and stormwater runoff), needs to be further explored. Plastic (re)mobilisation rates and patterns might differ depending on the flood characteristics and be further complexified in the case of compound (co-occurring) natural hazards. Coastal floods, when co-occurring with storm surges, hurricanes or tsunamis, can be hypothesised to generate new debris due to the coupled effects of sea surge and strong winds. Flash floods might be particularly effective in rapidly carrying (anthropogenic) debris over long distances. Taking into account the role of combined sewage overflows and stormwater runoff is also necessary to better estimate plastic waste sources and pathways in urban areas (Treilles *et al.*, 2020). Moreover, quantifying both deposition and remobilisation mechanisms of solid waste under fluvial flood conditions is necessary to ultimately determine whether plastic net transport is positive or negative during such events.

In the context of climate change, the expected increase in flood frequency and intensity is likely to result in higher plastic emissions into the oceans. However, the relation between the severity and frequency of floods, on the one hand, and the quantities of plastic transported and emitted, on the other hand, has not yet been resolved. In addition, plastic mobilisation during flood events might not necessarily immediately cause higher plastic emissions into the oceans, as plastic waste might be redeposited within rivers and on land multiple times before reaching the oceans (Roebroek *et al.*, 2021a). Catchment and river characteristics such as sinuosity (degree of meandering of a riverbed), valley morphology, soil types and land-use types of flood-prone areas are required for a holistic understanding of the diversity of plastic transfer modes during floods. Overall, research on plastic mobilisation during flood conditions is still a novel field and the influence of its control factors at different spatiotemporal scales has yet to be fully understood. The ubiquity of flood events worldwide and their likely increase due to climate change calls for further investigating this field of research.

3.3. *Plastic down a thousand rivers*

Global riverine plastic emission models are constantly evolving as more field data from plastic monitoring campaigns and new fundamental insights are being incorporated. A range of estimates have been made on global riverine plastic emissions into the oceans, each with varying results and levels of uncertainty. Here, we will provide a brief overview of the estimated global riverine plastic emissions into the oceans by different studies (Table 1) and explain where the differences in outcome between these studies come from.

One of the first estimates on global plastic emissions into the oceans comes from the study by Jambeck *et al.* (2015). They defined mismanaged plastic waste (MPW) as "material that is either littered or inadequately disposed" (Jambeck *et al.*, 2012). They used conversion rates from MPW to marine debris in order to estimate the mass of plastic waste entering the ocean, whereby they only considered the MPW generated by the population living within 50 km of the coast. Jambeck *et al.* (2012) provided a first rough estimate on plastic emissions into the oceans, however, their study did not look into the relative contribution of individual rivers. In 2017, Schmidt and colleagues were the first to consider the plastic emissions of individual rivers. They compared the amount of MPW generated in river basins (that drain into an ocean) with observational data on the plastic loads in these rivers. They defined the plastic

Table 1. Overview of studies from the past years that estimated the global riverine input of plastic into the oceans.

Study	Estimated global riverine plastic emission into the oceans (million metric tonnes/year)	Number of river basins considered	Global distribution of global riverine plastic emissions
Jambeck *et al.* (2015)	4.8–12.7	N/A	N/A
Schmidt *et al.* (2017)	0.47–2.75	1,494	Top 10 rivers account for 88–94%
Lebreton *et al.* (2017)	1.15–2.41	40,760	Top 20 rivers account for 67%
Mai *et al.* (2020)	0.06–0.27	1,518	Top 10 rivers account for 41%
Meijer *et al.* (2021)	0.8–2.7	100,887	1,656 rivers account for 80%

delivery ratio as the ratio between the plastic load observed in a river and the MPW generated in the catchment of that river, and found a value of ~2×10^{-6} for all plastics (microplastics and macroplastics). The results from Schmidt *et al.* (2017) suggest that only a small fraction of the MPW generated in a river basin is being transported in the river. In addition, they reported that the top 10 polluting rivers are responsible for 88–94% of the global riverine emissions into the oceans. In the same year, Lebreton and colleagues developed a global model that aims to predict the plastic emission from rivers into oceans based on geospatial data of population density, rates of MPW production, catchment runoff and artificial barriers (e.g., dams and weirs). They calibrated their model against field data measurements from 30 sampling studies across 13 rivers. Even though the estimations on the global riverine plastic emissions into the oceans from Lebreton *et al.* (2017) and Schmidt *et al.* (2017) are similar (Table 1), Lebreton *et al.* (2017) found that 67% of the emitted plastics come from the top 20 most polluting rivers. In 2020, Mai and colleagues applied a different approach to estimating the global riverine plastic emissions from rivers into the oceans. They used the Human Development Index (HDI) as a proxy for the amount of MPW that could potentially enter a river. Mai *et al.* (2020) found a global riverine plastic emission into the oceans of 0.06–0.27 million metric tonnes per year, which is considerably lower than the previous estimations from Jambeck *et al.* (2012), Schmidt *et al.*

(2017) and Lebreton *et al.* (2017). Most strikingly, Mai *et al.* (2020) found that only 41%, in contrast to 88–94% according to Schmidt *et al.* (2017), comes from the top 10 most polluting rivers. These discrepancies between different approaches emphasise the need for better fundamental understanding and additional observations to validate the global distribution of riverine plastic emissions.

To summarise, our understanding regarding the relative contribution of individual rivers to the global plastic input into the oceans has shifted from holding only a few large rivers accountable to holding many small- and medium-sized rivers accountable for the majority of the plastic emissions. This insight was even further reinforced by the results from the study by Meijer *et al.* (2021). They found that 1,656 of a total of 31,904 plastic leaking rivers are responsible for 80% of the global riverine plastic emissions into the oceans. Of these 1,656 rivers, 830 small rivers (discharges between 0.1 and 10 m^3/s) were found to account for 30%, 793 medium-sized rivers (discharges between 10 and 1,000 m^3/s) for 47% and 33 large rivers (discharges > 1,000 m^3/s) for 3% of the global riverine plastic emissions into the oceans. The model from Meijer *et al.* (2021) calculates the riverine plastic emission at the river mouth by multiplying the probability of MPW emission with the amount of MPW generated within the river basin. They found that the probability of MPW being emitted to the ocean is largely determined by precipitation rates, terrain characteristics (e.g., land use and slope) and the distance between MPW sources and nearest river rather than the size of the river. Insights like these imply that implementing mitigation measures in small- to medium-sized rivers will already have a large impact on the input of plastics from rivers into the oceans.

3.4. *Final destination?*

3.4.1. *Rivers as reservoirs: If it fits it sits*

Plastic research is often driven by the motivation to identify and quantify sources of marine plastic pollution. Pinpointing entry points for plastic waste into the ocean and estimating their respective contributions are crucial inputs for ocean plastic transport models (Kaandorp *et al.*, 2020; Onink *et al.*, 2021). To date, this has been a successful approach to explore the origin and fate of marine plastics. What is often neglected, however, is that only a small fraction of the total plastics leaked into the environment makes it into the marine environment. In global plastic export models, this is implicitly taken into account as a loss factor rather than considering the river basin also as a plastic reservoir. However, there is ample evidence that plastic accumulation in river systems is a key process that needs to be tackled in order to effectively solve the global plastic pollution challenge.

Once in a river, plastics can be deposited on riverbanks, on floodplains or in sediments (Figure 1). Factors that determine the rate of deposition include the river flow patterns, the shape and geometry of rivers, and hydraulic infrastructure (e.g., dams, sluices and groins). In river curves, plastics are often carried in the outer bend, increasing the likelihood of interaction with the riverbank or structures such as groins (van Calcar & van Emmerik, 2019). In such areas, increased riverbank plastic concentrations can be expected (Tramoy *et al.*, 2020a). Also, sideways-directed flow, for example, induced by navigation, can transport plastics towards the riverbank where they may deposit.

River plastics are also effectively retained by infrastructure or litter traps. For example, Gasperi *et al.* (2014) used the floating dams along the Seine to sample (plastic) litter. In the Rhine, Vriend *et al.* (2020) analysed plastic debris concentrated in a litter trap in Rotterdam to improve plastic mass transport estimates. Besides infrastructure, plastics can also be retained by dams (Castro-Jimenez *et al.*, 2019). In urban environments, plastic waste accumulates around urban drainage infrastructure where it effectively blocks waterways leading to increased flood risk (Honingh *et al.*, 2020).

Other natural plastic aggregators include floating aquatic vegetation and mangrove forests. Water hyacinths, an invasive species nowadays present in most tropical rivers, are shown to be very effective in entrapping macroplastic pollution. In some rivers, up to 80% of the total floating macroplastics are carried by these floating vegetation patches (Schreyers *et al.*, 2021). The entrapment is only of limited duration, as hyacinths disintegrate if the salinity and flow velocity become too high or when they are disturbed by anthropogenic factors such as navigation. Mangroves, on the other hand, can retain macroplastics for much longer periods and have been found to be an important ultimate sink for riverine, estuarine and coastal plastic pollution (Ivar do Sul *et al.*, 2014; Martin *et al.*, 2020).

Combining all this observational evidence suggests that rivers should be considered as plastic reservoirs rather than only sources of marine plastic debris. Several studies have quantified the fraction of total mismanaged plastic waste that enters the ocean each year. By combining macroplastic observations and a hydrodynamic model, the annual emission of the Jakarta rivers was estimated to be 2,100 metric tonnes/year (van Emmerik *et al.*, 2019c). This was found to be only 3% of the total plastic leaking into the river basins, suggesting that the other 97% is accumulating somewhere on land or within the river. At the global level, it has been estimated that only 1.5% of the annual mismanaged plastic waste is entering the oceans (Meijer *et al.*, 2021) equaling 1.2 million metric tonnes per year. The remaining almost 80 million metric tonnes is potentially accumulating somewhere within the river basin.

One of the main unknowns remains to discover plastic accumulation zones in terrestrial and riverine ecosystems. Several case studies provide observational evidence of accumulation on land, on riverbanks, in floodplains, around infrastructure and within vegetation. However, it remains unclear what factors determine the relative contributions of these factors on plastic accumulation. The next step in plastic transport modelling may include overland transport of plastics to also reveal terrestrial plastic accumulation zones (Mellink *et al.*, 2021). The retention of macroplastics in river systems also influences the degradation of plastics. Depending on whether plastics are accumulated in sediments, on riverbanks or around hydraulic infrastructure, the degree of weathering, biofouling and fragmentation may vary. As formulated by Tramoy *et al.* (2020a), although macroplastics are clearly retained for decades or even longer, this may imply that such accumulation zones are, in fact, also factories for microplastics and nanoplastics. Rapid identification of such locations is, therefore, absolutely crucial in minimizing the negative impact of macroplastics but also to rid the environment of sources of smaller plastic particles. For effective prevention and removal of plastics from the environment, better understanding and identification of plastic accumulation and retention in river systems are crucial.

3.4.2. *Estuaries: Will it stay or will it go?*

Land-based plastic waste is considered one of the main sources of marine plastic pollution (Meijer *et al.*, 2021). Reducing and preventing plastic pollution from entering the ocean, therefore, benefit from identifying entry sources and quantifying their emissions. One of the main assumptions in most model-based and observation-based plastic emission estimates is that plastics deposited on land, floodplains and riverbanks are conveyed by rivers to downstream regions (Schmidt *et al.*, 2017). Once at the river mouth, plastic debris is assumed to enter the marine environment, where it is transported by ocean currents to one of the world's gyres (van Emmerik & Schwarz, 2020). Water and, therefore, also plastics, however, do not simply flow into the ocean from river mouths. The tidal zone, or estuary, that connects freshwater systems to the sea is a complex system that is both influenced by tidal dynamics and river discharges (Savenije, 2012). These combined dynamics lead to diurnally changing water levels, flow velocities and flow direction. Neglecting these transport processes, therefore, introduces additional uncertainty in the estimates of both riverine plastic emissions into the ocean and the accumulation of plastics within estuaries and river systems (Figure 1).

Ample observational evidence shows that plastic debris accumulates at various timescales in estuaries. In the Rio de la Plata, Acha *et al.* (2003) quantified plastic debris concentration in sediment along the estuary. They found that plastic concentrations were always significantly higher on the riverside of the estuary compared to the oceanside of the estuary. The large difference in plastic concentrations clearly demonstrates plastic debris accumulation in estuarine sediments. Floating plastics have also been found to accumulate in river mouths. During flood tide, water and, therefore, plastics can flow in the upstream direction. For several rivers in the Philippines and Vietnam, the measured plastic flux was about the same in either flow direction (van Emmerik & Schwarz, 2020; van Emmerik *et al.*, 2020b). Especially during times of low freshwater discharge, this leads to zero net transport. In turn, this leads to the accumulation of plastics in the tidal zone, at least for certain periods of time. Floods may potentially (re) mobilise these plastics again, which may result in peak export of river plastics into the ocean (Liro *et al.*, 2020).

Riverbank macroplastic observations along the Dutch transects of the Rhine and Meuse rivers revealed a clear peak in the most downstream part of the delta, the Biesbosch National Park (van Emmerik *et al.*, 2020a). Here, around 50,000 macroplastic items/km of the riverbank were measured versus 2,000 items/km on average for these rivers. Estuaries are considered very biodiverse and productive ecosystems, making plastic accumulation and its associated negative effects even more problematic. The nature reserve is within the tidal area and has heavily vegetated riverbanks and floodplains. Similar results were found in the Amazon estuary where plastic litter retention was explained by dense vegetation and tidal influence (Gonçalves *et al.*, 2020). The combination of bidirectional flow, diurnally varying water level, vegetation and strong seasonality in discharge may explain the high accumulation rates in estuaries.

The impact of tidal dynamics on plastic deposition in estuaries has also been demonstrated by extensive work in the Seine, France. Tramoy *et al.* (2020a, 2020b) investigated both the short-term (year) and long-term (decades) transport dynamics. Macroplastic items experience a strong deposition and remobilisation cycle, which is influenced by hydrodynamics, tidal dynamics and river characteristics. Only occasionally, mainly during flood events, plastic debris may be emitted into the ocean. Under non-flood conditions, plastics do not travel far downstream and often accumulate on riverbanks. The residence time of macroplastics can, therefore, exceed decades and even centuries (Tramoy *et al.*, 2020a).

Examples from rivers globally show increased accumulation of floating plastics, plastics in sediment and plastic on riverbanks in estuaries compared to upstream locations in rivers, independent of the size range of the plastics considered. Despite the clear connection between tidal dynamics, hydrodynamics, river characteristics, and plastic transport and accumulation, the exact dependencies and relative contributions remain unresolved. It still remains unclear what determines the transport dynamics in estuaries. It is often hypothesised that plastics may accumulate in river systems or estuaries during periods of low discharge and can potentially be flushed into the ocean after floods or other extreme events (Roebroek *et al.*, 2021a). Yet, some observations have demonstrated that plastic loads in estuary wetlands remained high also after floods (Ryan & Perold, 2021). The retention time and response to extreme events could be related to the exact location where plastics accumulate. Future research should, therefore, explore what factors determine accumulation and remobilisation dynamics of plastics in riverine and estuarine systems. It also remains unclear whether marine plastics may actually enter estuaries during flood tide, making estuaries a potential sink for plastics originating from outside the system boundaries. The role of estuaries on plastic transfer between rivers and oceans remains largely unresolved. The increasing amount of observational evidence of short- and long-term plastic accumulation emphasises the need for improved understanding of estuarine plastic transport dynamics. These insights will play a key role in improving riverine emission estimates, and optimizing ocean plastic reduction and prevention strategies.

3.5. *You are what you eat*

Current research on plastic waste highlights the major role that rivers play in conveying plastic to the oceans (Meijer *et al.*, 2021). However, the exposure of freshwater ecosystems to plastic pollution remains largely undocumented with few studies estimating plastic concentrations within rivers and lakes. A common misconception seems to be that freshwater systems (especially rivers) do not retain many plastic particles and thus are not as exposed as the marine environment to their likely detrimental effects, such as fauna ingestion or entanglement.

Some plastic studies, however, tell a different story, with sinking mechanisms and large retention rates suggesting high levels of exposure of freshwater ecosystems to plastic pollution (van Emmerik *et al.*, 2019c; Meijer *et al.*, 2021). Microplastic concentrations in both riverine and marine water samples vary greatly (of several orders of magnitude) between study areas but were not found to be consistently higher in the marine environment than in rivers (Waldschläger *et al.*, 2020). The ingestion of microplastic by freshwater biota is of particular concern because of its pivotal role across the food chain. Given that fishes are a feeding resource to many terrestrial and aquatic organisms, the consequences of plastic ingestion by freshwater species reach far beyond the river and lakeshores. To date, the ingestion of plastic by freshwater organisms (from invertebrates to mammals) has been ascertained for more than 200 species worldwide (Azevedo-Santos *et al.*, 2021). In the Amazon, as much as 80% of the fish species examined had ingested plastic particles (Andrade *et al.*, 2019).

Another reason for concern stems from the fact that synthetic polymers either contain themselves highly toxic additives or adsorb contaminants present in the water. Industries and cities are the dominant sources of these contaminants, and their close locations to freshwater habitats likely cause leakage of these pollutants into the water. This cumulative pollution effect associated with plastics in freshwater systems is likely to generate even more adverse effects on freshwater biota. Measurements conducted on the shores of Lake Geneva showed that hazardous elements such as mercury, lead and bromine were found in plastic particles in higher concentrations than in beached marine plastics (Filella & Turner, 2018).

Many uncertainties remain on the linkages between plastic and freshwater ecosystems. These include robust quantifications of freshwater's exposure to both microplastic and macroplastic pollution. Baseline assessments of the amounts of plastic debris present in freshwater systems are lacking for most rivers and lakes of the world, especially in developing countries. This is the direct consequence of greater scientific focus on marine plastic pollution with close to 90% of scientific studies focusing on the marine environment (Blettler *et al.*, 2018). Determining the exposure of freshwater biota to plastics remains difficult also because the main sinks of plastic in lakes and rivers have not yet been identified with certainty. A better understanding of the exposure level of freshwater compartments to plastic accumulation is needed for assessing the overall risk associated with plastic toxicity in lakes and rivers.

4. Concluding Remarks

When we think about plastic pollution, we tend to picture ghost fishing gear along with daily single-use plastic waste items floating on the ocean's surface. However, plastic pollution is also widespread throughout terrestrial and freshwater ecosystems. Moreover, unlike what we might think, plastic pollution does not necessarily only consist of used items; each year, vast amounts of plastic pellets, the building materials for all new plastics, (accidentally) end up in the natural environment. Another common misconception is to consider low-income countries with poor waste management practices to be the major global plastic emitters. However, most global estimates on plastic emissions neglect the export of solid waste from high-income to low-income countries.

Once in the natural environment, plastic waste is distributed across different ecosystem compartments. In rivers, flow velocities and discharge mainly control the transport of riverine plastics. However, apart from these hydrometeorological variables, other factors such as material properties, e.g., size, density, shape and polymer composition, as well as processes such as biofouling, have also proven to play a major role in the transport mode and rate of plastics in aquatic environments. Compared to aquatic environments, the transport and accumulation of plastic waste in terrestrial environments have been studied to a lesser extent. Terrestrial plastic transport is assumed to be controlled by wind and precipitation-driven surface runoffs although this idea still lacks a strong empirical basis. This is one of the main future challenges in the field of plastic pollution: to determine the transport capacity of wind and surface runoffs for different types of plastics (e.g., bottles, bags and food wrappers) along terrestrial terrains with varying degrees of friction (e.g., urban, grasslands and forests). Another key challenge lies in identifying the relative role of rivers as plastic conduits and as plastic reservoirs. Although over a thousand rivers are found to be responsible for about 80% of the total river plastic emissions, observational evidence shows that large amounts of plastic waste do not make it into the oceans. Vast amounts of plastics are deposited on riverbeds, riverbanks and floodplains or become entrapped in aquatic vegetation (e.g., water hyacinths and mangrove forests), hydraulic structures (e.g., dams), urban drainage infrastructure or litter traps installed by man. During flood events, deposited plastic waste can be remobilised and emitted to the oceans. However, under normal conditions, diurnally changing flow directions and low river discharges may result in a low net transport of plastics from the river to the ocean leading to an accumulation of plastics in deltas and estuaries. Therefore, future efforts should focus on determining which factors influence the accumulation, retention and remobilisation dynamics of plastics in terrestrial and freshwater ecosystems.

At the moment, we still lack the fundamental understanding of the processes that lead to the distribution and accumulation of plastic waste in different environments. Additionally, a deeper knowledge on environmental plastic loads is necessary to determine the exposure of marine, freshwater and terrestrial biota and thus for a sound risk assessment of plastics in the environment. Hydrological flow models can help in this process, but in order to obtain the full benefit from those, we need more field observations and empirical experiments on the transport behaviour of plastics to calibrate and validate such models. These include measurements of (floating) plastic fluxes, accumulation zones along rivers, and observations before and after extreme events. Until then, global models on riverine plastic emissions to the oceans need to be treated with caution. To conclude this chapter, only with a comprehensive understanding of the sources, transport pathways and accumulation zones of plastics in the environment can the impacts on ecosystems be assessed and evaluated, and mitigation strategies be adopted to reduce plastic pollution.

Acknowledgements

The work of TvE is supported by the Veni research program The River Plastic Monitoring Project with project number 18211, which is (partly) financed by the Dutch Research Council (NWO). The work of LS is supported by NWO Open Mind grant 18127. The work of KW is supported by Investment Plan for strengthening the Technical Sciences at Wageningen University.

References

Acha, E. M., Mianzan, H. W., Iribarne, O., Gagliardini, D. A., Lasta, C., & Daleo, P. (2003). The role of the Rıo de la Plata bottom salinity front in accumulating debris. *Marine Pollution Bulletin*, *46*(2), 197–202.

Andrade, M. C., Winemiller, K. O., Barbosa, P. S., Fortunati, A., Chelazzi, D., Cincinelli, A., & Giarrizzo, T. (2019). First account of plastic pollution impacting freshwater fishes in the Amazon: Ingestion of plastic debris by piranhas and other serrasalmids with diverse feeding habits. *Environmental Pollution*, *244*, 766–773.

Araújo, M. C. B. & Costa, M. F. (2019). A critical review of the issue of cigarette butt pollution in coastal environments. *Environmental Research*, *172*, 137–149.

Asensio-Montesinos, F., Anfuso, G., & Williams, A. T. (2019). Beach litter distribution along the western Mediterranean coast of Spain. *Marine Pollution Bulletin*, *141*, 119–126.

Azevedo-Santos, V. M., Brito, M. F., Manoel, P. S., Perroca, J. F., Rodrigues-Filho, J. L., Paschoal, L. R., Gonçalves, G. R., Wolf, M. R., Blettler, M. C., Andrade, M. C., Nobile, A. B., Lima, F. P., Ruocco, A. M., Silva, C. V., Perbiche-Neves, G., Portinho, J. L., Giarrizzo, T., Arcifa, M. S., & Pelicice, F. M. (2021). Plastic pollution: A focus on freshwater biodiversity. *Ambio*, *50*(7), 1313–1324.

Belzagui, F., Buscio, V., Gutiérrez-Bouzán, C., & Vilaseca, M. (2021). Cigarette butts as a microfiber source with a microplastic level of concern. *Science of the Total Environment*, *762*, 144165.

Besseling, E., Quik, J. T., Sun, M., & Koelmans, A. A. (2017). Fate of nano-and microplastic in freshwater systems: A modeling study. *Environmental Pollution*, *220*, 540–548.

Bishop, G., Styles, D., & Lens, P. N. (2020). Recycling of European plastic is a pathway for plastic debris in the ocean. *Environment International*, *142*, 105893.

Blettler, M. C., Abrial, E., Khan, F. R., Sivri, N., & Espinola, L. A. (2018). Freshwater plastic pollution: Recognizing research biases and identifying knowledge gaps. *Water Research*, *143*, 416–424.

Bonanomi, G., Incerti, G., Cesarano, G., Gaglione, S. A., & Lanzotti, V. (2015). Cigarette butt decomposition and associated chemical changes assessed by 13 C CPMAS NMR. *PLoS One*, *10*(1), e0117393.

Bonanomi, G., Maisto, G., De Marco, A., Cesarano, G., Zotti, M., Mazzei, P., Libralato, G., Staropoli, A., Siciliano, A., De Filippis, F., La Storia, A., Piccolo, A., Vinale, F., Crasto, A., Guida, M., Ercolini, D., & Incerti, G. (2020). The fate of cigarette butts in different environments: Decay rate, chemical changes and ecotoxicity revealed by a 5-years decomposition experiment. *Environmental Pollution*, *261*, 114108.

Brooks, A. L., Wang, S., & Jambeck, J. R. (2018). The Chinese import ban and its impact on global plastic waste trade. *Science Advances*, *4*, eaat0131.

Castro-Jiménez, J., González-Fernández, D., Fornier, M., Schmidt, N., & Sempéré, R. (2019). Macro-litter in surface waters from the Rhone River: Plastic pollution and loading to the NW Mediterranean Sea. *Marine Pollution Bulletin, 146*, 60–66.

Chubarenko, I., Bagaev, A., Zobkov, M., & Esiukova, E. (2016). On some physical and dynamical properties of microplastic particles in marine environment. *Marine Pollution Bulletin*, *108*(1–2), 105–112.

Ciszewski, D. & Grygar, T. M. (2016). A review of flood-related storage and remobilization of heavy metal pollutants in river systems. *Water, Air, & Soil Pollution, 227*, 239.

Cole, G. & Sherrington, C. (2016). *Study to Quantify Pellet Emissions in the UK*. Bristol: Eunomia.

d'Ambrières, W. (2019). Plastics recycling worldwide: Current overview and desirable changes. *Field Actions Science Reports. The Journal of Field Actions*, (Special Issue 19), 12–21. Retrieved from https://journals.openedition.org/factsreports/4993.

Derraik, J. G. (2002). The pollution of the marine environment by plastic debris: A review. *Marine Pollution Bulletin*, *44*(9), 842–852.

Fazey, F. M. & Ryan, P. G. (2016). Biofouling on buoyant marine plastics: An experimental study into the effect of size on surface longevity. *Environmental Pollution, 210*, 354–360.

Fidra. (2021). Retrieved from https://www.nurdlehunt.org.uk/nurdle-finds.html. Accessed 27.04.2021.

Filella, M. & Turner, A. (2018). Observational study unveils the extensive presence of hazardous elements in beached plastics from Lake Geneva. *Frontiers in Environmental Science*, *6*, 1.

Gasperi, J., Dris, R., Bonin, T., Rocher, V., & Tassin, B. (2014). Assessment of floating plastic debris in surface water along the Seine River. *Environmental Pollution, 195*, 163–166.

Gonçalves, M., Schmid, K., Andrade, M. C., Andrades, R., Pegado, T., & Giarrizzo, T. (2020). Are the tidal flooded forests sinks for litter in the Amazonian estuary? *Marine Pollution Bulletin, 161*, 111732.

Haberstroh, C. J., Arias, M. E., Yin, Z., & Wang, M. C. (2021). Effects of hydrodynamics on the cross-sectional distribution and transport of plastic in an urban coastal river. *Water Environment Research, 93*(2), 186–200.

Hann, S., Sherrington, C., Jamieson, O., Hickman, M., Kershaw, P., Bapasola, A., & Cole, G. (2018). Investigating options for reducing releases in the aquatic environment of microplastics emitted by (but not intentionally added in) products. *Report for DG Environment of the European Commission*, 335.

Hartley, B. L., Pahl, S., Veiga, J., Vlachogianni, T., Vasconcelos, L., Maes, T., Doyle, T., d'Arcy Metcalfe, R., Öztürk, A. A., Di Berardo, M., & Thompson, R. C. (2018). Exploring public views on marine litter in Europe: Perceived causes, consequences and pathways to change. *Marine Pollution Bulletin*, *133*, 945–955.

Honingh, D., van Emmerik, T., Uijttewaal, W., Kardhana, H., Hoes, O., & van de Giesen, N. (2020). Urban river water level increase through plastic waste accumulation at a rack structure. *Frontiers in Earth Science*, *8*, 28.

Hurley, R., Woodward, J., & Rothwell, J. J. (2018). Microplastic contamination of river beds significantly reduced by catchment-wide flooding. *Nature Geoscience*, *11*(4), 251–257.

Ivar do Sul, J. A. , Costa, M. F., Silva-Cavalcanti, J. S., & Araújo, M. C. B. (2014). Plastic debris retention and exportation by a mangrove forest patch. *Marine Pollution Bulletin*, *78*(1–2), 252–257.

Jambeck, J. R., Geyer, R., Wilcox, C., Siegler, T. R., Perryman, M., Andrady, A., Narayan, R., & Law, K. L. (2015). Plastic waste inputs from land into the ocean. *Science*, *347*(6223), 768–771.

Kaandorp, M. L., Dijkstra, H. A., & van Sebille, E. (2020). Closing the Mediterranean marine floating plastic mass budget: Inverse modeling of sources and sinks. *Environmental Science & Technology*, *54*(19), 11980–11989.

Kaiser, D., Kowalski, N., & Waniek, J. J. (2017). Effects of biofouling on the sinking behavior of microplastics. *Environmental Research Letters*, *12*(12), 124003.

Karlsson, T. M., Arneborg, L., Broström, G., Almroth, B. C., Gipperth, L., & Hassellöv, M. (2018). The unaccountability case of plastic pellet pollution. *Marine Pollution Bulletin*, *129*(1), 52–60.

Khatmullina, L. & Isachenko, I. (2017). Settling velocity of microplastic particles of regular shapes. *Marine Pollution Bulletin*, *114*(2), 871–880.

Kooi, M., Nes, E. H. V., Scheffer, M., & Koelmans, A. A. (2017). Ups and downs in the ocean: Effects of biofouling on vertical transport of microplastics. *Environmental Science & Technology*, *51*(14), 7963–7971.

Kooi, M., Reisser, J., Slat, B., Ferrari, F. F., Schmid, M. S., Cunsolo, S., Brambini, R., Noble, K. D., Sirks, L-A., Linders, T., Schöneich-Argent, R., & Koelmans, A. A. (2016). The effect of particle properties on the depth profile of buoyant plastics in the ocean. *Scientific Reports*, *6*(1), 1–10.

Kowalski, N., Reichardt, A. M., & Waniek, J. J. (2016). Sinking rates of microplastics and potential implications of their alteration by physical, biological, and chemical factors. *Marine Pollution Bulletin*, *109*(1), 310–319.

Kumar, A. S. & Varghese, G.K. (2021). Microplastic pollution of Calicut beach — Contributing factors and possible impacts. *Marine Pollution Bulletin, 169,* 112492.

Kystverket. (16 June, 2020). Governmental environmental response to accidental release of plastic pellets. Retrieved from https://www.kystverket.no/en/EN_System/plastic-pellets-clean-up/. Accessed 27.04.2021.

Law, K. L., Starr, N., Siegler, T. R., Jambeck, J. R., Mallos, N. J., & Leonard, G. H. (2020). The United States' contribution of plastic waste to land and ocean. *Science Advances, 6*(44), eabd0288.

Lebreton, L. C., Slat, B., Ferrari, F., Sainte-Rose, B., Aitken, J., Marthouse, R., Hajbane, S., Cunsolo, S., Schwarz, A., Levivier, A., Noble, K., Debeljak, P., Maral, H., Schöneich-Argent, R., Brambini, R., & Reisser, J. (2018). Evidence that the Great Pacific Garbage Patch is rapidly accumulating plastic. *Scientific Reports*, *8*(1), 1–15.

Lebreton, L. C., van Der Zwet, J., Damsteeg, J. W., Slat, B., Andrady, A., & Reisser, J. (2017). River plastic emissions to the world's oceans. *Nature Communications*, *8*(1), 1–10.

Lechthaler, S., Waldschläger, K., Stauch, G., & Schüttrumpf, H. (2020). The way of macroplastic through the environment. *Environments*, *7*(10), 73.

Li, W. C., Tse, H. F., & Fok, L. (2016). Plastic waste in the marine environment: A review of sources, occurrence and effects. *Science of the Total Environment*, *566*, 333–349.

Liro, M., Emmerik, T. V., Wyżga, B., Liro, J., & Mikuś, P. (2020). Macroplastic storage and remobilization in rivers. *Water*, *12*(7), 2055.

Lobelle, D., Kooi, M., Koelmans, A. A., Laufkötter, C., Jongedijk, C. E., Kehl, C., & van Sebille, E. (2021). Global modeled sinking characteristics of biofouled microplastic. *Journal of Geophysical Research: Oceans*, 126(4), 1–15, e2020JC017098.

Mai, L., Sun, X. F., Xia, L. L., Bao, L. J., Liu, L. Y., & Zeng, E. Y. (2020). Global riverine plastic outflows. *Environmental Science & Technology*, *54*(16), 10049–10056.

Mani, T., Hauk, A., Walter, U., & Burkhardt-Holm, P. (2015). Microplastics profile along the Rhine River. *Scientific Reports*, *5*(1), 1–7.

Martin, C., Baalkhuyur, F., Valluzzi, L., Saderne, V., Cusack, M., Almahasheer, H., Krishnakumar, P. K., Rabaoui, L., Qurban, M. A., Arias-Ortiz, A., Masqué, P., & Duarte, C. M. (2020). Exponential increase of plastic burial in mangrove sediments as a major plastic sink. *Science Advances, 6*(44), eaaz5593.

Meijer, L. J. J., van Emmerik, T., van der Ent, R., Schmidt, C., & Lebreton, L. (2021). More than 1000 rivers account for 80% of global riverine plastic emissions into the ocean. *Science Advances*, *7*(18), eaaz5803.

Mellink, Y. A. M., van Emmerik, T., Kooi, M., Laufkötter, C., & Niemann, H. (2021). The trash-tracker: A macroplastic transport and fate model at river basin scale. *Earth.* ArXiv [Preprint]. https://doi.org/10.31223/X5303G.

Mihai, F.-C. (2018). Rural plastic emissions into the largest mountain lake of the Eastern Carpathians. *Royal Society Open Science*, *5*, 172396172396.
Moore, C. J., Lattin, G. L., & Zellers, A. F. (2011). Quantity and type of plastic debris flowing from two urban rivers to coastal waters and beaches of Southern California. *Journal of Integrated Coastal Zone Management*, *11*, 65–73.
NOLA. (2020). Tristan Baurick, August 21, 2020 With cleanup stalled for weeks, groups of volunteers tackle New Orleans nurdle spill on their own. Retrieved from https://www.nola.com/news/environment/article_81011dfa-e3d6-11ea-8739-776cd89891a0.html. Accessed 27.04.2021.
Novotny, T. E. & Slaughter, E. (2014). Tobacco product waste: An environmental approach to reduce tobacco consumption. *Current Environmental Health Reports*, *1*(3), 208–216.
Ockelford, A., Cundy, A. & Ebdon, J. E. (2020). Storm response of fluvial sedimentary microplastics. *Scientific Reports*, *10*, 1865.
Olivelli, A., Hardesty, B. D., & Wilcox, C. (2020). Coastal margins and backshores represent a major sink for marine debris: Insights from a continental-scale analysis. *Environmental Research Letters*, *15*(7), 074037.
Onink, V., Jongedijk, C., Hoffman, M., van Sebille, E., & Laufkötter, C. (2021). Global simulations of marine plastic transport show plastic trapping in coastal zones. *Environmental Research Letters*, *16*, 064053.
Piehl, S., Leibner, A., Löder, M. G., Dris, R., Bogner, C., & Laforsch, C. (2018). Identification and quantification of macro-and microplastics on an agricultural farmland. *Scientific Reports*, *8*(1), 1–9.
PlasticsEurope. (2018). *Operation Clean Sweep, Report 2018*. Belgium.
Rahman, M. T., Mohajerani, A., & Giustozzi, F. (2020). Possible recycling of cigarette butts as fiber modifier in bitumen for asphalt concrete. *Materials*, *13*(3), 734.
Rath, J. M., Rubenstein, R. A., Curry, L. E., Shank, S. E., & Cartwright, J. C. (2012). Cigarette litter: Smokers' attitudes and behaviors. *International Journal of Environmental Research and Public Health*, *9*(6), 2189–2203.
Roebroek, C. T., Harrigan, S., van Emmerik, T. H., Baugh, C., Eilander, D., Prudhomme, C., & Pappenberger, F. (2021a). Plastic in global rivers: Are floods making it worse? *Environmental Research Letters*, *16*(2), 025003.
Roebroek, C. T., Hut, R., Vriend, P., De Winter, W., Boonstra, M., & van Emmerik, T. H. (2021b). Disentangling variability in riverbank macrolitter observations. *Environmental Science & Technology*, *55*(8), 4932–4942.
Ryan, P. G. & Perold, V. (2021). Limited dispersal of riverine litter onto nearby beaches during rainfall events. *Estuarine, Coastal and Shelf Science*, *251*, 107186.
Ryberg, M. W., Hauschild, M. Z., Wang, F., Averous-Monnery, S., & Laurent, A. (2019). Global environmental losses of plastics across their value chains. *Resources, Conservation and Recycling*, *151*, 104459.

Savenije, H. H. G. (2012). *Salinity and Tides in Alluvial Estuaries*, 2nd edn. Delft, The Netherlands: Delft University of Technology. https://doi.org/10.1016/B978-044452107-1/50006-X.

Schmidt, C., Krauth, T., & Wagner, S. (2017). Export of plastic debris by rivers into the sea. *Environmental Science & Technology*, *51*(21), 12246–12253.

Schreyers, L., van Emmerik, T., Nguyen, T. L., Castrop, E., Phung, N.-A., Kieu-Le, T.-C., Strady, E., Biermann, L., & van der Ploeg, M. (2021). Plastic plants: The role of water hyacinths in plastic transport in tropical rivers. *Frontiers in Environmental Science*, *9*, 686334.

Schwarz, A. E., Ligthart, T. N., Boukris, E., & van Harmelen, T. (2019). Sources, transport, and accumulation of different types of plastic litter in aquatic environments: A review study. *Marine Pollution Bulletin*, *143*, 92–100.

Slaughter, E., Gersberg, R. M., Watanabe, K., Rudolph, J., Stransky, C., & Novotny, T. E. (2011). Toxicity of cigarette butts, and their chemical components, to marine and freshwater fish. *Tobacco Control*, *20*(Suppl 1), i25–i29.

Soares, J., Miguel, I., Venâncio, C., Lopes, I., & Oliveira, M. (2021). Public views on plastic pollution: Knowledge, perceived impacts, and pro-environmental behaviours. *Journal of Hazardous Materials*, *412*, 125227.

Suaria, G., Perold, V., Lee, J. R., Lebouard, F., Aliani, S., & Ryan, P. G. (2020). Floating macro-and microplastics around the Southern Ocean: Results from the Antarctic circumnavigation expedition. *Environment International*, *136*, 105494.

Takada, H., Mato, Y., Endo, S., Yamashita, R., & Zakaria, M. P. (2005). Pellet Watch: Global monitoring of persistent organic pollutants (POPs) using beached plastic resin pellets. In *The Plastic Debris Rivers to Sea Conference: Focusing on the Land-Based Sources of Marine Debris, Redondo Beach, CA, USA*.

Torkashvand, J., Farzadkia, M., Sobhi, H. R., & Esrafili, A. (2020). Littered cigarette butt as a well-known hazardous waste: A comprehensive systematic review. *Journal of Hazardous Materials*, *383*, 121242.

Tramoy, R., Gasperi, J., Colasse, L., & Tassin, B. (2020a). Transfer dynamic of macroplastics in estuaries — New insights from the Seine estuary: Part 1. Long term dynamic based on date-prints on stranded debris. *Marine Pollution Bulletin*, *152*, 110894.

Tramoy, R., Gasperi, J., Colasse, L., Silvestre, M., Dubois, P., Noûs, C., & Tassin, B. (2020b). Transfer dynamics of macroplastics in estuaries — New insights from the Seine estuary: Part 2. Short-term dynamics based on GPS-trackers. *Marine Pollution Bulletin*, *160*, 111566.

Treilles, R., Gasperi, J., Saad, M., Rabier, A., Breton, J., Rocher, V., Guérin, S., & Tassin, B. (2020). Macro and microplastics in stormwater and combined sewer overflows in Paris megacity. In Cocca M. *et al.* (Eds.), *Proceedings of the 2nd International Conference on Microplastic Pollution in the Mediterranean Sea. ICMPMS 2019*. Springer Water. Cham: Springer.

van Calcar, C. J. & van Emmerik, T. H. M. (2019). Abundance of plastic debris across European and Asian rivers. *Environmental Research Letters*, *14*(12), 124051.

van Emmerik, T. (2021). Macroplastic research in an era of microplastic. *Microplastics and Nanoplastics*, *1*, 4.

van Emmerik, T., Loozen, M., van Oeveren, K., Buschman, F., & Prinsen, G. (2019c). Riverine plastic emission from Jakarta into the ocean. *Environmental Research Letters*, *14*(8), 084033.

van Emmerik, T., Roebroek, C., De Winter, W., Vriend, P., Boonstra, M., & Hougee, M. (2020b). Riverbank macrolitter in the Dutch Rhine–Meuse delta. *Environmental Research Letters*, *15*(10), 104087.

van Emmerik, T. & Schwarz, A. (2020). Plastic debris in rivers. *Wiley Interdisciplinary Reviews: Water*, *7*(1), e1398.

van Emmerik, T., Strady, E., Kieu-Le, T. C., Nguyen, L., & Gratiot, N. (2019b). Seasonality of riverine macroplastic transport. *Scientific Reports*, *9*(1), 1–9.

van Emmerik, T., Tramoy, R., van Calcar, C., Alligant, S., Treilles, R., Tassin, B., & Gasperi, J. (2019a). Seine plastic debris transport tenfolded during increased river discharge. *Frontiers in Marine Science*, *6*, 642.

van Emmerik, T., van Klaveren, J., Meijer, L. J., Krooshof, J. W., Palmos, D. A. A., & Tanchuling, M. A. (2020a). Manila river mouths act as temporary sinks for macroplastic pollution. *Frontiers in Marine Science*, *7*, 770.

Vriend, P., van Calcar, C., Kooi, M., Landman, H., Pikaar, R., & van Emmerik, T. (2020). Rapid assessment of floating macroplastic transport in the Rhine. *Frontiers in Marine Science*, *7*, 10.

Wagner, S., Klöckner, P., Stier, B., Römer, M., Seiwert, B., Reemtsma, T., & Schmidt, C. (2019). Relationship between discharge and river plastic concentrations in a rural and an urban catchment. *Environmental Science & Technology*, *53*(17), 10082–10091.

Waldschläger, K., Lechthaler, S., Stauch, G., & Schüttrumpf, H. (2020). The way of microplastic through the environment — Application of the source-pathway-receptor model. *Science of the Total Environment*, *713*, 136584.

Woodall, L. C., Sanchez-Vidal, A., Canals, M., Paterson, G. L., Coppock, R., Sleight, V., Calafat, A., Rogers, A. D., Narayanaswamy, B. E., & Thompson, R. C. (2014). The deep sea is a major sink for microplastic debris. *Royal Society Open Science*, *1*(4), 140317.

Yadav, N. & Hakkarainen, M. (2021). Degradable or not? Cellulose acetate as a model for complicated interplay between structure, environment and degradation. *Chemosphere*, *265*, 128731.

https://doi.org/10.1142/9789811259111_0004

Chapter 3

Transport of Plastics Within the World's Oceans

Alethea Mountford

Department of Geography and Environmental Sciences, Northumbria University, Ellison Place, Newcastle upon Tyne NE1 8ST, UK

alethea.mountford@northumbria.ac.uk

1. Introduction

When thinking about plastics in the world's oceans, typically the "garbage patches" may be the first thing to come to a person's mind, with visions of vast islands of trash the size of Texas that you would be able to walk on. However, the reality of plastic pollution in the oceans is on a much smaller, more pervasive scale. Estimates suggest that less than 1% of the plastics that have entered the world's oceans are present at the sea surface (Mountford & Morales Maqueda, 2019; Wichmann *et al.*, 2019), and the processes that affect the distribution of plastics in the oceans are both complex and interconnected leaving no region untouched by plastic pollution. Once within the marine environment, plastics are subject to both horizontal and vertical transport. Horizontal transport in the form of currents and wind, for example, moves the plastics away from their point of entry over either short or long distances. Vertical transport, such as through turbulent mixing or sinking, moves the plastics through the water column. Processes such as fragmentation, biofouling, and ingestion can further influence both the horizontal and vertical transport of plastics of all types. The transport of plastics within the world's oceans is complex and interconnected but largely determined by the plastic itself. As discussed in previous chapters, the composition of a plastic, namely, the physico-chemical properties needed for its use and the additives that bring about these properties, give the plastic an inherent density. This density, whether it is smaller or greater than the density of the surrounding seawater, along with the plastic's size and shape, will determine its behaviour and transport within the marine environment. Broadly, a plastic can be less dense than seawater (positively buoyant), denser than seawater (negatively buoyant), or have a density close to that of seawater (neutrally buoyant). Over time, a plastic's density may change through processes such as mechanical degradation and fragmentation, ingestion and egestion from marine organisms, and biofouling. This chapter will discuss the oceanographic transport and biological processes which define the global distribution of plastics within the world's oceans, from the sea surface to the seafloor.

2. Beaching and Coastal Retention of Plastics

An estimated 80% of plastics enter the marine environment from land-based sources, for example, through direct littering, escaping waste management and via wastewater and sewer systems (Auta *et al.*, 2017). Given the ever-visible plastic pollution on beaches globally, it stands to reason that a portion of plastics will remain close to their point of entry at the coastlines. In fact, recent modelling work suggests that between 31% and 95% of positively buoyant plastic debris will beach at some point during its time in the marine environment, and at least 77% remains within 10 km of the shoreline (Onink *et al.*, 2021). In the coastal environment, plastics may become trapped in near-shore sediments, intertidal matrices (such as rockpools), on beaches (both on the surface and below) and within mangrove sediments (Martin *et al.*, 2020). While this may be the final resting place for some of these plastics, beaching may only be a temporary sink, leading to a cycle of beaching, resuspension and burial, on a short- or long-term basis. This continuous process of beaching and resuspension at the shorelines suggests that the coastal regions may be an important site for the generation of secondary microplastics. Mechanical degradation and fragmentation in the sea swash zone, through wave action against sand and/or pebbles, are thought to be the fastest form of degradation (compared to UV and oxidative degradation, for example), with new plastics almost entirely broken down into microplastics after 24 hours in laboratory experiments (Chubarenko *et al.*, 2020). Of course, these degradation processes would likely not occur in isolation in the natural environment (Corcoran *et al.*, 2009). Following formation, the secondary microplastics are at greater risk of being transported offshore than the mesoplastics and macroplastics that they originated from (Isobe *et al.*, 2014). Mesoplastics are preferentially transported onshore, as their larger size and increased buoyancy result in them residing in the upper layers of the sea surface, where wind, waves and Stokes drift transport the particles onshore more quickly (Isobe *et al.*, 2014). Microplastics, on the other hand, are free from near-shore trapping and most typically become suspended in the subsurface layers, where they are less under the influence of onshore, surface currents (Zhang, 2017).

With that being said, it is not only beaches in densely populated areas that are littered with plastic debris. Henderson Island is a remote and uninhabited island in the South Pacific, which has an estimated 37.7 million debris items on its shores, although this count is likely far greater now compared to when the study was conducted, given it is thought that over 20 items per metre are accumulating daily (Lavers & Bond, 2017). Instances of marine plastic litter on remote shorelines have been reported in the North and South Atlantic (Andrades *et al.*, 2018; Monteiro *et al.*, 2018), Indian Ocean (Duhec *et al.*, 2015), South Pacific (Perez-Venegas *et al.*, 2017), Arctic Ocean (Barnes, 2002; Bergmann *et al.*, 2017a) and Southern Ocean (Barnes, 2005). At-sea sources, for example, from fishing, tourist or research vessels as well as off-shore structures and aquaculture, may be a significant contributor to the plastics found in remote locations, but the role of long-range transport cannot be underestimated (Onink *et al.*, 2021).

3. Surface Transport and Accumulations of Plastics

Numerical modelling is an invaluable tool for exploring the transport pathways and eventual global distribution of marine plastic pollution. Marine plastic models are typically (although not exclusively) either Lagrangian or Eulerian, which accounts for the way in which the plastics are included in the model (see van Sebille *et al.*, (2018)) for a detailed discussion about the use of Lagrangian models). Lagrangian model frameworks consider the behaviour and transport of an individual plastic particle, allowing the tracking of each particle's trajectory within a two- or three-dimensional field. Eulerian formulations, on the other hand, consider plastic particles as a mass or volume concentration, which is then advected by ocean velocity and diffused by turbulence. This approach lends itself to looking at three-dimensional, global transport pathways of large concentrations of plastics (Mountford & Morales Maqueda, 2019). While both frameworks have advantages and disadvantages, they provide a complementary approach to marine plastic modelling, allowing for the exploration of both small- and large-scale behaviour and transport of plastics.

The inherent buoyancy of almost half of the plastics most readily produced globally (PlasticsEurope, 2020) suggests that much of the plastics that enter the oceans will float at the sea surface, where it is susceptible to transport over long distances (Ryan, 2015). Extensive sampling efforts and numerical modelling have confirmed the presence of "garbage patches" in five subtropical gyres: in the North and South Pacific, the North and South Atlantic, and the Indian Ocean. Contrary to popular belief, it is not possible to walk on "trash islands" as the vast majority of these pieces of plastic are classed as microplastics, forming more of a "plastic soup". These accumulations form as a result of Ekman currents at the sea surface and down to depths of around 100 m, which deflects the transport of water inwards, trapping plastics and other pollutants. Ekman transport, alongside geostrophic currents and Stokes drift, is thought to determine the eventual location of the garbage patches (Onink *et al.*, 2019) although their predicted location varies somewhat between numerical modelling studies due to differences in circulation, resolution and plastic inputs (Lebreton *et al.*, 2012; Maximenko *et al.*, 2015; van Sebille *et al.*, 2015; Mountford & Morales Maqueda, 2019; Wichmann *et al.*, 2019). The North Pacific garbage patch (also known as the Great Pacific Garbage Patch) is the most well-known and widespread of these accumulations, estimated to cover an area of approximately 1.6 million km^2, containing between 1.1 and 3.6 trillion pieces of plastic, of which 94% is believed to be microplastics (Lebreton *et al.*, 2018). The North Pacific garbage patch accumulates debris from both the North America to the east via the California current and Asia to the west. The Kuroshio Extension current system is thought to be the greatest transporter of debris from Asia to the patch, highlighted, in particular, following the 2011

Tohoku earthquake and subsequent tsunami, as debris from the tragedy was discovered within the gyre (Desforges *et al.*, 2014; Lebreton *et al.*, 2018). Long-range transport is of particular note for buoyant plastic that resides in the air-sea interface (such as Styrofoam and plastic bottles), as they are subject to the influence not only of sea surface currents but also by winds (Ko *et al.*, 2020). The greater the portion of plastic exposed from the sea surface, the greater the influence of windage, therefore this will be of more consequence to larger debris. Lebreton *et al.* (2018) estimated that over three-quarters of the mass of the debris found within the accumulation in the North Pacific was larger than 5 cm (considered to be macroplastics) but suggested that higher windage debris may pass through and exit the patch itself, with a greater likelihood of beaching. For particularly buoyant plastics, such as expanded polystyrene (the small spherical particles that comprise polystyrene cups, packaging and insulation, for instance), wind is the dominating transport mechanism with the potential of wind-driven drift speeds up to four times larger than current speeds and the probability that spherical particles will roll over the sea surface (Chubarenko *et al.*, 2016). This is in contrast to denser polymers, such as polyvinyl chloride, which will sink through the water column, where they will be transported by turbulent mixing processes, before they reach the seafloor to be transported largely by near-bottom currents (Mountford & Morales Maqueda, 2019; Kvale *et al.*, 2020b), as will be discussed later in this chapter. This suggests that for larger plastics, in particular, density is likely a factor in determining the dominant transport mechanism.

The epicentre of the North Atlantic garbage patch is in the Western North Atlantic, with small microplastics (categorised as 25−1,000 μm in the following study by Poulain *et al.* (2019)) thought to dominate within the accumulation zone, with between 60 million and 4.5 billion particles km^{-2} small microplastics compared to 13,000–188,000 particles km^{-2} of large microplastics (categorised as 1–5 mm) (Poulain *et al.*, 2019). This high proportion of small microplastics compared to larger plastics is thought to partly explain why previous studies reported no observed change in plastic abundance between 1986 and 2008 (Law *et al.*, 2010). The combination of these smaller particles likely being overlooked in early studies and the continuous influx and breakdown of larger plastics may account for the lack of observed change. However, differences in sampling locations and protocols may also contribute to these observational differences (Law *et al.*, 2010; Wilcox *et al.*, 2019). The South Atlantic and South Pacific garbage patches (and the Southern Hemisphere as a whole) are largely understudied and undersampled. Although surface drifter models predicted the location of the patches in the South Atlantic and South Pacific, the presence of the accumulations was not confirmed by observational data until 2014 (Ryan, 2014) and 2013, respectively (Eriksen *et al.*, 2013). In 2013, the average abundance of microplastics in the South Pacific subtropical gyre was reported to be 26,898 particles km^{-2} (Eriksen *et al.*, 2013) although, since this initial study, there have been no further observational estimates of the plastic abundance in the South Pacific. While not focused specifically on the South Atlantic gyre, Pabortsava and Lampitt (2020) assessed the levels of plastic pollution along a 10,000 km North–South transect in the Atlantic Ocean. They found the lowest concentrations of plastics at the sampling stations within the South Atlantic gyre, although polyethylene and polypropylene particles were found in subsurface samples in the remote South Atlantic, including comparably high concentrations around South Georgia (up to 2,553 and 726 particles m^{-3}, respectively). There is significant work to be done in the Southern Hemisphere, in particular, in order to gain a more thorough understanding of the significance of these accumulations.

The garbage patch in the Indian Ocean is the least well defined and studied despite large coastal and riverine inputs. The Indian Ocean gyre is thought to be the weakest attractor of surface plastic pollution, due to its complex and interconnected surface circulation, and it has been argued that the plastic accumulation in the Southern Indian Ocean does not exist in the same sense as those in the North Pacific and North Atlantic (van der Mheen *et al.*, 2019). Despite the vast majority of plastic pollution entering the Indian Ocean in the Northern Hemisphere, the patch itself resides in the Southern Hemisphere; due to a lack of a subtropical gyre in the Northern Indian Ocean, plastics that remain in the north end up beached along the rim (Pattiaratchi *et al.*, 2021). The interconnectivity between the Southern Indian Ocean and both the South Atlantic Ocean and South Pacific Ocean results in a "leaky" garbage patch. Stokes drift, as well as surface ocean circulation and wind-driven transport, has been shown to play an important role in the transport of plastic debris into the South Atlantic, as the influence of Stokes drift causes the epicentre of the accumulation to drift towards the west of the basin (Dobler *et al.*, 2019; van der Mheen *et al.*, 2019). Plastics are transported into the South Pacific via a "superconvergence pathway" (Maes *et al.*, 2018) following the Leeuwin Current along the west coast of Australia (Waite *et al.*, 2007) before joining the Tasman Leakage (van Sebille *et al.*, 2014), which connects the Southern Indian and South Pacific.

Aside from the garbage patches, the Mediterranean is believed to be one of the regions most impacted by plastic pollution. As a semi-enclosed sea with an estimated coastal population of around 150 million, plastic waste inputs are thought to amount to approximately 100,000 tonnes per year, much of which remains trapped within the Mediterranean region itself (Cincinelli *et al.*, 2019; Liubartseva *et al.*, 2018). van Sebille *et al.* (2015) estimated a total mass of microplastics trapped within the Mediterranean between 4.8 and 30.3 Mt (based on three ocean circulation models), with an estimated particle load of between 3.2 and 28.2×10^{12} particles. While the total estimated mass in other basins exceeded that of the Mediterranean, the numerical particle count here was the greatest of all the basins in all of the model outcomes, with the exception of the North Pacific in one of the ocean circulation models, suggesting a higher proportion of smaller particles and/or less dense polymer types.

4. Surface Transport to Remote Regions

Surface-dwelling plastics are also making their way north into the Arctic, with a proposed garbage patch developing in the Barents Sea (van Sebille *et al.*, 2012) (although the presence of this garbage patch has not been observed), as well as high concentrations in coastal areas. As population density north of 60°N is comparatively low, local sources are not thought to be a significant contributor to plastic pollution, suggesting the majority of plastics found in the Arctic must be from further afield. Thermohaline circulation (Cózar *et al.*, 2017) and Stokes drift (Onink *et al.*, 2018) have both been suggested as important transport vectors of plastic pollution from the North Atlantic into the Arctic. The Greenland, Barents and Kara seas are thought to be particularly affected by floating plastics (Bergmann *et al.*, 2016; Tošić *et al.*, 2020), acting somewhat as a trapping zone for the buoyant plastics, leading to claims that the Arctic is a "dead end" for floating plastic pollution (Cózar *et al.*, 2017). Deep-water formation in the Arctic, such as in the Labrador Sea, can lead to the transport of microplastics through the water column towards the seafloor. Cózar *et al.* (2017) hypothesise that the Arctic seafloor will be the final resting place for much of the plastic pollution that enters the Arctic region, and while substantial quantities of macroplastics and microplastics have been recorded on the seafloor (e.g., Bergmann *et al.*, 2017b; Tekman *et al.*, 2020), microplastics have also been discovered within Arctic sea ice (Peeken *et al.*, 2018; Kanhai *et al.*, 2020) at concentrations comparable to those within subtropical gyres (Obbard *et al.*, 2014). Although the process of microplastic incorporation into sea ice is not yet fully understood, it is believed that there may be a preferential uptake and enrichment of small, low-density, irregularly shaped microplastics (Obbard *et al.*, 2014; Geilfus *et al.*, 2019). The seasonal cycle of sea ice growth and retreat means that while sea ice may not be a permanent sink for microplastics, it could be an important temporal sink for microplastic pollution (Obbard *et al.*, 2014). Once trapped in sea ice, microplastics can be transported from the Eurasian basin (where the majority of microplastics are entrained within sea ice) across the Arctic basin via the Transpolar Drift towards the Fram Strait, where microplastic-laden meltwater may be exported from the Arctic once more (Peeken *et al.*, 2018).

Antarctica was once considered to be the last pristine region on Earth, as the Antarctic Circumpolar Current and Polar Front systems have long been thought to protect the Antarctic Continent from debris travelling southwards due to the dominant northward Ekman transport. Waller *et al.* (2017) provided the first review of the state of microplastic research in the Southern Ocean and surrounding the Antarctic continent, and highlighted the need for more observational data. While concentrations of floating macroplastics and microplastics in the Southern Ocean have been shown to be the lowest globally (Suaria *et al.*, 2020), there is evidence of the presence of plastics all around the Antarctic continent with reports of microplastics present at the sea surface (Cincinelli *et al.*, 2017), within Antarctic sea ice (Kelly *et al.*, 2020), and within sediments (Munari *et al.*, 2017; Reed *et al.*, 2018; Cunningham *et al.*, 2020). The Pacific Ocean sector of the Southern Ocean has been suggested to be the most prevalent source of buoyant microplastics to the waters surrounding Antarctica although deep-water transport of neutrally buoyant plastics or local sources, such as scientific bases and increased ship traffic, may be a more likely source of microplastics (Cincinelli *et al.*, 2017; Munari *et al.*, 2017; Mountford & Morales Maqueda, 2021). A thorough review of plastics in polar samples can be found in Tirelli, Suaria and Lusher (2020) as well as in the "Plastics in the Antarctic and Southern Ocean" chapter.

5. Vertical Transport of Plastics

Despite the majority of plastics being produced and, as a consequence, entering the world's oceans being less dense than seawater, estimates of plastics floating at the sea surface account for less than 1% of the total amount of plastics believed to have entered the oceans. The location of the "missing plastics" has been the focus of much research in recent years.

The sinking behaviour of plastics (aside from entrainment in marine snow and under the effects of biofouling) is determined by their inherent density, size and, perhaps most importantly, its shape. Macroplastics and microplastics alike are heterogeneous in terms of their shape, but this difference in shape and its impact on sinking are particularly relevant for microplastics. The most frequently encountered forms of microplastics are fibres, films, fragments, foams and beads with secondary microplastics typically having a more irregular morphology and primary microplastics which have not yet been weathered or degraded having a more regular (and often spherical) shape (Chubarenko *et al.*, 2016; Khatmullina & Isachenko, 2017). Weathering and degradation (through UV radiation, mechanical or biological degradation) can lead to changes to the shape, surface texture and weight of a particle, and fragmentation from larger into smaller plastics is likely to modify a plastic's rise or sink velocity. Within the water column, plastics may settle out when they reach an area of similar density, for example, at a thermocline or pycnocline (Zobkov *et al.*, 2019). The water column may be an important repository for neutrally buoyant plastics, which remain at depths of equal buoyancy unless they are drawn towards the seafloor by biofouling, turbulence, deep-water formation or downwelling. While possible residence times for plastics in the water column under biofouling conditions have been reported, with polyethylene fibres estimated to spend 6–8 months in the euphotic zone before sinking (Chubarenko *et al.*, 2016), residence times for neutrally buoyant plastics and a wider range of fouled buoyant plastics are still lacking.

Through processes such as biofouling, fragmentation and ingestion/egestion, floating plastics may lose their buoyancy and hence sink out of the sea surface layer. There has been an observed paucity of smaller plastics (i.e., microplastics) at the sea surface (Cózar *et al.*, 2014), as with decreasing size there is also decreasing buoyancy, with small microplastics and nanoplastics hypothesised to be essentially neutrally buoyant (Kooi *et al.*, 2017). Evidence of fallout of positively buoyant plastics from the North Pacific Garbage Patch has been recorded with over half of the particles (52%)

detected in the water column being smaller than 1.5 mm in size (Egger *et al.*, 2020). These smaller particles are more susceptible to processes such as turbulence and downwelling and are able to be drawn down into the water column. Pabortsava and Lampitt (2020) similarly observed elevated concentrations of microplastics in the Atlantic Ocean, with the top 200 m estimated to contain between 11.6 and 21.1 Mt of microplastics (32–651 μm) of only the three most littered plastics (polyethylene, polypropylene and polystyrene). Wind-driven mixing has been shown to mix plastic particles of positive buoyancy downward through the upper water column, both through observational data (Reisser *et al.*, 2015) and modelling experiments (Kukulka *et al.*, 2012). As well as these wind-driven processes, heat-driven turbulence is also a potential mechanism, whereby buoyant particles are also transported below the sea surface. Kukulka *et al.* (2016) observed a significant increase in surface plastics during surface heating and a significant decrease during surface cooling and hypothesise that the transport of buoyant plastics out of the sea surface layer is enhanced at night.

The formation of a biofilm occurs within seconds after submersion, with colonising species of bacteria adhering to the surface of the material and releasing a conditioning matrix of extracellular polymeric substances (EPS). This initial biofilm encourages the settlement of further microorganisms, such as bacteria and diatoms. Biofouling succession eventually leads to the potential adhesion of macroalgae and organisms such as mussels, polychaetes and bryozoans to larger debris. Biofouling, whether it is on a microscale or macroscale, affects the density of plastics and can cause positively and neutrally buoyant plastics to become negatively buoyant. After these particles sink, they may settle out at depth once they reach an area of equal buoyancy, most likely out of the euphotic zone. This lack of light, in combination with grazing, can lead to "defouling", in which fouling organisms detach, and density decreases (Ye & Andrady, 1991; Andrady, 2011). This cycle of biofouling and defouling can lead to an oscillation effect of sinking and resuspension of plastics over time (Kooi *et al.*, 2017). The density changes and timescales of this biofouling/defouling cycle are dependent upon the initial density and size of the particle. Modelling work by Kooi *et al.* (2017) explores the trade-off between particle size (radius) and the surface area to volume ratio. The results suggest that larger particles (that is, with a larger radius) have the potential to lose buoyancy more rapidly than smaller particles as they are more likely to collide with algae. However, smaller particles need fewer algal cells to increase their density to the point of sinking.

The presence of biofouling material on the surface of microplastics may encourage grazing (and thus ingestion) in species that use chemoreception to detect prey; this has been observed in a number of copepod species (Cole *et al.*, 2013) although the opposite was observed in some hard coral species (Allen *et al.*, 2017). Ingestion of both clean and biofouled plastics has been observed on both a macroscale and microscale, and while the digestive tracts of organisms are a temporary sink for the plastics, the subsequent egestion can influence a plastic particle's vertical distribution within the water column. This is particularly true of microplastics, which become incorporated within faecal pellets. Microplastics can alter the density and sinking rates of the faecal pellets, and vice versa, with faecal pellets believed to be a vector for transporting buoyant plastics away from the sea surface and into the ocean interior (Cole *et al.*, 2016). Faecal pellets, and the microplastics contained within, form aggregates with detritus, particulate organic matter (POM), phytoplankton, extrapolymeric substances (EPS) and other matter forming marine snow, an essential component of the biological carbon pump, and the export of organic matter to the deep ocean (Coyle *et al.*, 2020; Zhao *et al.*, 2017). Kvale *et al.* (2020a) found that if pollution were to stop, marine snow and aggregates would be able to transport most existing surface microplastics to the seafloor within two years, suggesting that marine snow may be a significant removal mechanism of floating microplastics.

Plastics that are inherently denser than seawater (or those that lose buoyancy over time, due to processes such as biofouling or aggregation) that reach the seafloor, similar to plastics that remain in coastal regions, will either become trapped within sediments where they settle initially or be subject to horizontal transport by bottom currents. As discussed previously, coastal sediments are a likely temporary sink for plastics although there is the potential for them to be sequestered within coastal sediments over longer timescales. Those plastics that do not become trapped within near-shore sediments are then subject to bottom or near-bottom currents, which transport them away from their point of entry. Far less is known about the distribution of plastic pollution on the seafloor compared to at the sea surface. Despite this, instances of both macrolitter and microlitter have been recorded on the seafloor and within seafloor sediments in all major ocean basins, albeit with a focus on sampling in coastal regions, European seas, the Western Pacific and North America, as detailed by Canals *et al.* (2020). As mentioned earlier in this chapter, the Mediterranean is thought to be one of the most affected regions, with floating plastic concentrations to rival those of the garbage patches (Suaria *et al.*, 2016). It stands to reason, therefore, that the Mediterranean seafloor is also a likely hotspot of plastic pollution. While there is no comprehensive review of the state of plastic pollution on the Mediterranean seafloor at present, the region has been the focus of studies since the 1990s (Galgani *et al.*, 1995, 1996), and the MEDiterranean International bottom Trawl Survey (MEDITS) has provided the opportunity for monitoring the state of seafloor litter across the Mediterranean (e.g., García-Rivera *et al.*, 2018; Alomar *et al.*, 2020; Garofalo *et al.*, 2020, among others). In the Indian Ocean, it has been estimated that the sediments may contain up to 4 billion plastic fibres per square kilometre, which equates to approximately 1.4 Bt of plastic (Woodall *et al.*, 2014; Coppock *et al.*, 2017). Once on the seafloor, bioturbating organisms, such as mussels (Van Colen *et al.*, 2020), clams and polychaetes (Näkki *et al.*, 2017), are able to transport small plastics into deeper layers of marine sediments.

6. Plastic Accumulation in the Deep Sea

Despite the focus on the coastal seafloor, in part due to the relative ease of collecting observational data in these regions, abyssal and hadal plains and trenches are thought to be the "ocean's ultimate trashcan", as once plastics enter these deep-ocean areas, the exit is near-impossible (Peng *et al.*, 2020). Turbidity currents are believed to be an important mechanism for transporting microplastics away from the continental shelf and into submarine canyons, where a portion of the plastics may be buried, in particular, fibres, which are preferentially removed from suspension by these currents (Pohl *et al.*, 2020). Once off the continental slope, thermohaline currents are able to pick up microplastics deposited by turbidity currents and transport them further offshore leading to microplastic "hotspots" in the deep sea, with the reported highest concentration of microplastics ever found in sediments, of up to 3,800 microplastics kg^{-1} (Kane *et al.*, 2020). Finally, funnelling mechanisms along submarine canyons and hadal trenches concentrate plastics into these regions, with little opportunity for escape due to steep slopes (Peng *et al.*, 2020). Microplastics and macroplastics have been found down to depths of 10,927 m in the Challenger Deep, and in other abyssal plains and hadal trenches in the Western Pacific Ocean (Chiba *et al.*, 2018; Street, 2019; Peng *et al.*, 2020). The concentrations that microplastic pollution reaches these depths is not insignificant, reaching abundances of over 100 items per kg of the dry weight of sediment. Many of the fibres recovered from these deep-ocean locations were polymers of a lower density than seawater (Peng *et al.*, 2020), suggesting that they were either small enough to have reached neutral buoyancy and been drawn down into the deep ocean through turbulent mixing or downwelling processes, or perhaps more likely, may have lost buoyancy through biofouling or incorporation into marine snow, for example.

Large accumulations of plastics on the seafloor, such as the large debris dumps discovered in Xisha Trough, South China Sea, have the potential to provide a new ecosystem for organisms, resulting in biodiversity hotspots, particularly for sessile organisms (Song *et al.*, 2021). Since bottom currents are essential for transporting the nutrients and oxygen that support biodiversity, it is not surprising that (micro)plastics are transported along the same pathways, and flourishing ecosystems emerge on plastic accumulations (Kane *et al.*, 2020). However, plastics accumulating on the seafloor may be releasing high concentrations of additives, as has been observed in Sagami Bay, Japan, and the West Pacific Ocean, in which over 1,500 kg of additives (dibutyl phthalate butylated hydroxytoluene, bis (2-ethylhexyl) phthalate and 1,2,4-trichlorobenzene) were estimated to be present on the seafloor under the Kuroshio Extension and its recirculation gyre (Yamauchi *et al.*, 2021). The release of these chemicals could have damaging effects on local ecosystems and may also be transported away from the region via ocean currents.

7. Conclusion

The transport and ultimate fate of plastics in the world's oceans are largely defined by its inherent buoyancy, that is, whether it is more, less or equally as dense as the surrounding seawater. The coastlines represent an important temporary (or potentially permanent) sink for marine plastics of all densities, although cycling or beaching followed by resuspension can lead to the generation of secondary microplastics, which are preferentially transported away from the coastal regions. Broadly, many buoyant plastics find their way into the subtropical garbage patches and are also readily transported northwards into the Arctic, thought to be a potential "dead end" for plastic pollution. Floating plastics that lose their buoyancy and small microplastics and nanoplastics are largely assumed to have neutral buoyancy and are at the mercy of ocean currents transporting them vertically through the water column. The water column itself is likely a large repository for microplastic pollution, although processes such as biofouling and aggregation can result in the export of these particles towards the seafloor. The seafloor, in particular, abyssal and hadal plains and trenches, is thought to be the final resting place for the majority of plastics that enter the world's oceans.

References

Allen, A. S., Seymour, A. C., & Rittschof, D. (2017). Chemoreception drives plastic consumption in a hard coral. *Marine Pollution Bulletin, 124*(1), 198–205.

Alomar, C., Compa, M., Deudero, S., & Guijarro, B. (2020). Spatial and temporal distribution of marine litter on the seafloor of the Balearic Islands (western Mediterranean Sea). *Deep Sea Research Part I: Oceanographic Research Papers, 155*, 103178.

Andrades, R., Santos, R. G., Joyeux, J. C., Chelazzi, D., Cincinelli, A., & Giarrizzo, T. (2018). Marine debris in Trindade Island, a remote island of the South Atlantic. *Marine Pollution Bulletin, 137*, 180–184.

Andrady, A. L. (2011). Microplastics in the marine environment. *Marine Pollution Bulletin, 62*(8), 1596–1605.

Barnes, D. K. (2002). Invasions by marine life on plastic debris. *Nature, 416*(6883), 808–809.

Barnes, D. K. (2005). Remote islands reveal rapid rise of southern hemisphere sea debris. *The Scientific World Journal, 5*, 915–921.

Bergmann, M., Lutz, B., Tekman, M. B., & Gutow, L. (2017a). Citizen scientists reveal: Marine litter pollutes Arctic beaches and affects wild life. *Marine Pollution Bulletin, 125*(1–2), 535–540.

Bergmann, M., Sandhop, N., Schewe, I., & D'Hert, D. (2016). Observations of floating anthropogenic litter in the Barents Sea and Fram Strait, Arctic. *Polar Biology, 39*, 553–560.

Bergmann, M., Wirzberger, V., Krumpen, T., Lorenz, C., Primpke, S., Tekman, M. B., & Gerdts, G. (2017b). High quantities of microplastic in Arctic deep-sea sediments from the HAUSGARTEN observatory. *Environmental Science & Technology, 51*(19), 11000–11010.

Canals, M., Pham, C. K., Bergmann, M., Gutow, L., Hanke, G., van Sebille, E., Angiolillo, M., Buhl-Mortensen, L., Cau, A., Ioakeimidis, C., & Kammann, U. (2020). The quest for seafloor macrolitter: A critical review of background knowledge, current methods and future prospects. *Environmental Research Letters, 16*, 023001.

Chubarenko, I., Efimova, I., Bagaeva, M., Bagaev, A., & Isachenko, I. (2020). On mechanical fragmentation of single-use plastics in the sea swash zone with different types of bottom sediments: Insights from laboratory experiments. *Marine Pollution Bulletin, 150*, 110726.

Cincinelli, A., Martellini, T., Guerranti, C., Scopetani, C., Chelazzi, D., & Giarrizzo, T. (2019). A potpourri of microplastics in the sea surface and water column of the Mediterranean Sea. *TrAC Trends in Analytical Chemistry*, *110*, 321–326.

Cole, M., Lindeque, P., Fileman, E., Halsband, C., Goodhead, R., Moger, J., & Galloway, T.S. (2013). Microplastic ingestion by zooplankton. *Environmental Science & Technology*, *47*(12), 6646–6655.

Cole, M., Lindeque, P. K., Fileman, E., Clark, J., Lewis, C., Halsband, C., & Galloway, T. S. (2016). Microplastics alter the properties and sinking rates of zooplankton faecal pellets. *Environmental Science & Technology*, *50*(6), 3239–3246.

Coyle, R., Hardiman, G., & O'Driscoll, K. (2020). Microplastics in the marine environment: A review of their sources, distribution processes, uptake and exchange in ecosystems. *Case Studies in Chemical and Environmental Engineering*, *2*, 100010.

Cózar, A., Echevarría, F., González-Gordillo, J. I., Irigoien, X., Úbeda, B., Hernández-León, S., Palma, Á. T., Navarro, S., García-de-Lomas, J., Ruiz, A., & Fernández-de-Puelles, M. L. (2014). Plastic debris in the open ocean. *Proceedings of the National Academy of Sciences*, *111*(28), 10239–10244.

Cózar, A., Martí, E., Duarte, C. M., García-de-Lomas, J., van Sebille, E., Ballatore, T. J., Eguíluz, V. M., González-Gordillo, J. I., Pedrotti, M. L., Echevarría, F., & Troublè, R. (2017). The Arctic Ocean as a dead end for floating plastics in the North Atlantic branch of the thermohaline circulation. *Science Advances*, *3*(4), e1600582.

Cunningham, E. M., Ehlers, S. M., Dick, J. T., Sigwart, J. D., Linse, K., Dick, J. J., & Kiriakoulakis, K. (2020). High abundances of microplastic pollution in deep-sea sediments: Evidence from Antarctica and the Southern Ocean. *Environmental Science & Technology*, *54*(21), 13661–13671.

Desforges, J. P. W., Galbraith, M., Dangerfield, N., & Ross, P. S. (2014). Widespread distribution of microplastics in subsurface seawater in the NE Pacific Ocean. *Marine Pollution Bulletin*, *79*(1–2), 94–99.

Dobler, D., Huck, T., Maes, C., Grima, N., Blanke, B., Martinez, E., & Ardhuin, F. (2019). Large impact of stokes drift on the fate of surface floating debris in the South Indian Basin. *Marine Pollution Bulletin*, *148*, 202–209.

Duhec, A. V., Jeanne, R. F., Maximenko, N., & Hafner, J. (2015). Composition and potential origin of marine debris stranded in the Western Indian Ocean on remote Alphonse Island, Seychelles. *Marine Pollution Bulletin*, *96*(1–2), 76–86.

Eriksen, M., Maximenko, N., Thiel, M., Cummins, A., Lattin, G., Wilson, S., Hafner, J., Zellers, A., & Rifman, S. (2013). Plastic pollution in the South Pacific subtropical gyre. *Marine Pollution Bulletin, 68*(1–2), 71–76.

Galgani, F., Brien, A. S. O., Weis, J., Ioakeimidis, C., Schuyler, Q., Makarenko, I., Griffiths, H., Bondareff, J., Vethaak, D., Deidun, A., & Sobral, P. (2021). Are litter, plastic and microplastic quantities increasing in the ocean? *Microplastics and Nanoplastics*, *1*(1), 1–4.

Galgani, F., Jaunet, S., Campillo, A., Guenegen, X., & His, E. (1995). Distribution and abundance of debris on the continental shelf of the north-western Mediterranean Sea. *Marine Pollution Bulletin, 30*(11), 713–717.

Galgani, F., Souplet, A., & Cadiou, Y. (1996). Accumulation of debris on the deep sea floor off the French Mediterranean coast. *Marine Ecology Progress Series*, *142*, 225–234.

García-Rivera, S., Lizaso, J. L. S., & Millán, J. M. B. (2018). Spatial and temporal trends of marine litter in the Spanish Mediterranean seafloor. *Marine Pollution Bulletin, 137*, 252–261.

Garofalo, G., Quattrocchi, F., Bono, G., Di Lorenzo, M., Di Maio, F., Falsone, F., Gancitano, V., Geraci, M. L., Lauria, V., Massi, D., & Scannella, D. (2020). What is in our seas? Assessing anthropogenic litter on the seafloor of the central Mediterranean Sea. *Environmental Pollution, 266*, 115213.

Geilfus, N.-X., Munson, K., Sousa, J., Germanov, Y., Bhugaloo, S., Babb, D., & Wang, F. (2019). Distribution and impacts of microplastic incorporation within sea ice. *Marine Pollution Bulletin, 145*, 463–473.

Hinata, H., Mori, K., Ohno, K., Miyao, Y., & Kataoka, T. (2017). An estimation of the average residence times and onshore-offshore diffusivities of beached microplastics based on the population decay of tagged meso-and macrolitter. *Marine Pollution Bulletin, 122*(1–2), 17–26.

Isobe, A., Kubo, K., Tamura, Y., Nakashima, E., & Fujii, N. (2014). Selective transport of microplastics and mesoplastics by drifting in coastal waters. *Marine Pollution Bulletin, 89*(1–2), 324–330.

Kane, I. A., Clare, M. A., Miramontes, E., Wogelius, R., Rothwell, J. J., Garreau, P., & Pohl, F. (2020). Seafloor microplastic hotspots controlled by deep-sea circulation. *Science, 368*(6495), 1140–1145.

Kanhai, L. D. K., Gardfeldt, K., Krumpen, T., Thompson, R. C., & O'Connor, I. (2020). Microplastics in sea ice and seawater beneath ice floes from the Arctic ocean. *Scientific Reports*, *10*, 1–11.

Kelly, A., Lannuzel, D., Rodemann, T., Meiners, K., & Auman, H. (2020). Microplastic contamination in east Antarctic sea ice. *Marine Pollution Bulletin, 154*, 111130.

Khatmullina, L. & Isachenko, I. (2017). Settling velocity of microplastic particles of regular shapes. *Marine Pollution Bulletin, 114*(2), 871–880.

Kooi, M., van Nes, E. H., Scheffer, M., & Koelmans, A. A. (2017). Ups and downs in the ocean: Effects of biofouling on the vertical transport of microplastics. *Environmental science & technology*, *51*(14), 7963–7971.
Kukulka, T., Law, K. L., & Proskurowski, G. (2016). Evidence for the influence of surface heat fluxes on turbulent mixing of microplastic marine debris. *Journal of Physical Oceanography*, *46*(3), 809–815.
Kukulka, T., Proskurowski, G., Morét-Ferguson, S., Meyer, D. W., & Law, K. L. (2012). The effect of wind mixing on the vertical distribution of buoyant plastic debris. *Geophysical Research Letters*, *39*(7), 1–6.
Kvale, K. F., Friederike Prowe, A. E., & Oschlies, A. (2020a). A critical examination of the role of marine snow and zooplankton fecal pellets in removing ocean surface microplastic. *Frontiers in Marine Science*, *6*, 808.
Kvale, K., Prowe, A. E. F., Chien, C. T., Landolfi, A., & Oschlies, A. (2020b). The global biological microplastic particle sink. *Scientific Reports*, *10*(1), 1–12.
Lavers, J. L. & Bond, A. L. (2017). Exceptional and rapid accumulation of anthropogenic debris on one of the world's most remote and pristine islands. *Proceedings of the National Academy of Sciences*, *114*(23), 6052–6055.
Law, K. L., Morét-Ferguson, S., Maximenko, N. A., Proskurowski, G., Peacock, E. E., Hafner, J., & Reddy, C. M. (2010). Plastic accumulation in the North Atlantic subtropical gyre. *Science*, *329*(5996), 1185–1188.
Lebreton, L. C.-M., Greer, S. D., & Borrero, J. C. (2012). Numerical modelling of floating debris in the world's oceans. *Marine Pollution Bulletin*, *64*(3), 653–661.
Lebreton, L. C.-M., Slat, B., Ferrari, F., Sainte-Rose, B., Aitken, J., Marthouse, R., Hajbane, S., Cunsolo, S., Schwarz, A., Levivier, A., & Noble, K. (2018). Evidence that the Great Pacific Garbage Patch is rapidly accumulating plastic. *Scientific Reports*, *8*(1), 1–15.
Liubartseva, S., Coppini, G., Lecci, R., & Clementi, E. (2018). Tracking plastics in the Mediterranean: 2D Lagrangian model. *Marine Pollution Bulletin*, *129*(1), 151–162.
Martin, C., Baalkhuyur, F., Valluzzi, L., Saderne, V., Cusack, M., Almahasheer, H., Krishnakumar, P. K., Rabaoui, L., Qurban, M. A., Arias-Ortiz, A., & Masqué, P. (2020). Exponential increase of plastic burial in mangrove sediments as a major plastic sink. *Science Advances*, *6*(44), eaaz5593.
Maximenko, N., Hafner, J., & Niiler, P. (2012). Pathways of marine debris derived from trajectories of Lagrangian drifters. *Marine Pollution Bulletin*, *65*(1–3), 51–62.
Monteiro, R. C., do Sul, J. A. I., & Costa, M. F. (2018). Plastic pollution in islands of the Atlantic Ocean. *Environmental Pollution*, *238*, 103–110.
Mountford, A. S. & Morales Maqueda, M. A. (2019). Eulerian modeling of the three-dimensional distribution of seven popular microplastic types in the

global ocean. *Journal of Geophysical Research: Oceans*, *124*(12), 8558–8573.

Mountford, A. S. & Morales Maqueda, M. A. (2021). Modeling the accumulation and transport of microplastics by sea ice. *Journal of Geophysical Research: Oceans*, *126*, e2020JC016826. https://doi.org/10.1029/2020JC016826.

Munari, C., Infantini, V., Scoponi, M., Rastelli, E., Corinaldesi, C., & Mistri, M. (2017). Microplastics in the sediments of Terra Nova Bay (Ross Sea, Antarctica). *Marine Pollution Bulletin*, *122*(1–2), 161–165.

Näkki, P., Setälä, O., & Lehtiniemi, M. (2017). Bioturbation transports secondary microplastics to deeper layers in soft marine sediments of the northern Baltic Sea. *Marine Pollution Bulletin*, *119*(1), 255–261.

Obbard, R. W., Sadri, S., Wong, Y. Q., Khitun, A. A., Baker, I., & Thompson, R. C. (2014). Global warming releases microplastic legacy frozen in Arctic Sea ice. *Earth's Future*, *2*, 315–320.

Onink, V., Jongedijk, C., Hoffman, M., van Sebille, E., & Laufkötter, C. (2021). Global simulations of marine plastic transport show plastic trapping in coastal zones. *Environmental Research Letters*. In Press: https://doi.org/10.1088/1748-9326/abecbd.

Onink, V., Wichmann, D., Delandmeter, P., & van Sebille, E. (2019). The role of Ekman currents, geostrophy, and stokes drift in the accumulation of floating microplastic. *Journal of Geophysical Research: Oceans*, *124*, 1474–1490. https://doi.org/10.1029/2018JC014547.

Pabortsava, K. & Lampitt, R. S. (2020). High concentrations of plastic hidden beneath the surface of the Atlantic Ocean. *Nature Communications*, *11*(1), 1–11.

Pattiaratchi, C., van der Mheen, M., Schlundt, C., Narayanaswamy, B. E., Sura, A., Hajbane, S., White, R., Kumar, N., Fernandes, M., & Wijeratne, S. (2021). Plastics in the Indian Ocean–sources, fate, distribution and impacts. *Ocean Science Ocean Science*, 18(1), 1–28.

Peeken, I., Primpke, S., Beyer, B., Gütermann, J., Katlein, C., Krumpen, T., Bergmann, M., Hehemann, L., & Gerdts, G. (2018). Arctic sea ice is an important temporal sink and means of transport for microplastic. *Nature Communications*, *9*, 1505.

Peng, G., Bellerby, R., Zhang, F., Sun, X., & Li, D. (2020). The ocean's ultimate trashcan: Hadal trenches as major depositories for plastic pollution. *Water Research*, *168*, 115121.

Perez-Venegas, D., Pavés, H., Pulgar, J., Ahrendt, C., Seguel, M., & Galbán-Malagón, C. J. (2017). Coastal debris survey in a remote island of the Chilean Northern Patagonia. *Marine Pollution Bulletin*, *125*(1–2), 530–534.

PlasticsEurope. (2020). Plastics — The facts 2020: An analysis of European plastics production, demand and waste data. Retrieved from https://www.plasticseurope.org/resources/market-data.

Pohl, F., Eggenhuisen, J. T., Kane, I. A., & Clare, M. A. (2020). Transport and burial of microplastics in deep-marine sediments by turbidity currents. *Environmental Science & Technology*, *54*(7), 4180–4189.

Poulain, M., Mercier, M. J., Brach, L., Martignac, M., Routaboul, C., Perez, E., Desjean, M. C., & Ter Halle, A. (2018). Small microplastics as a main contributor to plastic mass balance in the North Atlantic Subtropical Gyre. *Environmental Science & Technology*, *53*(3), 1157–1164.

Reed, S., Clark, M., Thompson, R., & Hughes, K. A. (2018). Microplastics in marine sediments near Rothera research station, Antarctica. *Marine Pollution Bulletin*, *133*, 460–463.

Reisser, J., Slat, B., Noble, K., Du Plessis, K., Epp, M., Proietti, M., de Sonneville, J., Becker, T., & Pattiaratchi, C. (2015). The vertical distribution of buoyant plastics at sea: An observational study in the North Atlantic Gyre. *Biogeosciences*, *12*(4), 1249–1256.

Ryan, P. G. (2014). Litter survey detects the South Atlantic 'garbage patch'. *Marine Pollution Bulletin*, *79*(1–2), 220–224.

Ryan, P. G. (2015). Does size and buoyancy affect the long-distance transport of floating debris? *Environmental Research Letters*, *10*(8), 084019.

Song, X., Lyu, M., Zhang, X., Ruthensteiner, B., Ahn, I. Y., Pastorino, G., Wang, Y., Gu, Y., Ta, K., Sun, J., & Liu, X. (2021). Large plastic debris dumps: New biodiversity hot spots emerging on the deep-sea floor. *Environmental Science & Technology Letters*, *8*(2), 148–154.

Street, F. (2019). Deepest ever dive finds 'plastic bag' at bottom of Mariana Trench. Retrieved from https://edition.cnn.com/travel/article/victor-vescovo-deepest-dive-pacific/index.html.

Suaria, G., Perold, V., Lee, J. R., Lebouard, F., Aliani, S., & Ryan, P. G. (2020). Floating macro-and microplastics around the Southern Ocean: Results from the Antarctic circumnavigation expedition. *Environment International*, *136*, 105494.

Tekman, M. B., Wekerle, C., Lorenz, C., Primpke, S., Hasemann, C., Gerdts, G., & Bergmann, M. (2020). Tying up loose ends of microplastic pollution in the Arctic: Distribution from the sea surface through the water column to deep-sea sediments at the HAUSGARTEN observatory. *Environmental Science & Technology*, *54*(7), 4079–4090.

Tirelli, V., Suaria, G., & Lusher, A. L. (2020). Microplastics in polar samples. In Rocha-Santos, T., Costa, M., & Mouneyrac, C. (Eds.), *Handbook of Microplastics in the Environment*. Cham: Springer. https://doi.org/10.1007/978-3-030-10618-8_4-1.

Tošić, T. N., Vruggink, M., & Vesman, A. (2020). Microplastics quantification in surface waters of the Barents, Kara and White Seas. *Marine Pollution Bulletin*, *161*, 111745.

Van Colen, C., Moereels, L., Vanhove, B., Vrielinck, H., & Moens, T. (2021). The biological plastic pump: Evidence from a local case study using blue mussel and infaunal benthic communities. *Environmental Pollution, 274*, 115825.

van der Mheen, M., Pattiaratchi, C., & van Sebille, E. (2019). Role of Indian Ocean dynamics on accumulation of buoyant debris. *Journal of Geophysical Research: Oceans, 124*(4), 2571–2590.

van Sebille, E., Griffies, S. M., Abernathey, R., Adams, T. P., Berloff, P., Biastoch, A., Blanke, B., Chassignet, E. P., Cheng, Y., Cotter, C. J., & Deleersnijder, E. (2018). Lagrangian ocean analysis: Fundamentals and practices. *Ocean Modelling, 121*, 49–75.

van Sebille, E., Sprintall, J., Schwarzkopf, F. U., Sen Gupta, A., Santoso, A., England, M. H., Biastoch, A., & Böning, C. W. (2014). Pacific-to-Indian Ocean connectivity: Tasman leakage, Indonesian throughflow, and the role of ENSO. *Journal of Geophysical Research: Oceans, 119*(2), 1365–1382.

van Sebille, E., Wilcox, C., Lebreton, L., Maximenko, N., Hardesty, B. D., van Franeker, J. A., Eriksen, M., Siegel, D., Galgani, F., & Law, K. L. (2015). A global inventory of small floating plastic debris. *Environmental Research Letters, 10*(12), 124006.

Waite, A. M., Thompson, P. A., Pesant, S., Feng, M., Beckley, L. E., Domingues, C. M., Gaughan, D., Hanson, C. E., Holl, C. M., Koslow, T., & Meuleners, M. (2007). The Leeuwin current and its eddies: An introductory overview. *Deep Sea Research Part II: Topical Studies in Oceanography, 54*(8–10), 789–796.

Wichmann, D., Delandmeter, P., & van Sebille, E. (2019). Influence of near-surface currents on the global dispersal of marine microplastic. *Journal of Geophysical Research: Oceans, 124*(8), 6086–6096.

Wilcox, C., Hardesty, B. D., & Law, K. L. (2019). Abundance of floating plastic particles is increasing in the Western North Atlantic Ocean. *Environmental Science & Technology, 54*(2), 790–796.

Yamauchi, T., Nakajima, R., Tsuchiya, M., Yabuki, A., Kitahashi, T., Nagano, Y., Isobe, N., & Nakata, H. (2021). Plastic additives in deep-sea debris collected from the western North Pacific and estimation for their environmental loads. *Science of the Total Environment, 768*, 144537.

Ye, S. & Andrady, A. L. (1991). Fouling of floating plastic debris under Biscayne Bay exposure conditions. *Marine Pollution Bulletin, 22*(12), 608–613.

Zhang, H. (2017). Transport of microplastics in coastal seas. *Estuarine, Coastal and Shelf Science, 199*, 74–86.

Zhao, S., Danley, M., Ward, J. E., Li, D., & Mincer, T. J. (2017). An approach for extraction, characterization and quantitation of microplastic in natural marine snow using Raman microscopy. *Analytical Methods, 9*(9), 1470–1478.

Zobkov, M. B., Esiukova, E. E., Zyubin, A. Y., & Samusev, I. G. (2019). Microplastic content variation in water column: The observations employing a novel sampling tool in stratified Baltic Sea. *Marine Pollution Bulletin, 138*, 193–205.

https://doi.org/10.1142/9789811259111_0005

Chapter 4

Plastic Pollution and Marine Megafauna: Recent Advances and Future Directions

Sarah E. Nelms[*,¶], Bethany L. Clark[*,†], Emily M. Duncan[*], Elitza Germanov[‡], Brendan J. Godley[*], Kristian J. Parton[*], Christopher K. Pham[§] and Yasmina Rodríguez[§]

[*]*Centre for Ecology and Conservation, University of Exeter, Cornwall, UK*

[†]*BirdLife International, The David Attenborough Building, Pembroke Street, Cambridge, UK*

[‡]*Marine Megafauna Foundation, Truckee, CA, USA*

[§]*Okeanos R&D Centre, University of the Azores, Horta, Portugal*

[¶]*s.nelms@exeter.ac.uk*

1. Introduction

The widespread and pervasive nature of plastic pollution has resulted in a growing body of evidence documenting the detrimental effects of anthropogenic waste on aquatic wildlife (Provencher *et al.*, 2017; Kühn & van Franeker, 2020). Over the last two decades, the number of marine species known to ingest or become entangled in debris, of which the majority is plastic, has more than trebled from 267 (Laist, 1997) to 914 (Kühn & van Franeker, 2020).

Of particular concern are marine megafauna, namely, marine mammals (cetaceans, pinnipeds, sirenians and the polar bear), elasmobranchs (sharks and rays), marine turtles and seabirds. These large marine vertebrates play key roles in the functioning and maintenance of marine habitats and are often considered indicators of marine ecosystem health (Alves *et al*., 2016; Pimiento *et al*., 2020). They also provide economic, social and cultural benefits and serve as flagship species for enhancing public engagement in marine conservation issues (Frazier, 2005; Germanov *et al*., 2018). Despite this, many marine megafauna species face a range of anthropogenic pressures, including fisheries (direct take and by-catch), climate change, habitat loss and pollution (Pimiento *et al*., 2020; Nelms *et al*., 2021), one-third are considered vulnerable to extinction as a result (Pimiento *et al*., 2020).

The recently emerged threat of plastic pollution presents an additional and omnipresent issue for these species; as the volume of anthropogenic debris circulating within the global ocean increases, so too does the likelihood of its interaction with marine wildlife. Indeed, plastic pollution is now so ubiquitous that even species residing in remote habitats, such as the polar regions, are exposed to it (Bergmann *et al*., 2017; Bessa *et al*., 2019). Currently, all seven marine turtle species, 70% of marine mammal species, and 55% of seabird species are known to interact with anthropogenic debris through ingestion and entanglement (Kühn & van Franeker, 2020). These numbers are likely to grow further still due to the recent increase in attention from the scientific community and general public in plastic pollution. Notably, elasmobranchs have been missing from previous assessments but are beginning to receive greater attention as evidence of interactions with plastic pollution emerges (Abreo *et al*., 2019; Germanov *et al*., 2019b, Mucientes & Queiroz, 2019; Parton *et al*., 2019, 2020).

For individual animals, ingestion of or entanglement in plastic pollution poses a risk of injury or mortality, as well as sub-lethal impacts that may affect their health, resource acquisition and reproductive success (Senko *et al*., 2020). Plastic pollution can also damage and degrade key marine megafauna habitats in both terrestrial and aquatic environments that are important for foraging, breeding and resting. The effects of plastic interactions on individuals may also lead to population-level impacts for some species (Lavers *et al*., 2014), particularly those that are less resilient due to other anthropogenic stressors, but evidence for this is limited to date (Vegter *et al*., 2014; Senko *et al*., 2020).

In this chapter, we (i) outline the impacts of plastic pollution on marine megafauna, (ii) explore recent advances in our understanding of this field of research, and (iii) highlight knowledge gaps and recommend priority future work.

2. Plastic Ingestion by Marine Megafauna

Plastic pollution in the marine environment occurs on a continuous size scale from macroplastics and mesoplastics (>5 mm) to microplastics (<5 mm) and nanoplastics (<1 μm), and the smaller the size, the more bioavailable it is to a greater variety of species. Recent estimates indicate that more than 700 species of marine vertebrates are known to ingest plastic, including 44% of seabird species ($n = 409$), 56% of marine mammal species ($n = 123$), and 100% of sea turtle species ($n = 7$; Kühn & van Franeker, 2020). It is not known how many elasmobranch species ingest plastic due to a lack of assessment.

2.1. *Plastic ingestion pathways*

Depending on foraging behaviours, the uptake of plastic items by marine megafauna may occur through direct (i.e., active and passive ingestion) and/or indirect pathways (i.e., trophic and intergenerational transfer); Figure 1.

2.1.1. *Direct pathways*

Active selection, where animals purposefully target plastic items for consumption, is most likely to occur in species whose foraging strategy is driven by sensory cues, such as visual or olfactory signals (Vegter *et al.*, 2014; Savoca *et al.*, 2016; Pfaller *et al.*, 2020). The degree of similarity between the plastic item characteristics (shape, size, colour and smell) and potential prey items is likely a key driver of selectivity (Schuyler *et al.*, 2014). For example, leatherback turtles (*Dermochelys coriacea*) have the propensity to ingest plastic items with visual similarities to their primary prey of jellyfish-like organisms (Mrosovsky *et al.*, 2009), and green turtles (*Chelonia mydas*) are known to target linear sheet and thread-like items that are black, clear and green in colour, therefore resembling seagrass and algae (Duncan *et al.*, 2019). It should be noted, however, that the colour of plastic may change once ingested, colour classification is subjective and there are clear discrepancies among studies, prompting a

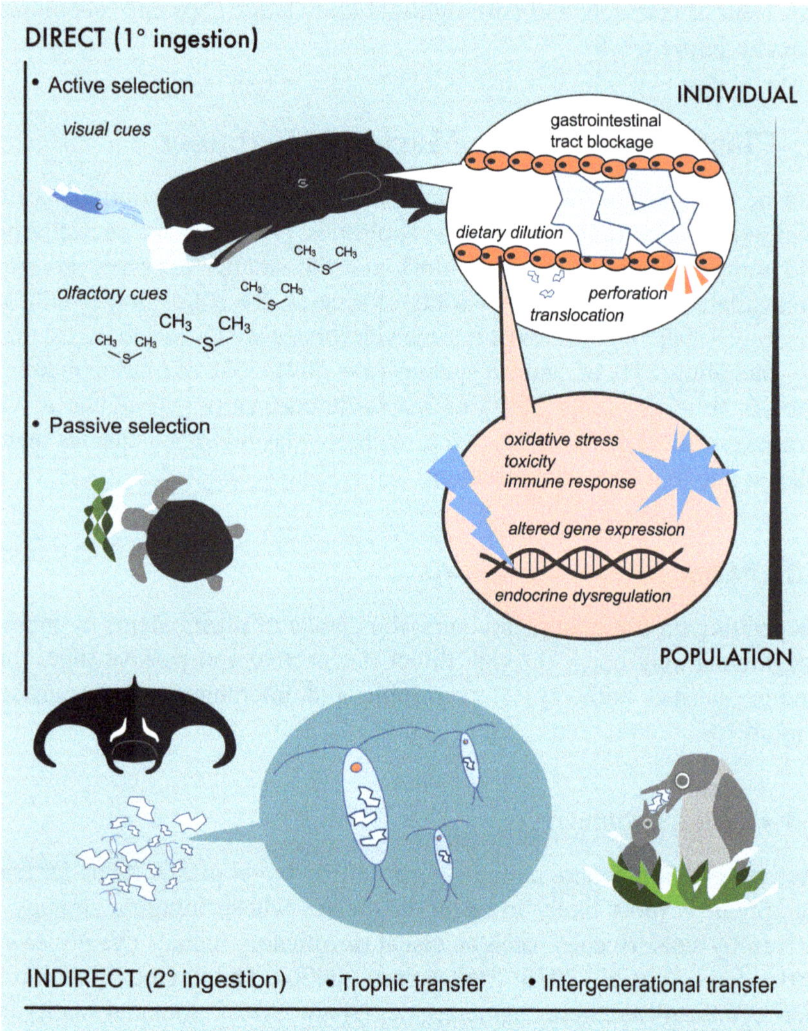

Figure 1. The pathways of plastic ingestion and potential impacts.

Note: Artwork by Elitza Germanov.

call for using standardised colour schemes (Martí *et al.*, 2020). An additional factor influencing ingestion rates of transparent plastic may be polarised vision, which enables animals to distinguish and interpret polarised light and, therefore, detect transparent gelatinous prey (Kagel *et al.*,

2021). A study by Kagel *et al.* (2021) found that transparent plastic bags had higher polarisation contrast than gelatinous zooplankton and a similar radiance contrast, which may be linked to the high ingestion rates of transparent plastic by marine animals.

The hypothesis of olfactory attraction to plastic is based on the potential for dimethyl sulphide (DMS), an odoriferous, airborne and waterborne molecule produced by various marine organisms including phytoplankton, to emit from plastic items as a result of biofouling (Savoca *et al.*, 2016; Pfaller *et al.*, 2020). The DMS signature can be detected by marine organisms, such as some seabird species and turtles, and may be interpreted as an indicator of a food source (Endres & Lohmann, 2012; Dell'Ariccia *et al.*, 2014; Pfaller *et al.*, 2020). Though this theory is difficult to examine in a wild setting and has been questioned for seabirds (Dell'Ariccia *et al.*, 2017), Pfaller *et al.* (2020) employed an experimental set-up and found that turtles detected and responded to airborne odorants emanating from biofouled plastic in the same way they respond to food odours.

Passive ingestion of plastic may occur due to non-selective feeding strategies, such as filter- or suction-feeding marine mammals and elasmobranchs (e.g., Besseling *et al.*, 2015; Germanov *et al.*, 2019b, Parton *et al.*, 2020). This is particularly likely to occur when small plastic pieces are mixed with food items and cannot be distinguished or excluded during the intake of seawater.

2.1.2. *Indirect pathways*

Trophic transfer occurs when prey containing plastic, usually microplastics, are consumed by predators. While there is some evidence of this occurring in invertebrates and small vertebrates, such as crabs and fish (Setälä *et al.*, 2014; Watts *et al.*, 2014; Miller *et al.*, 2020), the number of studies demonstrating its occurrence in marine megafauna is limited and most research to date has focused on pinnipeds. For example, Eriksson and Burton (2003) surmised that wild fur seals (*Arctocephalus* spp.) acquired the small plastic pieces present in their scats through the consumption of contaminated prey (pelagic fish species, *Electrona subaspera*). Subsequently, Nelms *et al.* (2018) empirically demonstrated this process by attributing the presence of microplastics in the scats of captive grey seals (*Halichoerus grypus*) to trophic transfer from the wild-caught mackerel (*Scomber scombrus*) they were fed upon. Filter- and suction-feeding marine mammal and elasmobranch species, and perhaps some turtle species, are likely to be exposed to plastic ingestion both directly from seawater or sediment, and indirectly through ingesting contaminated prey (Casale *et al.*, 2008; Germanov *et al.*, 2018; Parton *et al.*, 2020; Zantis *et al.*, 2022).

Intergenerational transfer occurs when plastic items ingested by adult animals are inadvertently fed to their young during food provisioning. For example, seabird parents regurgitate their stomach contents to feed nestlings with prey items, and sometimes plastic. Procellariiform seabirds (i.e., albatrosses and petrels) that accumulate plastics in their gastrointestinal tract, and regurgitate almost exclusively to feed their chicks, appear to be the most affected (e.g., Carey, 2011; Rodríguez *et al.*, 2012). Microplastics found in human placenta samples indicate maternal transmission of plastic contamination may be possible in marine mammals (Ragusa *et al.*, 2021). Intergenerational transfer could also possibly occur in mammals where very small plastic particles (i.e., nanoplastics) are passed to suckling young during lactation, although this pathway is yet to be investigated.

2.1.3. *Factors affecting the likelihood of plastic ingestion*

Understanding the ecological drivers of plastic ingestion is a key concern for determining which species may be most at risk, and several variables, such as spatial distribution, foraging strategy and diet, can influence the probability of plastic ingestion in marine megafauna (Vegter *et al.*, 2014; Critchell *et al.*, 2019). For example, animals whose foraging ranges overlap with plastic pollution hotspots, namely, oceanic gyres and coastal areas in proximity to dense human population centres, seem to ingest higher amounts of plastic items (e.g., sea turtles (Schuyler *et al.*, 2016; Pham *et al.*, 2017), seabirds (Franeker & Law, 2015; Roman *et al.*, 2019b), and filter feeders (Germanov *et al.*, 2018; Zantis *et al.*, 2022). In addition, differences in plastic ingestion can be influenced by the foraging strategy and the type of habitats in which the animals feed (Di Beneditto & Awabdi, 2014). For example, surface seizing, diving and filtering seabirds are the most affected by plastic ingestion due to their mixed surface-feeding strategies, specifically Procellariiformes (Bourne, 1976). In contrast, pursuit feeders and species foraging below the surface are comparatively at the lowest risk (Roman *et al.*, 2019a). In sea turtles, plastic ingestion rates vary according to life stage and species. For example, immature oceanic stage sea turtles are at greater risk from plastic ingestion than older benthic feeders, and, in general, green, hawksbill (*Eretmochelys imbricata*) and leatherback turtles have been noted to ingest greater quantities and volumes of plastic litter than other turtle species (Schuyler *et al.*, 2014; Lynch, 2018). In marine mammals, deep-diving cetaceans, such as beaked and sperm whales, seem to have a high propensity of ingesting and being compromised by ingesting macroplastic and subsequently being compromised, perhaps due to their reliance on echo-location, rather than vision, to identify prey (Abreo *et al.*, 2016; Alexiadou *et al.*, 2019; Zantis *et al.*, 2021). It is likely that filter-feeding baleen whales consume high numbers of microplastics, but the levels of exposure will depend on the contamination load of seawater and prey within their feeding areas (Zantis *et al.*, 2022).

2.2. *Impact*

2.2.1. *Lethal effects*

Though it is often difficult to categorically identify the main cause of death, the number of reports documenting plastic ingestion-induced mortality in marine megafauna species has increased in the last decade (Senko *et al.*, 2020). Injuries to the digestive tract, such as ulcerations, perforations and lesions, can be caused by hard and sharp pieces of plastic or flexible linear items. Obstructions can also result from large volumes of compacted debris (Di Beneditto & Awabdi, 2014; Nelms *et al.*, 2015). If left untreated, these injuries will most likely lead to ongoing morbidity and could eventually result in death (Levy *et al.*, 2009; Guebert-Bartholo *et al.*, 2011).

Turtles are at risk of fatal injury from plastic ingestion due to their inability to regurgitate indigestible items. The spikes in their oesophagus, called papillae, force consumed items into their digestive tract. Similarly, though many marine mammals can regurgitate and expel indigestible items, mortality from ingestion of large plastic items has been documented in several species (Fossi *et al.*, 2018), particularly sperm whales (*Physeter macrocephalus*; Stamper *et al.*, 2006; Jacobsen *et al.*, 2010; de Stephanis *et al.*, 2013; Unger *et al.*, 2016) and those living in urbanised settings, including harbour seals, porpoises and manatees (Reinert *et al.*, 2017; Unger *et al.*, 2017). Little information on lethal impacts on elasmobranchs exist; however, a stomach perforation caused by a hardened drinking straw was thought to be the cause of death in a whale shark (*Rhincodon typus*) in one of three currently available reports (Haetrakul *et al.*, 2009; Sampaio *et al.*, 2018; Abreo *et al.*, 2019). The digestive tracts of seabirds can also become obstructed or damaged by ingestion of sharp plastics (Dau *et al.*, 2009; Seif *et al.*, 2018); however, direct lethal effects are rarely reported (often because it is not possible to determine the cause of mortality, particularly for dead-stranded animals) but may be underestimated (Roman *et al.*, 2020; Senko *et al.*, 2020).

In all marine megafauna taxa, the population-level effects of mortality from plastic ingestion are poorly understood (Senko *et al.*, 2020). This is an important consideration if we are to understand the true scale of harm caused by plastic pollution (Marn *et al.*, 2020).

2.2.2. *Sub-lethal effects*

In general, plastic ingestion by marine megafauna, especially microplastics and nanoplastics, is not likely to directly or immediately lead to mortality. In most cases, plastic ingestion will have slower-acting implications through sub-lethal effects, such as dietary dilution, malnutrition and disease that will reduce general health condition (e.g., Santos *et al.*, 2015; Fossi *et al.*, 2018), with the potential for impacts on reproductive fitness (Puskic *et al.*, 2020; Senko *et al.*, 2020). For example, gastrointestinal tract obstruction from plastic in Laysan albatross (*Phoebastria immutabilis*) chicks causes false satiation and malnourishment, which can lead to individuals becoming less resistant to other stressors, such as lead poisoning and avian poxvirus (Auman *et al.*, 2014). Gut dysbiosis (bacterial imbalance within the digestive tract) is another sub-lethal effect of plastic ingestion and a potentially chronic condition that may be caused by infections from pathogens carried on plastics (Oberbeckmann *et al.*, 2015) or interference with host immune system functioning when the gut microbiota is chronically exposed to plastics (Fackelmann & Sommer, 2019).

Several experimental systems have indicated that microplastics and nanoplastics (Cole & Galloway, 2015), and their associated compounds, such as plasticisers, stabilisers, colourants and flame retardants (Hahladakis *et al.*, 2018), can translocate through the gut wall, enter the blood-stream, and permeate cell membranes (Browne *et al.*, 2013; Galloway 2015). Once internalised, some plastic polymers (e.g., polyethylene), monomers (e.g., styrenes) and additives are known to cause oxidative stress, toxicity, immunological responses and endocrine dysregulation, potentially impacting reproductive fitness (Rochman *et al.*, 2014; Jeong *et al.*, 2017; Alimba & Faggio, 2019; Karbalaei *et al.*, 2021).

In addition to chemicals added during production, plastic can adsorb chemical pollutants, such as polychlorinated biphenyls (PCBs), polycyclic aromatic hydrocarbons (PAHs) and pesticides, dichlorodiphenyltrichloroethane (DDT) and derivatives, and heavy metals from the marine environment (Rios *et al.*, 2007, 2010; Ogata *et al.*, 2009; Cole *et al.*, 2011), and concentrate them by up to one millionfold of those in the surrounding water (Mato & Isobe, 2001). Upon ingestion, these chemicals can leach into biological tissues (Teuten *et al.*, 2009; Rochman *et al.*, 2013).

Necropsies and non-lethal sampling of marine megafauna species have detected plastic additives and associated compounds in liver and skin tissues, yet strong linkages between the observed contaminant loads and plastic exposure are still to be made (Fossi *et al.*, 2014; Savoca *et al.*, 2018).

2.3. *Methods of investigating plastic ingestion in marine megafauna*

Various approaches and protocols are available to assess the extent and risk of plastic ingestion in marine megafauna (López-Martínez *et al.*, 2020). Although recommendations for cohesive practices are increasing (e.g., Provencher *et al.*, 2017; Bessa *et al.*, 2019; Zantis *et al.*, 2021), much of the current information is not directly comparable due to the absence of standardised techniques (López-Martínez *et al.*, 2020; Zantis *et al.*, 2021). To date, necropsies have served as surveillance tools to investigate and monitor plastic ingestion in marine vertebrates (e.g., Bond *et al.*, 2014; Duncan *et al.*, 2019; Nelms *et al.*, 2019a). This method allows researchers to identify plastic accumulated along different sections of the gastrointestinal tract, investigate sub-lethal effects, and provides the opportunity to collect additional data, such as age, sex or body condition (Provencher *et al.*, 2017). There are challenges, however, in irrefutably linking mortality to plastic ingestion as many marine megafauna are under several confounding anthropogenic and natural pressures. Accordingly, researchers have explicitly proposed standardised guidelines for necropsy protocols for determining the cause of death (Moore *et al.*, 2013; Fossi *et al.*, 2018).

Additionally, the sampling of dead animals may introduce bias as animals that died from ill-health (as opposed to trauma) do not accurately represent the population as a whole (Casale *et al.*, 2016; Rodríguez *et al.*, 2018; Nelms *et al.*, 2019a). Non-lethal and non-invasive methodologies are also used to investigate plastic ingestion in live animals. Seabird and pinniped populations are more easily studied than other vertebrates due to their breeding and resting periods on land. Plastic ingestion in seabirds is commonly studied by analysing pellets or boluses (e.g., Lindborg *et al.*, 2012), namely, the natural regurgitations composed of the non-digestible parts of the diet, such as fish bones and otoliths, squid beaks and plastics (Ryan, 1987). For example, albatross chicks regurgitate a bolus immediately before fledging, which offers an opportunity to evaluate the differences in plastic ingestion between locations and examine the survival of new generations (Young *et al.*, 2009). Furthermore, pellets regurgitated by adults can also be found in nesting areas of several seabird species, representing an opportunistic but effective opportunity to obtain extended time-series data and track temporal variations of plastic pollution in remote areas (Perold *et al.*, 2020).

Faecal analysis is another non-invasive technique to study plastic ingestion (e.g., Nelms *et al.*, 2018, 2019b; Le Guen *et al.*, 2020). For example, data collected from guano in seabird nests may be useful to estimate the quantity of plastic that is excreted and concentrated around the colony (Provencher *et al.*, 2018). Pinniped scats are used to monitor the exposure of populations to plastics across geographic scales (Donohue *et al.*, 2019) and investigate the role of trophic transfer in dietary microplastic exposure (Nelms *et al.*, 2018, 2019b). Similar methods are being trialled for opportunistic in-water collection of faecal matter from cetaceans (Harlacher, 2020; Zantis *et al.*, 2022) and elasmobranchs (Germanov *et al.*, 2019b).

Other, more invasive, techniques involve inducing regurgitations in live animals using chemical emetics (e.g., Bond & Lavers, 2013) or stomach flushing using water (e.g., Lavers *et al.*, 2014). These techniques require careful ethical and practical considerations, particularly as some individuals may be harmed when plastics are forced out of the alimentary canal, or the animals may lose an energetically costly meal. Additionally, the full amount of ingested litter may not be expelled during the process, thus leading to underestimation of the plastic load (Provencher *et al.*, 2019). It is a beneficial technique, however, when investigating seabird species that do not produce pellets.

3. Entanglement in Plastic Pollution

Entanglement in plastic pollution is now recognised as a major threat to many marine species, with almost all marine taxa impacted in some form. The levels of impact vary greatly across species and are influenced by many factors related to the nature of the entangling debris and the life histories, behaviour and morphology of the species (Gall & Thompson, 2015; Kühn & van Franeker, 2020).

Though land-based debris, such as packing straps, plastic bags, balloons and clothing (Colmenero *et al.*, 2017; Duncan *et al.*, 2017; Parton *et al.*, 2019), is known to entangle marine species, by far, the most common form of entangling debris is abandoned, lost or otherwise discarded fishing gear (ALDFG; Richardson *et al.*, 2019; Stelfox *et al.*, 2020; Gilman *et al.*, 2021). Made of synthetic materials that do not biodegrade, this "ghost gear" (nets, ropes, lines and floats) can passively drift over large distances, indiscriminately "fishing" marine organisms, entrapping and killing organisms over years (Wilcox *et al.*, 2013).

It is important to note that entanglement and bycatch are often conflated, but distinguishing between the two issues is key to unravelling their impacts; bycatch refers to non-target species caught by active fishing gear (Davies *et al.*, 2009), whereas entanglement is the passive entrapment of marine species within the debris, including derelict fishing gear (Ryan, 2018; Tulloch *et al.*, 2019). Bycatch is one of the biggest threats to marine megafauna worldwide, and the number of marine species killed as a result of bycatch is likely significantly greater than that of entanglement (Lewison *et al.*, 2004). However, confusion between the two issues potentially means the incidences of entanglement in plastic (including derelict fishing gear) is significantly underreported (Tulloch *et al.*, 2019). Additionally, differences in how active and derelict fishing gears are monitored mean that incidences of bycatch are more likely to be recorded than entanglement. For example, active gear is frequently checked by fishers (to recover their catch) and on-board observers, and any trapped marine animals are more likely to be encountered before they decay or are predated. Drifting ghost gear can travel large distances into remote areas, such as the open ocean, where evidence of entanglement may not be observed unless through chance encounters (Stelfox *et al.*, 2019). Entanglement incidents are also less likely to be published within the scientific literature, particularly incidents featuring only single individuals (Nelms *et al.*, 2020), meaning that many entanglement observations go unreported, are only provided as anecdotes with minimal descriptions or are held in databases that are not widely accessible (Parton *et al.*, 2019).

Marine megafauna species may become entangled in debris because they are attracted to it out of curiosity or while seeking shelter in open water (Derraik, 2002; Galgani *et al.*, 2018), and because biofouling by marine organisms aggregates grazers and, subsequently predators (Wilcox *et al.*, 2013). Additionally, collisions could be due to animals failing to detect debris. For example, entanglement in fishing lines and nets is common for manta rays whose laterally placed eyes likely limit their binocular field of view (Deakos *et al.*, 2011; Germanov *et al.*, 2019a).

At an individual level, entanglement can cause serious wounds, as well as increased drag, restricted movement and breathing or choking (Votier *et al.*, 2011; Lawson *et al.*, 2015; Colmenero *et al.*, 2017). For example, derelict fishing gear in Svalbard has been observed wrapped around the body of bearded (*Erignathus barbatus*) and harbour seals (*Phoca vitulina*), that, if left untreated, could lead to death through wound infection or strangulation (Bergmann *et al.*, 2017). Entanglement of the limbs of marine turtles results in necrotic processes and amputation (e.g., Barreiros & Raykov, 2014). Entanglement can also affect the fitness of individuals by compromising their ability to capture and digest food, reproduce, migrate and escape from predators (Linnebjerg *et al.*, 2021). For example, Wegner and Cartamil (2012) observed a shortfin mako shark (*Isurus oxyrinchus*) entangled in a biofouled fishing rope, which showed scoliosis of the spine and severe undernourishment.

Many megafauna species display traits, such as spatial distribution patterns, morphological traits or behaviours, that predispose them to entanglement in plastic pollution. For example, although all turtle species across all oceans basins become entangled in plastic pollution, juvenile pelagic turtles may be impacted at greater levels than adults due to naivety relating to floating debris (Duncan *et al.*, 2017). Additionally, higher entanglement rates for olive ridley turtles (*Lepidochelys olivacea*) may occur because they forage along major oceanic fronts, which aggregate high concentrations of marine debris, and their generalist feeding behaviour potentially attracts them to forage on biofouled items (Stelfox *et al.*, 2016; Duncan *et al.*, 2019).

Elasmobranchs, a taxon for which one of the earliest entanglement reports was recorded (unidentified Carcharhinid shark entangled in a rubber vehicle tyre; Gudger & Hoffmann, 1931), are relatively understudied in relation to entanglement (Stelfox *et al.*, 2016). Nonetheless, a recent study revealed over 70% of entanglement reports for sharks could be attributed to ghost fishing gear (Parton *et al.*, 2019). Many elasmobranch species must continuously swim forwards to pass oxygen across their gills through a process known as ram ventilation. Entanglement can restrict their movement and, therefore, oxygen intake which may lead to suffocation (Parton *et al.*, 2019). Additionally, certain elasmobranch species display morphological features that could predispose them to entanglement. For example, the elongated and serrated rostrum of sawfish can easily become entwined in monofilament line and other ghost fishing gear (Parton *et al.*, 2019).

Marine mammals are arguably one of the most studied and perhaps worst impacted marine species groups for entanglement (Kühn & van Franeker, 2020). Though cetaceans are known to become entangled in debris (Baulch & Perry, 2014), pinnipeds appear to be more susceptible; approximately 67% of species (n = 22 of 33) have been recorded with entanglements, the majority of which involved ghost fishing gear (Jepsen & de Bruyn, 2019; Allyn & Scordino, 2020). The spatial overlap of species distributions with large aggregations of debris, such as the North Pacific Gyre ("Great Pacific Garbage Patch"), increases their proximity to plastic pollution and, therefore, the risk of entanglement. For example, the endangered Hawaiian monk seal (*Monachus schauinslandi*) has one of, if not the highest, entanglement rates of any pinniped species (Henderson, 2001; Lowry *et al.*, 2011), and marine debris is considered a significant threat to the species (www.iucnredlist.org/species/13654/45227978; last accessed on 19 April 2021).

In seabirds, 27.4–36% of species have been documented as entangled in plastic pollution, and 83% of records reported the entangling material to be fishing gear (Ryan, 2018; Kühn & van Franeker, 2020). Auks, which perform long dives to forage, are particularly vulnerable to entanglement in ghost fishing nets (e.g., Camphuysen, 2000). Plunge-diving species, such as sulids and terns, are most prone to becoming entangled in plastic bags, particularly around their necks, presumably as a result of pursuing fish sheltering beneath the debris (Ryan, 2018).

4. Habitat Implications

4.1. *Direct impacts*

Plastic debris can significantly alter habitat characteristics, which can in turn affect the marine animals that use such habitats. Marine turtles spend most of their lives at sea but rely on sandy beaches for nesting, and many studies have shown that some key beaches are hotspots of plastic debris (Beckwith & Fuentes, 2018; Duncan *et al.*, 2018; Garrison & Fuentes, 2019). Entanglement in plastic debris presents a risk to both females emerging to lay their eggs as well as their hatchlings (Triessnig, 2012). High debris abundance increases the frequency of false crawls (when females crawl on the beach but do not deposit a clutch of eggs) and disturbs turtle nest site selection (Fujisaki & Lamont, 2016; Gündoğdu *et al.*, 2019). Additionally, debris within beach sediments may prevent hatchlings from reaching the surface after hatching from eggs (Nelms *et al.*, 2015). In green turtles, the percentage of hatchlings reaching the sea is negatively correlated with the quantities of plastic debris present on the beach (Özdilek *et al.*, 2006). This is likely because the time needed for them to cross heavily polluted beaches is significantly higher (Aguilera *et al.*, 2018). Another potential impact of debris on turtle reproduction is the alteration of the incubating environment for eggs. Plastic can alter the physical properties of the sediment, including substrate temperature (Carson *et al.*, 2011; Lavers *et al.*, 2021). For example, Lavers *et al.* (2021) found that plastic debris on the beaches of two islands in the South Pacific was associated with fluctuations in sediment temperature (e.g., increases in daily maximum temperatures of 2.45°C and decreases of daily minimum by −1.50°C) compared to unaffected beaches. Such changes may affect embryo development, sex determination and hatching success (Yntema & Mrosovsky, 1980; Ackerman, 1997). In Hawksbill turtles, a 1.8°C difference in incubation temperature causes a shift from 100% males to 100% females (Mrosovsky & Provancha, 1992). Considering that a plastic concentration of only 1.5% can decrease sediment temperature by 0.75°C (Carson *et al.*, 2011), sex ratios in turtles could be significantly affected by the presence of plastic pollution. Furthermore, increased permeability caused by heavy plastic concentration in beach sediment may result in the desiccation of the eggs due to reduced humidity (Carson *et al.*, 2011). Plastics can also release toxic compounds, such as phthalates, into the sand, which have been found in the eggs (shell, yolk and albumen) of loggerhead turtle nests in the Mediterranean (Savoca *et al.*, 2021). Such chemicals are known to cause endocrine and reproductive effects on animals and could potentially impact hatchling development, but further research is needed on this topic (Savoca *et al.*, 2021).

Seabirds can become entangled in plastic debris they use to build their nests (Ryan, 2018; Jagiello *et al.*, 2019), which can lead to injury or death for both chicks (Robinson *et al.*, 2012) and adults (Votier *et al.*, 2011). Surveys have recorded plastic in the nests of at least 27 seabird species, primarily gannets, boobies, cormorants, gulls and terns (Table 1). Yet, not much is known about the scale of the impact, with many studies not reporting injury or mortality and only one study to date (to the authors' knowledge) has examined population-level mortality rates, which were found not to be significant to the population (Votier *et al.*, 2011).

Plastic in seabird nests consists mostly of ropes, lines and netting from fisheries (Norman *et al.*, 1995; Votier *et al.*, 2011; Ryan, 2018), although domestic plastics can be the primary source (Garcia-Cegarra *et al.*, 2020), especially near urban areas (Lopes *et al.*, 2020). The availability of fishing debris can influence the scale of debris incorporation; a substantially higher proportion of northern gannet (*Morus bassanus*) nests contained plastic at the same colony in years with more gillnet fishing (Bond *et al.*, 2012). Furthermore, the colony for which no fishing took place nearby had far fewer nests containing debris and a lower proportion of debris originating from fisheries (Bond *et al.*, 2012). Visiting colonies after the adults have departed to release any remaining entangled individuals is an effective way to reduce mortality and document the potential impact (Votier *et al.*, 2011). However, the survival and recruitment rates of these freed fledglings are unknown. Even if all plastic pollution was eliminated from the oceans, there would be a legacy of it in nests used over decades by the same pairs and over many generations by colony members. Plastics could be removed from nests, but this may have unintended consequences via disturbance.

4.2. *Indirect impacts*

Many megafauna species are dependent on a suite of sensitive ecosystems which are themselves impacted by anthropogenic stressors, including

Table 1. Species for which studies of the frequency of occurrence of plastic debris in seabird nests have been recorded.

Species name	Location	*N*	FO%	Source
Sulids				
Northern Gannet *Morus bassanus*	Canada, Faroes, France, Germany, Iceland, Ireland, Italy, Norway, UK	8,570	0–100	Montevecchi (1991), Votier *et al.* (2011), Bond *et al.* (2012), Merlino *et al.* (2018), O'Hanlon *et al.* (2019)
Brown Booby *Sula leucogaster*	Atlantic, Australia, Brazil, Cayman Islands, Pacific	2,611	0–100	Lavers *et al.* (2013), Verlis *et al.* (2014), Tavares *et al.* (2016), Grant *et al.* (2018), Brentano *et al.* (2020)
Australasian Gannet *Morus serrator*	Australia	338	28	Norman *et al.* (1995)
Cormorants				
Great Cormorant *Phalacrocorax carbo*	Senegal	68	40	Tavares *et al.* (2019)
Double-crested Cormorant *Phalacrocorax auritus*	USA	497	37	Podolsky and Kress (1988)
European Shag *Phalacrocorax aristotelis*	Scotland	170	80	Thompson *et al.* (2020)
Great Cormorant *Phalacrocorax carbo*	Scotland	98	24	Thompson *et al.* (2020)
White-breasted Cormorant *Phalacrocorax lucidus*	South Africa	121	26	Tavares *et al.* (2020)
Red-legged Cormorant *Phalacrocorax gaimardi*	Chile	18	100	Garcia-Cegarra *et al.* (2020)
Gulls				
Black-legged Kittiwake *Rissa tridactyla*	Denmark	777	39–57	Clemens and Hartwig (1993), Hartwig *et al.* (2007)
Hartlaub's Gull *Larus hartlaubii*	South Africa	265	1	Tavares *et al.* (2020)
Sooty Gull *Larus hemprichii*	UAE	258	11	Yaghmour and Al (2020)

Table 1. (*Continued*)

Species name	Location	N	FO%	Source
Yellow-legged Gull *Larus michahellis*	Portugal	204	33	Lopes *et al.* (2020)
Kelp Gull *Larus dominicanus*	South Africa	630	36	Witteveen *et al.* (2017)
Herring Gull *Larus argentatus*	Scotland	1,022	36	Thompson *et al.* (2020)
Lesser Black-backed Gull *Larus fuscus*	Scotland	221		Thompson *et al.* (2020)
Great Black-backed Gull *Larus marinus*	Scotland	86		Thompson *et al.* (2020)
Terns				
Grey Noddy *Anous albivitta*	Salas y Gomez, E Pacific	6	100	Luna-Jorquera *et al.* (2019)
Brown Noddy *Anous stolidus*	SE Pacific, S Atlantic			Ryan (2020)
Sooty Tern *Onychoprion fuscatus*	Brazil	1,800	3	de Souza Petersen *et al.* (2016)
Caspian Tern *Hydroprogne caspia*	Senegal	529	15	Tavares *et al.* (2019)
Others				
Great Frigatebirds *Fregata minor*	Salas y Gomez, E Pacific	35	100	Luna-Jorquera *et al.* (2019)
Flesh-footed Shearwater *Ardenna carneipes*	Australia	67	1	Paterson and Dunlop (2018)
African Penguin *Spheniscus demersus*	South Africa	100	3	Tavares *et al.* (2020)
Great White Pelican *Pelecanus onocrotalus*	South Africa	78	6	Tavares *et al.* (2020)
Polynesian Storm-petrel *Nesofregetta fuliginosa*	Salas y Gomez, E Pacific	4	100	Luna-Jorquera *et al.* (2019)
Yellow-legged Gull *Larus michahellis*	Spain	80	33	Delgado *et al.* (2020)

Note: *N* = number of nests surveyed across all studies and FO% = the percentage of nests containing debris in each study (frequency of occurrence). There have also been individual reports of species not yet surveyed: Arctic tern, Atlantic puffin, Brown pelican, Galapagos cormorant, Gentoo penguin, Silver gull, Yelkouan shearwater and Yellow-eyed penguin (birdsanddebris.com).

plastic debris (Vegter *et al.*, 2014). Tropical coral reefs provide functional benefits to a wide variety of megafauna, including turtles, elasmobranchs and cetaceans (Wabnitz *et al.*, 2010; Roff *et al.*, 2016). However, plastic pollution represents an important stressor for many coral reef systems across the globe, notably with the presence of derelict fishing gear (Mulochau *et al.*, 2020; Patterson Edward *et al.*, 2020) but also by elevated concentrations of microplastics (Huang *et al.*, 2021). The ingestion of microplastics or its passive adhesion on coral surfaces has numerous effects on habitat-forming corals, affecting feeding, physiology, energetics, growth and promoting diseases (e.g., Allen *et al.*, 2017; Hankins *et al.*, 2018, 2021; Martin *et al.*, 2019; Huang *et al.*, 2021). Larger plastic debris causes suffocation, disease, tissue abrasion, bleaching and mortality of habitat-forming corals (Lamb *et al.*, 2018; Beneli *et al.*, 2020; Mueller & Schupp, 2020). Entanglement in ALDFG and other plastic items causes severe damage to corals with evident signs of tissue damage (Asoh *et al.*, 2004) and a high probability of mortality (Yoshikawa & Asoh, 2004). Accordingly, coral mortality is higher in areas where ALDFG is abundant (Beneli *et al.*, 2020).

Decline in coral cover promotes novel community dynamics, directly affecting the numerous megafauna that are dependent upon the reefs (Bellwood *et al.*, 2004). Studies have reported more degraded reefs to have reduced shark abundance compared to healthy reefs (Sandin *et al.*, 2008; Espinoza *et al.*, 2014). Conversely, it is possible that for spongivorous turtles, such as some hawksbill turtle populations, decline in coral covers results in increased sponge abundance that could be beneficial for turtle populations (Hawkes *et al.*, 2009). Vegetated coastal habitats, such as seagrass meadows, macroalgae beds, mangrove forests or saltmarshes, are also key habitats for some marine megafauna (Short *et al.*, 2011; Sievers *et al.*, 2019). These ecosystems can also act as traps for particulate matter and more recently as important plastic sinks (de los Santos *et al.*, 2021; Luo *et al.*, 2021). For example, there is mounting evidence that mangroves, vital coastal systems in tropical oceans, are acting as a plastic trap (Riascos *et al.*, 2019; Luo *et al.*, 2021). Although mangrove trees are relatively resilient when partially covered by plastic, prolonged suffocation can result in significant leaf loss (van Bijsterveldt *et al.*, 2021) with potentially severe consequences for the animals that depend on such ecosystems for food and shelter (Makowski *et al.*, 2006; Amorocho & Reina, 2007).

5. Discussion

In this chapter, we outlined the various impacts of plastic pollution (via ingestion, entanglement and habitat degradation) on marine megafauna, namely, elasmobranchs, turtles, seabirds and marine mammals. There are a number of overarching themes requiring further research, which are worthy of discussion and are outlined in the following.

5.1. *Population-level effects*

Across the studies on marine megafauna species around the world, there are significant knowledge gaps concerning whether population-level impacts occur as a result of interaction with plastic pollution, and in understanding the relative contribution to mortality plastic has compared to, and in synergy with, other threats such as climate change, ocean acidification, invasive species and emerging diseases. There is an inherent lack of evidence of this in recent studies, with actual population-level impacts on species rarely assessed (Parton *et al.*, 2019; Senko *et al.*, 2020). The potential for this impact is present; however, without additional data, it remains undocumented. While there may not currently be the cause for concern at population level, the injuries caused by macroplastic ingestion and entanglement undoubtedly fall under the category of an animal welfare issue for any individuals that are impacted (Duncan *et al.*, 2017, 2019). With increasing amounts of plastic pollution entering the global ocean every year (Jambeck *et al.*, 2015), the likelihood of this issue becoming a greater problem is of profound concern. Future studies should seek to obtain more data at population level for marine species and explore the potential impacts of plastic with other environmental stressors to understand the cascading effects for megafauna. Impacts on population levels will also require studies to investigate the additional pressure that plastic pollution might pose to already threatened populations or life stages under heightened anthropogenic pressure (Puskic *et al.*, 2020; Rice *et al.*, 2021).

5.2. *Threat hotspots*

The threat of plastic pollution is global, but it is not evenly distributed. According to global models, there are hotspots of high plastic densities in enclosed sea regions (Mediterranean, Black Sea, Southeast Asia, China and Japan) and ocean gyres (North Pacific, South Atlantic and South Indian Ocean; van Sebille *et al.*, 2015), and areas of low plastic densities in upwelling regions (Humboldt and Canary Current), high latitudes, (particularly the Southern Ocean) and equatorial mid-ocean areas (Atlantic, Pacific and Indian; van Sebille *et al.*, 2015). The distributions of marine megafauna species are also not evenly spread around the oceans, and there has been increasing research into how they overlap with areas of plastic pollution levels. Wilcox *et al.* (2015) assessed the overlap between seabird range maps and plastic pollution, but recent technological advancements in animal tracking and species niche modelling allow for more accurate observations of spatio-temporal data of marine megafauna distributions (Jeffers & Godley, 2016; Schofield *et al.*, 2019). Overlaying databases of megafauna distributions and model outputs identifying areas of high plastic density, alongside areas where megafauna interact with plastic pollution at higher rates, can help to identify threat hotspots (Wilcox *et al.*, 2013; Germanov *et al.*, 2018). Matching these data to jurisdictions could facilitate the development of mitigation strategies and policies to reduce its impact (Hays *et al.*, 2019).

5.3. *Indicator species*

Using internationally agreed indicator species to investigate plastic pollution prevalence and impacts across a range of marine megafauna taxa would be beneficial. Currently, OSPAR (the mechanism by which 15 governments and the European Union cooperate to protect the marine environment of the North-East Atlantic) monitors the abundance of plastic particles in the stomachs of northern fulmars (*Fulmarus glacialis*) as one of its indicators of environmental quality (van Franeker *et al.*, 2011). The loggerhead turtle has also been identified as a suitable indicator species for monitoring plastic ingestion by marine animals in the Mediterranean basin due to its propensity to ingest marine debris and cosmopolitan distribution in the region, involving all range states (Matiddi *et al.*, 2017).

Seabird nests could be good indicators for the prevalence of plastic debris in the environment. Nests can be monitored by sampling them after breeding or with quick, low-impact, cheap methods such as direct observation (Bond *et al.*, 2012) and photography (Luna-Jorquera *et al.*, 2019; Ryan, 2020) and can also be carried out by citizen scientists, massively increasing the potential for data collection (see birdsanddebris.com; last accessed on 1 June 2021). However, seabird nest data must be interpreted carefully. Seabirds are selective, particularly targeting items that resemble natural nest-building materials, such as floating seaweeds and terrestrial plant materials (Lavers *et al.*, 2013; Ryan, 2020). Furthermore, other factors influence the prevalence of plastic in nests (Delgado *et al.*, 2020), which may include interspecific differences in preferences, natural nesting material availability, competition with conspecifics, and the immediate environment around the nest. For the North Atlantic, northern gannet nests are suitable indicators for debris from fisheries that have been surveyed throughout their range at 29 sites across eight countries (O'Hanlon *et al.*, 2019). For the tropics, the brown booby (*Sula leucogaster*) has been proposed as an indicator species (Tavares *et al.*, 2020), as it has a pantropical distribution and the proportion of plastic in their nests has been quantified for 18 sites from Australia, American Samoa, Palmyra Atoll, the Caribbean and Eastern Brazil, with the frequency of plastic occurrence ranging from 0% to 100% (Table 1). There are clear advantages of using standardised techniques across studies (Provencher *et al.*, 2017), so there is a need for protocols to be agreed to make this comparable worldwide. Additionally, if the detection of plastic additives and pollutants can be quantified in relation to plastic ingestion (see Ecotoxicology section in the following), marine megafauna species could serve as sentinels for plastic prevalence in wider marine ecosystems (Fossi & Panti, 2018; Germanov *et al.*, 2018; Fossi *et al.*, 2020).

5.4. *Ecotoxicology*

Currently, the ability to establish the extent of toxicological risks posed by plastic ingestion and its associated chemicals to individuals and populations is extremely limited (Koelmans *et al.*, 2017; Burns & Boxall, 2018; Bucci *et al.*, 2020). The extent to which plastic ingestion exposes organisms to pollutants compared to other routes, such as via water and their prey, remains unclear, and elucidating the potential role of plastic as a pollutant vector in various settings needs clarification. In some areas, where the concentration of plastic is very high relative to prey (Collignon *et al.*, 2012), or where higher levels of pollutants are adsorbed to plastic relative to other sources (Chen *et al.*, 2018), plastic ingestion might be a significant contributor of pollutants. Future toxicology investigations will also need to define and separate the effects (e.g., impacts on animal health) from the cocktail of chemicals and compounds that organisms are exposed to through plastic ingestion. Novel techniques, such as cell-based bioassays, make it possible to assess the toxicity of chemicals from ingested plastics, and it may be possible to identify biomarkers in tissue or blood, to indicate levels of plastic ingestion (Finlayson *et al.*, 2020).

5.5. *Data gaps*

Morbidity and mortality have been linked to plastic ingestion by marine megafauna but are difficult to directly attribute, especially as veterinary diagnostic testing is infrequently used to support a diagnosis (Orós *et al.*, 2016). Therefore, it would be advantageous to work alongside veterinarians and other specialists (i.e., veterinary pathologists) for access to veterinary diagnostic testing services where possible (Stamper *et al.*, 2009; Poppi *et al.*, 2012).

Recently, the use of data from social media platforms has helped in identifying the disparity between entanglement reports in the scientific literature and actual entanglement rates for species around the world (Hart *et al.*, 2018; Parton *et al.*, 2019). In addition, descriptive reporting of interactions between wildlife and fishing gear (i.e., entangling material, degree of biofouling, presence of other dead animals and their condition) will allow for rates of both bycatch and entanglement to be more accurately assessed (Wilcox *et al.*, 2015; Duncan *et al.*, 2019; Parton *et al.*, 2019; Nelms *et al.*, 2021).

Data regarding the absence of plastic pollution interactions and impacts are lacking due to a bias towards positive results. The few existing studies reporting negative results relate mainly to populations of pinnipeds. For instance, no plastic particles were found in scats collected on colonies of fur seals (*Arctocephalus* spp.) from the South Atlantic and South-west Indian Ocean (Ryan *et al.*, 2016), and similar results were reported in different life stages on three seal species (*Phoca hispida, Erignathus barbatus* and *Phoca vitualina*) from the Eastern Canadian Arctic (Bourdages *et al.*, 2020).

Furthermore, observations of interactions with plastic by a single or few individuals are often not published in peer-reviewed literature, but there is a need for such records to be stored (i.e., a database) and made available for inclusion in assessments and analyses, particularly for rare or cryptic species.

Continued monitoring and reporting of negative results relating to plastic interactions or impacts, as well as observations from small sample sizes, will provide a more accurate representation of occurrence rates and the extent of effects on individuals, populations and across species (Kühn & van Franeker, 2020; Puskic *et al.*, 2020).

6. Conclusion

Plastic pollution research is a rapidly evolving field and researchers are racing to determine its impacts on species and habitats in line with the growing pace of plastic production and leakage into the environment. Marine megafauna species are complex in both their life histories and their spatial distribution around the world. Consequently, their interaction with plastic pollution differs, not only among species but also across life stages, geographic location and habitat type. These complexities require robust monitoring, particularly for marine megafauna that are already threatened by a range of anthropogenic factors. In a field of research that is continually changing, scientists must continue to collaborate and share data in order to identify and fill the existing knowledge gaps that have been mentioned in detail here. Alongside this, working closely with policymakers, stakeholders and the general public will be critical for overcoming the challenges marine species will face as a result of plastic pollution in the future.

References

Abreo, N. A. S., Blatchley, D., & Superio, M. D. (2019). Stranded whale shark (*Rhincodon typus*) reveals vulnerability of filter-feeding elasmobranchs to marine litter in the Philippines. *Marine Pollution Bulletin, 141*, 79–83.

Abreo, N. A. S., Macusi, E. D., Blatchley, D. D., & Cuenca-Ocay, G. (2016). First evidence of plastic ingestion by the rare Deraniyagala's beaked whale (*Mesoplodon hotaula*). *IAMURE International Journal of Ecology and Conservation, 19*, 16–36.

Ackerman, R. (1997). The nest environment and the embryonic development of sea turtles. In *The Biology of Sea Turtles*, pp. 83–106. Boca Raton, New York, London: CRC press.

Aguilera, M., Medina-Suárez, M., Pinós, J., Liria-Loza, A., & Benejam, L. (2018). Marine debris as a barrier: Assessing the impacts on sea turtle hatchlings on their way to the ocean. *Marine Pollution Bulletin, 137*, 481–487.

Alexiadou, P., Foskolos, I., & Frantzis, A. (2019). Ingestion of macroplastics by odontocetes of the Greek Seas, Eastern Mediterranean: Often deadly! *Marine Pollution Bulletin, 146*, 67–75.

Alimba, C. G. & Faggio, C. (2019). Microplastics in the marine environment: Current trends in environmental pollution and mechanisms of toxicological profile. *Environmental Toxicology and Pharmacology, 68*, 61–74.

Allen, A. S., Seymour, A. C., & Rittschof, D. (2017). Chemoreception drives plastic consumption in a hard coral. *Marine Pollution Bulletin, 124*, 198–205.

Allyn, E. M. & Scordino, J. J. (2020). Entanglement rates and haulout abundance trends of Steller (*Eumetopias jubatus*) and California (*Zalophus californianus*) sea lions on the north coast of Washington state. *PLoS One, 15*, 1–21.

Alves, L. M. F., Nunes, M., Marchand, P., Le Bizec, B., Mendes, S., Correia, J. P. S., Lemos, M. F. L., & Novais, S. C. (2016). Blue sharks (*Prionace glauca*) as bioindicators of pollution and health in the Atlantic Ocean: Contamination levels and biochemical stress responses. *Science of the Total Environment, 563–564*, 282–292.

Amorocho, D. & Reina, R. (2007). Feeding ecology of the East Pacific green sea turtle *Chelonia mydas agassizii* at Gorgona National Park, Colombia. *Endangered Species Research, 3*, 43–51.

Asoh, K., Yoshikawa, T., Kosaki, R., & Marschall, E. A. (2004). Damage to cauliflower coral by monofilament fishing lines in Hawaii. *Conservation Biology, 18*, 1645–1650.

Auman, H. J., Ludwig, J. P., Giesy, J. P., & Colborn, T. (2014). Plastic ingestion by Laysan Albatross chicks on Sand Island, Midway Atoll, in 1994 and 1995. In Robertson, G., & Gales, R. (Eds.), *Albatross Biology and Conservation*, pp. 239–344. Chipping Norton: Surrey Beatty.

Barreiros, J. P. & Raykov, V. S. (2014). Lethal lesions and amputation caused by plastic debris and fishing gear on the loggerhead turtle *Caretta caretta* (Linnaeus, 1758). Three case reports from Terceira Island, Azores (NE Atlantic). *Marine Pollution Bulletin, 86*, 518–22.

Baulch, S. & Perry, C. (2014). Evaluating the impacts of marine debris on cetaceans. *Marine Pollution Bulletin, 80*, 210–21.

Beckwith, V. K. & Fuentes, M. M. P. B. (2018). Microplastic at nesting grounds used by the northern Gulf of Mexico loggerhead recovery unit. *Marine Pollution Bulletin, 131*, 32–37.

Bellwood, D. R., Hughes, T. P, Folke, C., & Nyström, M. (2004). Confronting the coral reef crisis. *Nature, 429*, 827–833.

Beneli, T. M., Pereira, P. H. C., Nunes J. A. C. C., & Barros, F. (2020). Ghost fishing impacts on hydrocorals and associated reef fish assemblages. *Marine Environmental Research, 161*, 105129.

Bergmann, M., Lutz, B., Tekman, M. B., & Gutow, L. (2017). Citizen scientists reveal: Marine litter pollutes Arctic beaches and affects wildlife. *Marine Pollution Bulletin, 125*, 535–540.

Bessa, F., Ratcliffe, N., Otero, V., Sobral, P., Marques, J. C., Waluda, C. M., Trathan, P. N., & Xavier, J. C. (2019). Microplastics in gentoo penguins from the Antarctic region. *Scientific Reports, 9*, 1–7.

Besseling, E., Foekema, E. M., van Franeker, J. A., Leopold, M. F., Kühn, S., Rebolledo, E. L. B., Heße, E., Mielke, L., Ijzer, J., Kamminga, P., & Koelmans, A. A. (2015). Microplastic in a macro filter feeder: Humpback whale *Megaptera novaeangliae*. *Marine Pollution Bulletin, 95*, 248–252.

Bond, A. L. & Lavers, J. L. (2013). Effectiveness of emetics to study plastic ingestion by Leach's Storm-petrels (*Oceanodroma leucorhoa*). *Marine Pollution Bulletin, 70*, 171–175.

Bond, A. L., Montevecchi, W. A., Guse, N., Regular, P. M., Garthe, S., & Rail, J. F. (2012). Prevalence and composition of fishing gear debris in the nests of northern gannets (*Morus bassanus*) are related to fishing effort. *Marine Pollution Bulletin, 64*, 907–911.

Bond, A. L., Provencher, J. F., Daoust, P.-Y., & Lucas, Z. N. (2014). Plastic ingestion by fulmars and shearwaters at Sable Island, Nova Scotia, Canada. *Marine Pollution Bulletin, 87*, 68–75.

Bourdages, M. P. T., Provencher, J. F., Sudlovenick, E., Ferguson, S. H., Young, B. G., Pelletier, N., Murphy, M. J. J., Addario, A. D., Vermaire, J. C., D'Addario, A., & Vermaire, J. C. (2020). No plastics detected in seal (Phocidae) stomachs harvested in the eastern Canadian Arctic. *Marine Pollution Bulletin, 150*, 110772.

Bourne, W. R. P. (1976). Seabirds and pollution. In Johnson, R. (Ed.), *Marine Pollution*, pp. 403–502. London: Academic Press.

Brentano, R., de Brum, A. C., Montone, R. C., & Petry, M. V. (2020). Incidence of anthropogenic material in Sula leucogaster nests in a distant archipelago of Brazil. *Marine Pollution Bulletin, 151*, 110815.

Browne, M. A., Niven, S. J., Galloway, T. S., Rowland, S. J., & Thompson, R. C. (2013). Microplastic moves pollutants and additives to worms, reducing functions linked to health and biodiversity. *Current Biology, 23*, 2388–2392.

Bucci, K., Tulio, M., & Rochman, C. M. (2020). What is known and unknown about the effects of plastic pollution: A meta-analysis and systematic review. *Ecological Applications, 30*(2), 1–16.

Burns, E. E. & Boxall, A. B. A. (2018). Microplastics in the aquatic environment: Evidence for or against adverse impacts and major knowledge gaps. *Environmental Toxicology and Chemistry, 37*, 2776–2796.

Camphuysen, K. C. J. (2000). Short notes: Seabirds drowned in fishing nets off Jan Mayen (Greenland Sea). *Atlantic Seabirds, 2*, 5–7.

Carey, M. J. (2011). Intergenerational transfer of plastic debris by Short-tailed Shearwaters (*Ardenna tenuirostris*). Emu, *111*, 229–234.

Carson, H. S., Colbert, S. L., Kaylor, M. J., & McDermid, K. J. (2011). Small plastic debris changes water movement and heat transfer through beach sediments. *Marine Pollution Bulletin, 62*, 1708–1713.

Casale, P., Abbate, G., Freggi, D., Conte, N., Oliverio, M., & Argano, R. (2008). Foraging ecology of loggerhead sea turtles Caretta caretta in the central Mediterranean Sea: Evidence for a relaxed life history model. *Marine Ecology Progress Series, 372*, 265–276.

Casale, P., Freggi, D., Paduano, V., & Oliverio, M. (2016). Biases and best approaches for assessing debris ingestion in sea turtles, with a case study in the Mediterranean. *Marine Pollution Bulletin, 110*, 238–249.

Chen, Q., Reisser, J., Cunsolo, S., Kwadijk, C., Kotterman, M., Proietti, M., Slat, B., Ferrari, F. F., Schwarz, A., Levivier, A., Yin, D., Hollert, H., & Koelmans, A. A. (2018). Pollutants in plastics within the North Pacific Subtropical Gyre. *Environmental Science & Technology, 52*, 446–456.

Clemens, V. T. & Hartwig, E. (1993). Müll als Nistmaterial von Dreizehenmöwen Untersuchung einer Brutkolonie an der Jammerbucht, Dänemark. *Seevöge, 14*, 6–7.

Cole, M. & Galloway, T. S. (2015). Ingestion of nanoplastics and microplastics by Pacific Oyster Larvae. *Environmental Science & Technology, 49*, 14625–14632.

Cole, M., Lindeque, P., Halsband, C., & Galloway, T. S. (2011). Microplastics as contaminants in the marine environment: A review. *Marine Pollution Bulletin, 62*, 2588–2597.

Collignon, A., Hecq, J. H., Glagani, F., Voisin, P., Collard, F., & Goffart, A. (2012). Neustonic microplastic and zooplankton in the North Western Mediterranean Sea. *Marine Pollution Bulletin, 64*, 861–864.

Colmenero, A. I., Barría, C., Broglio, E., & García-barcelona, S. (2017). Plastic debris straps on threatened blue shark *Prionace glauca. Marine Pollution Bulletin, 115*, 436–438.

Critchell, K., Hamann, M., Wildermann, N., & Grech, A. (2019). Predicting the exposure of coastal species to plastic pollution in a complex island archipelago. *Environmental Pollution, 252*, 982–991.

Dau, B. K., Gilardi, K. V. K., Gulland, F. M., Higgins, A., Holcomb, J. B., Leger, J. S., & Ziccardi, M. H. (2009). Fishing gear–related injury in California marine wildlife. *Journal of Wildlife Diseases, 45*, 355–362.

Davies, R. W. D., Cripps, S. J., Nickson, A., & Porter, G. (2009). Defining and estimating global marine fisheries bycatch. *Marine Policy, 33*, 661–672.

de los Santos, C. B., Krång, A. S., & Infantes, E. (2021). Microplastic retention by marine vegetated canopies: Simulations with seagrass meadows in a hydraulic flume. *Environmental Pollution, 269*, 116050.

de Stephanis, R., Giménez, J., Carpinelli, E., Gutierrez-Exposito, C., & Cañadas, A. (2013). As main meal for sperm whales: Plastics debris. *Marine Pollution Bulletin, 69*, 206–214.

Deakos, M. H., Baker, J. D., & Bejder, L. (2011). Characteristics of a manta ray *Manta alfredi* population off Maui, Hawaii, and implications for management. *Marine Ecology Progress Series, 429*, 245–260.

Delgado, S., Zorrozua, N., & Arizaga, J. (2020). Marginal presence of plastic in nests of yellow–legged gulls (*Larus michahellis*) in the southeastern bay of biscay. *Animal Biodiversity and Conservation, 43*, 191–195.

Dell'Ariccia, G., Célérier, A., Gabirot, M., Palmas, P., Massa, B., & Bonadonna, F. (2014). Olfactory foraging in temperate waters: Sensitivity to dimethylsulphide of shearwaters in the Atlantic Ocean and Mediterranean Sea. *The Journal of Experimental Biology, 217*, 1701–1709.

Dell'Ariccia, G., Phillips, R. A., van Franeker, J. A., Gaidet, N., Catry, P., Granadeiro, J. P., Ryan, P. G., & Bonadonna, F. (2017). Comment on "Marine plastic debris emits a keystone infochemical for olfactory foraging seabirds" by Savoca *et al. Science Advances, 3*, e1700526.

Derraik, J. G. B. (2002). The pollution of the marine environment by plastic debris: A review. *Marine Pollution Bulletin, 44*, 842–852.

de Souza Petersen, E., Krüger, L., Dezevieski, A., Petry, M., & Carmela, R. (2016). Incidence of plastic debris in Sooty Tern nests: A preliminary study on Trindade Island, a remote area of Brazil. *Marine Pollution Bulletin, 105*, 373–376.

Di Beneditto, A. P. M., & Awabdi, D. R. (2014). How marine debris ingestion differs among megafauna species in a tropical coastal area. *Marine Pollution Bulletin, 88*, 86–90.

Donohue, M. J., Masura, J., Gelatt, T., Ream, R., Baker, J. D., Faulhaber, K., & Lerner, D. T. (2019). Evaluating exposure of northern fur seals, *Callorhinus ursinus*, to microplastic pollution through fecal analysis. *Marine Pollution Bulletin, 138*, 213–221.

Duncan, E. M., Arrowsmith, J., Bain, C., Broderick, A. C., Lee, J., Metcalfe, K., Pikesley, S. K., Snape R. T. E., van Sebille, E., & Godley, B. J. (2018). The true depth of the Mediterranean plastic problem: Extreme microplastic pollution on marine turtle nesting beaches in Cyprus. *Marine Pollution Bulletin, 136*, 334–340.

Duncan, E. M., Botterell, Z. L., Broderick, A. C., Galloway, T. S., Lindeque, P. K., Nuno, A., & Godley, B. J. (2017). A global review of marine turtle entanglement in anthropogenic debris: A baseline for further action. *Endangered Species Research, 34*, 431–448.

Duncan, E. M., Broderick, A. C., Fuller, W. J., Galloway, T. S., Godfrey, M. H., Hamann, M., Limpus, C. J., Lindeque, P. K., Mayes, A. G., Omeyer, L. C. M., Santillo, D., Snape, R. T. E., & Godley, B. J. (2019). Microplastic ingestion ubiquitous in marine turtles. *Global Change Biology, 25*, 744–752.

Endres, C. S. & Lohmann, K. J. (2012). Perception of dimethyl sulfide (DMS) by loggerhead sea turtles: A possible mechanism for locating high-productivity oceanic regions for foraging. *The Journal of Experimental Biology, 215*, 3535–3538.

Eriksson, C. & Burton, H. (2003). Origins and biological accumulation of small plastic particles in fur seals from Macquarie Island. *Ambio, 32*, 380–384.

Espinoza, M., Cappo, M., Heupel, M. R., Tobin, A. J., & Simpfendorfer, C. A. (2014). Quantifying shark distribution patterns and species-habitat associations: Implications of Marine Park zoning. *PLoS One*, *9*(9), e106885.

Fackelmann, G. & Sommer, S. (2019). Microplastics and the gut microbiome: How chronically exposed species may suffer from gut dysbiosis. *Marine Pollution Bulletin, 143*, 193–203.

Finlayson, K. A., Madden Hof, C. A., & van de Merwe, J. P. (2020). Development and application of species-specific cell-based bioassays to assess toxicity in green sea turtles. *Science of the Total Environment, 747*, 142095.

Fossi, M. C., Baini, M., & Simmonds, M. P. (2020). Cetaceans as ocean health indicators of marine litter impact at global scale. *Frontiers in Environmental Science, 8*, 1–7.

Fossi, M. C., Baini, M., Panti, C., & Baulch, S. (2018). Impacts of marine litter on cetaceans: A focus on plastic pollution. In Maria Cristina Fossi, C. P. (Ed.), *Marine Mammal Ecotoxicology: Impacts of Multiple Stressors on Population Health*, pp. 147–184. Cambridge: Academic Press.

Fossi, M. C., Coppola, D., Baini, M., Giannetti, M., Guerranti, C., Marsili, L., Panti, C., de Sabata, E., & Clò, S. (2014). Large filter feeding marine organisms as indicators of microplastic in the pelagic environment: The case studies of the Mediterranean basking shark (*Cetorhinus maximus*) and fin whale (*Balaenoptera physalus*). *Marine Environmental Research, 100*, 17–24.

Fossi, M. C. & Panti, C. (2018). *Marine Mammal Ecotoxicology: Impacts of Multiple Stressors on Population Health.* Cambridge: Academic Press.

Frazier, J. (2005). Marine turtles: The role of flagship species in interactions between people and the sea. *Montana: The Magazine of Western History, 3*, 5–38.

Fujisaki, I. & Lamont, M. M. (2016). The effects of large beach debris on nesting sea turtles. *Journal of Experimental Marine Biology and Ecology, 482*, 33–37.

Galgani, F., Pham, C. K., Claro, F., & Consoli, P. (2018). Marine animal forests as useful indicators of entanglement by marine litter. *Marine Pollution Bulletin, 135*, 735–738.

Gall, S. C. & Thompson, R. C. (2015). The impact of debris on marine life. *Marine Pollution Bulletin, 92*(1–2), 170–179.

Galloway, T. S. (2015). Micro- and nano-plastics and human health. In Bergmann, M., Gutow, L., & Klages, M. (Eds.), *Marine Anthropogenic Litter*, pp. 343–366. Berlin: Springer.

Garcia-Cegarra, A. M., Ramirez, R., & Orrego, R. (2020). Red-legged cormorant uses plastic as nest material in an artificial breeding colony of Atacama Desert coast. *Marine Pollution Bulletin, 160*, 111632.

Garrison, S. R. & Fuentes, M. M. P. B. (2019). Marine debris at nesting grounds used by the Northern Gulf of Mexico loggerhead recovery unit. *Marine Pollution Bulletin, 139*, 59–64.

Germanov, E. S., Bejder, L., Chabanne, D. B., Dharmadi, D., Hendrawan, I. G., Marshall, A. D., Pierce, S. J., van Keulen, M., & Loneragan, N. R. (2019a). Contrasting habitat use and population dynamics of reef manta rays within the Nusa Penida Marine Protected Area, Indonesia. *Frontiers in Marine Science*, *6*, 1–17.

Germanov, E. S., Marshall, A. D., Bejder, L., Fossi, M. C., & Loneragan, N. R. (2018). Microplastics: No small problem for filter-feeding megafauna. *Trends in Ecology & Evolution, 33*(4), 227–232.

Germanov, E. S., Bejder, L., Chabanne, D. B., Dharmadi, D., Hendrawan, I. G., Marshall, A. D., Pierce, S. J., van Keulen, M., & Loneragan, N. R. (2019a). Contrasting habitat use and population dynamics of reef manta rays within the Nusa Penida Marine Protected Area, Indonesia. *Frontiers in Marine Science*, *6*, 1–17.

Germanov, E. S., Marshall, A. D., Hendrawan, I. G., Admiraal, R., Rohner, C. A., Argeswara, J., Wulandari, R., Himawan, M. R., & Loneragan, N. R. (2019b). Microplastics on the menu: Plastics pollute Indonesian manta ray and whale shark feeding grounds. *Frontiers in Marine Science*, *6*, 1–21.

Gilman, E., Musyl, M., Suuronen, P., Chaloupka, M., Gorgin, S., Wilson, J., & Kuczenski, B. (2021). Highest risk abandoned, lost and discarded fishing gear. *Scientific Reports*, *11*(1), 1–11.

Grant, M. L., Lavers, J. L., Stuckenbrock, S., Sharp, P. B., & Bond, A. L. (2018). The use of anthropogenic marine debris as a nesting material by brown boobies (*Sula leucogaster*). *Marine Pollution Bulletin, 137*, 96–103.

Gudger, E. & Hoffmann, W. (1931). A shark encircled with a rubber automobile tire. *The Scientific Monthly*, *46*, 281–285.

Guebert-Bartholo, F., Barletta, M., Costa, M., & Monteiro-Filho, E. (2011). Using gut contents to assess foraging patterns of juvenile green turtles *Chelonia mydas* in the Paranagua Estuary, Brazil. *Endangered Species Research, 13*, 131–143.

Gündoğdu, S., Yeşilyurt, İ. N., & Erbaş, C. (2019). Potential interaction between plastic litter and green turtle Chelonia mydas during nesting in an extremely polluted beach. *Marine Pollution Bulletin, 140*, 138–145.

Haetrakul, T., Munanansup, S., Assawawongkasem, N., & Chansue, N. (2009). A case report: Stomach foreign object in whaleshark (*Rhincodon typus*) stranded in Thailand. In *Proceedings of the 4th International Symposium on SEASTAR2000 and Asian Bio-logging Science (The 8th SEASTAR2000 workshop) (2009)*, pp. 83–85.

Hahladakis, J. N., Velis, C. A., Weber, R., Iacovidou, E., & Purnell, P. (2018). An overview of chemical additives present in plastics: Migration, release, fate and environmental impact during their use, disposal and recycling. *Journal of Hazardous Materials*, *344*, 179–199.

Hankins, C., Duffy, A., & Drisco, K. (2018). Scleractinian coral microplastic ingestion: Potential calcification effects, size limits, and retention. *Marine Pollution Bulletin*, *135*, 587–593.

Hankins, C., Moso, E., & Lasseigne, D. (2021). Microplastics impair growth in two atlantic scleractinian coral species, *Pseudodiploria clivosa* and *Acropora cervicornis*. *Environmental Pollution*, *275*, 116649.

Harlacher, J. (2020). Whale, what do we have here? Evidence of microplastics in top predators: Analysis of two populations of resident killer whale fecal samples. Master's thesis, University of Washington.

Hart, A. G., Carpenter, W. S., Hlustik-Smith, E., Reed, M., & Goodenough, A. E. (2018). Testing the potential of Twitter mining methods for data acquisition: Evaluating novel opportunities for ecological research in multiple taxa. *Methods in Ecology and Evolution*, *9*, 2194–2205.

Hartwig, E., Clemens, T., & Heckroth, M. (2007). Plastic debris as nesting material in a Kittiwake- (*Rissa tridactyla*) — Colony at the Jammerbugt, Northwest Denmark. *Marine Pollution Bulletin*, *54*, 595–597.

Hawkes, L., Broderick, A., Godfrey, M., & Godley, B. (2009). Climate change and marine turtles. *Endangered Species Research*, *7*, 137–154.

Hays, G. C., Bailey, H., Bograd, S. J., Bowen, W. D., Campagna, C., Carmichael, R. H., Casale, P., Chiaradia, A., Costa, D. P., & Cuevas, E. (2019). Translating marine animal tracking data into conservation policy and management. *Trends in Ecology & Evolution*, 34(5), 459–473.

Henderson, J. R. (2001). A pre- and post-MARPOL annex V summary of Hawaiian monk seal entanglements and marine debris accumulation in the Northwestern Hawaiian Islands, 1982–1998. *Marine Pollution Bulletin*, *42*, 584–589.

Huang, W., Chen, M., Song, B., Deng, J., Shen, M., Chen, Q., Zeng, G., & Liang, J. (2021). Microplastics in the coral reefs and their potential impacts on corals: A mini-review. *Science of the Total Environment*, *762*, 143112.

Jacobsen, J. K., Massey, L., & Gulland, F. (2010). Fatal ingestion of floating net debris by two sperm whales (*Physeter macrocephalus*). *Marine Pollution Bulletin*, *60*, 765–767.

Jagiello, Z., Dylewski, Ł., Tobolka, M., & Aguirre, J. I. (2019). Life in a polluted world: A global review of anthropogenic materials in bird nests. *Environmental Pollution*, *251*, 717–722.

Jambeck, J. R., Geyer, R., Wilcox, C., Siegler, T. R., Perryman, M., Andrady, A., Narayan, R., & Law, K. L. (2015). Plastic waste inputs from land into the ocean. *Science*, *347*, 768–771.

Jeffers, V. F. & Godley, B. J. (2016). Satellite tracking in sea turtles: How do we find our way to the conservation dividends? *Biological Conservation*, *199*, 172–184.

Jeong, C., Kang, H., Lee, M., Kim, D., Han, J., Hwang, D., Souissi, S., Lee, S., Shin, K., Park, H. G., & Lee, J. (2017). Adverse effects of microplastics and oxidative stress-induced MAPK/mechanisms in the marine copepod Paracyclopina nana. Nature Publishing Group, 1–11.

Jepsen, E. M. & de Bruyn, P. J. N. J. N. (2019). Pinniped entanglement in oceanic plastic pollution: A global review. *Marine Pollution Bulletin*, *145*, 295–305.

Kagel, S. M., Garcia, M., Cummings, M. E., Gruev, V., & Brady, C. (2021). Comparison of the polarization contrast of gelatinous zooplankton and a transparent single-use plastic bag — Implications for marine animals. *Marine Pollution Bulletin*, *168*, 112438.

Karbalaei, S., Hanachi, P., Rafiee, G., Seifori, P., & Walker, T. R. (2021). Toxicity of polystyrene microplastics on juvenile *Oncorhynchus mykiss* (rainbow trout) after individual and combined exposure with chlorpyrifos. *Journal of Hazardous Materials*, *403*, 123980.

Koelmans, A. A., Besseling, E., Foekema, E., Kooi, M., Mintenig, S., Ossendorp, B. C., Redondo-Hasselerharm, P. E., Verschoor, A., van Wezel, A. P., & Scheffer, M. (2017). Risks of plastic debris: Unravelling fact, opinion, perception, and belief. *Environmental Science & Technology*, *51*, 11513–11519.

Kühn, S. & van Franeker, J. A. (2020). Quantitative overview of marine debris ingested by marine megafauna. *Marine Pollution Bulletin*, *151*, 110858.

Laist, D. W. (1997). Impacts of marine debris: entanglement of marine life in marine debris including a comprehensive list of species with entanglement and ingestion records. In Coe, J. M., & Rogers, D. B. (Eds.), *Marine Debris: Sources, Impacts and Solutions*, pp. 99–139. New York: Springer-Verlag Inc.

Lamb, J. B., Willis, B. L., Fiorenza, E. A., Couch, C. S., Howard, R., Rader, D. N., True, J. D., Kelly, L. A., Ahmad, A., & Jompa, J. (2018). Plastic waste associated with disease on coral reefs. *Science*, *2010*, 26–29.

Lavers, J. L., Bond, A. L., & Hutton, I. (2014). Plastic ingestion by Flesh-footed Shearwaters (*Puffinus carneipes*): Implications for fledgling body condition and the accumulation of plastic-derived chemicals. *Environmental Pollution*, *187*, 124–129.

Lavers, J. L., Hodgson, J. C., & Clarke, R. H. (2013). Prevalence and composition of marine debris in Brown Booby (*Sula leucogaster*) nests at Ashmore Reef. *Marine Pollution Bulletin*, *77*, 320–324.

Lavers, J. L., Rivers-Auty, J., & Bond, A. L. (2021). Plastic debris increases circadian temperature extremes in beach sediments. *Journal of Hazardous Materials*, *416*, 126140.

Lawson, T. J., Wilcox, C., Johns, K., Dann, P., & Hardesty, B. D. (2015). Characteristics of marine debris that entangle Australian fur seals (*Arctocephalus pusillus doriferus*) in Southern Australia. *Marine Pollution Bulletin*, *98*, 354–357.

Le Guen, C., Suaria, G., Sherley, R. B., Ryan, P. G., Aliani, S., Boehme, L., & Brierley, A. S. (2020). Microplastic study reveals the presence of natural and synthetic fibres in the diet of King Penguins (*Aptenodytes patagonicus*) foraging from South Georgia. *Environment International*, *134*, 105303.

Levy, A. M., Brenner, O., Scheinin, A., Morick, D., Ratner, E., Goffman, O., & Kerem, D. (2009). Laryngeal snaring by ingested fishing net in a common bottlenose dolphin (*Tursiops truncatus*) off the Israeli shoreline. *Journal of Wildlife Diseases*, *45*, 834–838.

Lewison, R. L., Crowder, L. B., Read, A. J., & Freeman, S. A. (2004). Understanding impacts of fisheries bycatch on marine megafauna. *Trends in Ecology & Evolution*, *19*, 598–604.

Lindborg, V. A., Ledbetter, J. F., Walat, J. M., & Moffett, C. (2012). Plastic consumption and diet of Glaucous-winged Gulls (*Larus glaucescen*) Study Area. *Marine Pollution Bulletin*, *64*, 2351–2356.

Linnebjerg, J. F., Baak, J. E., Barry, T., Gavrilo, M. V., Mallory, M. L., Merkel, F. R., Price, C., Strand, J., Walker, T. R., & Provencher, J. F. (2021). Review of plastic pollution policies of Arctic countries in relation to seabirds. *Facets*, *6*, 1–25.

Lopes, C. S., de Faria, J. P., Paiva, V. H., & Ramos, J. A. (2020). Characterization of anthropogenic materials on yellow-legged gull (*Larus michahellis*) nests breeding in natural and urban sites along the coast of Portugal. *Environmental Science and Pollution Research*, *27*, 36954–36969.

López-Martínez, S., Morales-Caselles, C., Kadar, J., & Rivas, M. L. (2020). Overview of global status of plastic presence in marine vertebrates. *Global Change Biology*, *27*(4), 728–737.

Lowry, L. F., Laist, D. W., Gilmartin, W. G., & Antonelis, G. A. (2011). Recovery of the Hawaiian Monk Seal (*Monachus schauinslandi*): A review of conservation efforts, 1972–2010, and thoughts for the future. *Aquatic Mammals*, *37*, 397–419.

Luna-Jorquera, G., Thiel, M., Portflitt-Toro, M., & Dewitte, B. (2019). Marine protected areas invaded by floating anthropogenic litter: An example from the South Pacific. *Aquatic Conservation: Marine and Freshwater Ecosystems*, *29*, 245–259.

Luo, Y. Y., Not, C., & Cannicci, S. (2021). Mangroves as unique but understudied traps for anthropogenic marine debris: A review of present information and the way forward. *Environmental Pollution*, *271*, 116291.

Lynch, J. M. (2018). Quantities of marine debris ingested by sea turtles: Global meta-analysis highlights need for standardized data reporting methods and reveals relative risk. *Environmental Science & Technology*, *52*, 12026–12038.

Makowski, C., Seminoff, J. A., & Salmon, M. (2006). Home range and habitat use of juvenile Atlantic green turtles (*Chelonia mydas L.*) on shallow reef habitats in Palm Beach, Florida, USA. *Marine Biology*, *148*, 1167–1179.

Marn, N., Jusup, M., Kooijman, S. A. L. M., & Klanjscek, T. (2020). Quantifying impacts of plastic debris on marine wildlife identifies ecological breakpoints. *Ecology Letters*, *23*, 1479–1487.

Martí, E., Martin, C., Galli, M., Echevarría, F., Duarte, C. M., & Cózar, A. (2020). The colors of the ocean plastics. *Environmental Science & Technology*, *54*, 6594–6601.

Martin, C., Corona, E., Mahadik, G. A., & Duarte, C. M. (2019). Adhesion to coral surface as a potential sink for marine microplastics. *Environmental Pollution*, *255*, 113281.

Matiddi, M., Hochsheid, S., Camedda, A., Baini, M., Cocumelli, C., Serena, F., Tomassetti, P., Travaglini, A., Marra, S., Campani, T., Scholl, F., Mancusi, C., Amato, E., Briguglio, P., Maffucci, F., Fossi, M. C., Bentivegna, F., & de Lucia, G. A. (2017). Loggerhead sea turtles (*Caretta caretta*): A target species for monitoring litter ingested by marine organisms in the Mediterranean Sea. *Environmental Pollution*, *230*, 199–209.

Mato, Y. & Isobe, T. (2001). Plastic resin pellets as a transport medium for toxic chemicals in the marine environment. *Environmental Science & Technology*, *35*, 318–324.

Merlino, S., Abbate, M., Pietrelli, L., Canepa, P., & Varella, P. (2018). Marine litter detection and correlation with the seabird nest content. *Rendiconti Lincei*, *29*, 867–875.

Miller, M. E., Hamann, M., & Kroon, F. J. (2020). Bioaccumulation and biomagnification of microplastics in marine organisms: A review and meta-analysis of current data. *PLoS One*, *15*, 1–25.

Montevecchi, W. A. (1991). Incidence and types of plastic in gannets' nests in the northwest Atlantic. *Canadian Journal of Zoology*, *69*, 295–297.

Moore, E. M. J., van Der Hoop, J., Barco, S. G., Costidis, A. M., Gulland, F. M., Jepson, P. D., Moore, K. T., Raverty, S., & Mclellan, W. A. (2013). Criteria and case definitions for serious injury and death of pinnipeds and cetaceans caused by anthropogenic trauma. *Diseases of Aquatic Organisms*, *103*, 229–264.

Mrosovsky, N. & Provancha, J. (1992). Sex ratio of hatchling loggerhead sea turtles: Data and estimates from a 5-year study. *Canadian Journal of Zoology*, *70*, 530–538.

Mrosovsky, N., Ryan, G. D., & James, M. C. (2009). Leatherback turtles: The menace of plastic. *Marine Pollution Bulletin*, *58*, 287–289.

Mucientes, G. & Queiroz, N. (2019). Presence of plastic debris and retained fishing hooks in oceanic sharks. *Marine Pollution Bulletin*, *143*, 6–11.

Mueller, J. S. & Schupp, P. J. (2020). Shading by marine litter impairs the health of the two Indo-Pacific scleractinian corals *Porites rus* and *Pavona cactus*. *Marine Pollution Bulletin*, *158*, 111429.

Mulochau, T., Lelabousse, C., & Séré, M. (2020). Estimations of densities of marine litter on the fringing reefs of Mayotte (France – South Western Indian Ocean) — Impacts on coral communities. *Marine Pollution Bulletin*, *160*, 111643.

Nelms, S., Alfaro-Shigueto, J., Arnould, J., Avila, I., Bengtson Nash, S., Campbell, E., Carter, M., Collins, T., Currey, R., Domit, C., Franco-Trecu, V., Fuentes, M., Gilman, E., Harcourt, R., Hines, E., Hoelzel, A., Hooker, S., Johnston, D., Kelkar, N., Kiszka, J., Laidre, K., Mangel, J., Marsh, H., Maxwell, S., Onoufriou, A., Palacios, D., Pierce, G., Ponnampalam, L., Porter, L., Russell, D., Stockin, K., Sutaria, D., Wambiji, N., Weir, C., Wilson, B., & Godley, B. (2021). Marine mammal conservation: over the horizon. *Endangered Species Research*, *44*, 291–325.

Nelms, S. E., Barnett, J., Brownlow, A., Davison, N. J., Deaville, R., Galloway, T. S., Lindeque, P. K., Santillo, D., & Godley, B. J. (2019a). Microplastics in marine mammals stranded around the British coast: ubiquitous but transitory? *Scientific Reports*, *9*, 1075.

Nelms, S. E., Duncan, E. M., Broderick, A. C., Galloway, T. S., Godfrey, M. H., Hamann, M., Lindeque, P. K., & Godley, B. J. (2015). Plastic and marine turtles: A review and call for research. *ICES Journal of Marine Science*, *73*, 165–181.

Nelms, S. E., Duncan, E. M., Patel, S., Badola, R., Bhola, S., Chakma, S., Chowdhury, G. W., Godley, B. J., Haque, A. B., Johnson, J. A., Khatoon, H., Kumar, S., Napper, I. E., Niloy, M. D. N. H., Akter, T., Badola, S., Dev, A., Rawat, S., Santillo, D., Sarker, S., Sharma, E., & Koldewey, H. (2020). Riverine plastic pollution from fisheries: Insights from the Ganges River system. *Science of the Total Environment*, *756*, 143305.

Nelms, S. E., Galloway, T. S., Godley, B. J., Jarvis, D. S., & Lindeque, P. K. (2018). Investigating microplastic trophic transfer in marine top predators. *Environmental Pollution*, *238*, 999–1007.

Nelms, S. E., Parry, H. E., Bennett, K. A., Galloway, T. S., Godley, B. J., Santillo, D., & Lindeque, P. K. (2019b) What goes in, must come out: Combining scat-based molecular diet analysis and quantification of ingested microplastics in a marine top predator. *Methods in Ecology and Evolution, 10*, 1712–1722.

Norman, F. I., Menkhorst, P. W., & Hurley, V. G. (1995). Plastics in nests of Australasian gannets *Morns serrator* in Victoria, Australia. Emu, *95*, 129–133.

Oberbeckmann, S., Löder, M. G. J., & Labrenz, M. (2015). Marine microplastic-associated biofilms — A review. *Environmental Chemistry, 12*, 551–562.

Ogata, Y., Takada, H., Mizukawa, K., Hirai, H., Iwasa, S., Endo, S., Mato, Y., Saha, M., Okuda, K., Nakashima, A., Murakami, M., Zurcher, N., Booyatumanondo, R., Pauzi, M., Quang, L., Gordon, M., Miguez, C., Suzuki, S., Moore, C., Karapanagioti, H. K., Weerts, S., Mcclurg, T., Burres, E., Smith, W., Van Velkenburg, M., Selby, J., Lang, R. C., Laursen, D., Danner, B., Stewardson, N., & Thompson, R. C. (2009). International Pellet Watch: Global monitoring of persistent organic pollutants (POPs) in coastal waters. 1. Initial phase data on PCBs, DDTs, and HCHs. *Marine Pollution Bulletin, 58*, 1437–1446.

O'Hanlon, N. J., Bond, A. L., Lavers, J. L., Masden, E. A., & James, N. A. (2019). Monitoring nest incorporation of anthropogenic debris by Northern Gannets across their range. *Environmental Pollution, 255*, 113152.

Orós, J., Montesdeoca, N., Camacho, M., Arencibia, A., & Calabuig, P. (2016). Causes of stranding and mortality, and final disposition of loggerhead sea turtles (*Caretta caretta*) admitted to a wildlife rehabilitation center in Gran Canaria Island, Spain (1998-2014): A long-term retrospective study. *PLoS One, 11*, 1–14.

Özdilek, H. G., Yalçin-Özdilek, Ş,, Ozaner, F. S., & Sönmez, B. (2006). Impact of accumulated beach litter on *Chelonia mydas L.* 1758 (green turtle) hatchlings of the Samandağ coast, Hatay, Turkey. *Fresenius Environmental Bulletin, 15*, 1–9.

Parton, K. J., Galloway, T. S., & Godley, B. J. (2019). A global review of shark and ray entanglement in anthropogenic marine debris. *Endangered Species Research, 39*, 173–190.

Parton, K. J., Godley, B. J., Santillo, D., Tausif, M., Omeyer, L. C. M., & Galloway, T. S. (2020). Investigating the presence of microplastics in demersal sharks of the North-East Atlantic. *Scientific Reports, 10*, 1–11.

Paterson, H. L. & Dunlop, J. N. (2018). Minimal plastic in Flesh-footed Shearwater *Ardenna carneipes* burrows at southwestern Australia colonies. *Marine Ornithology, 46*, 165–167.

Patrício, A., Hawkes, L., Monsinjon, J., Godley, B., & Fuentes, M. (2021). Climate change and marine turtles: recent advances and future directions. *Endangered Species Research, 44*, 363–395.

Patterson Edward, J. K., Mathews, G., Raj, K. D., Laju, R. L., Bharath, M. S., Kumar, P. D., Arasamuthu, A., & Grimsditch, G. (2020). Marine debris

— An emerging threat to the reef areas of Gulf of Mannar, India. *Marine Pollution Bulletin, 151*, 110793.

Perold, V., Schoombie, S., & Ryan, P. G. (2020). Decadal changes in plastic litter regurgitated by albatrosses and giant petrels at sub-Antarctic Marion Island. *Marine Pollution Bulletin, 159*, 111471.

Perrault, J., R., Levin, M., Mott, C. R., Bovery, C. M., Bresette, M. J., Chabot, R. M., Gregory, C. R., Guertin, J. R., Hirsch, S. E., Ritchie, B. W., Weege, S. T., Welsh, R. C., Witherington, B. E., & Page-Karjian, A. (2021). Insights on immune function in free-ranging green sea turtles (*Chelonia mydas*) with and without fibropapillomatosis. *Animals, 11*, 1–20.

Pfaller, J. B., Goforth, K. M., Gil, M. A., Savoca, M. S., & Lohmann, K. J. (2020). Odors from marine plastic debris elicit foraging behavior in sea turtles. *Current Biology, 30*, R213–R214.

Pham C. K., Rodríguez, Y., Dauphin, A., Carriço, R., Frias, J. P. G. L., Vandeperre, F., Otero, V., Santos, M. R., Martins, H. R., Bolten, A. B., & Bjorndal, K. A. (2017). Plastic ingestion in oceanic-stage loggerhead sea turtles (*Caretta caretta*) off the North Atlantic subtropical gyre. *Marine Pollution Bulletin, 121*, 222–229.

Pimiento, C., Leprieur, F., Silvestro, D., Lefcheck, J. S., Albouy, C., Rasher, D. B., Davis, M., Svenning, J. C., & Griffin, J. N. (2020). Functional diversity of marine megafauna in the Anthropocene. *Science Advances, 6* (16), eaay7650.

Podolsky, R. H., & Kress, S. W. (1988). Plastic debris incorporated into double-crested cormorant nests in the Gulf of Maine. *Journal of Field Ornithology, 60*, 248–250.

Poppi, L., Zaccaroni, A., Pasotto, D., Dotto, G., Marcer, F., Scaravelli, D., & Mazzariol, S. (2012). Post-mortem investigations on a leatherback turtle *Dermochelys coriacea* stranded along the Northern Adriatic coastline. *Diseases of Aquatic Organisms, 100*, 71–76.

Provencher, J. F., Bond, A. L., Avery-Gomm, S., Borrelle, S. B., Bravo Rebolledo, E. L., Hammer, S., Kühn, S., Lavers, J. L., Mallory, M. L., Trevail, A., & van Franeker, J. A. (2017). Quantifying ingested debris in marine megafauna: A review and recommendations for standardization. *Analytical Methods, 9*, 1454–1469.

Provencher, J. F., Borrelle, S. B., Bond, A. L., Lavers, J. L., & van Franeker, J. A. (2019). Recommended best practices for plastic and litter ingestion studies in marine birds: Collection, processing, and reporting. *Facets, 4*(1), 111–130.

Provencher, J. F., Vermaire, J. C., Avery-Gomm, S., Braune, B. M., & Mallory, M. L. (2018). Garbage in guano? Microplastic debris found in faecal precursors of seabirds known to ingest plastics. *Science of the Total Environment, 644*, 1477–1484.

Puskic, P. S., Lavers, J. L., & Bond, A. L. (2020). A critical review of harm associated with plastic ingestion on vertebrates. *Science of the Total Environment*, 743, 140666.

Ragusa, A., Svelato, A., Santacroce, C., Catalano, P., Notarstefano, V., Carnevali, O., Papa, F., Rongioletti, M. C. A., Baiocco, F., Draghi, S., D'Amore, E., Rinaldo, D., Matta, M., & Giorgini, E. (2021). Plasticenta: First evidence of microplastics in human placenta. *Environment International*, *146*, 106274.

Reinert, T. R., Spellman, A. C., & Bassett, B. L. (2017). Entanglement in and ingestion of fishing gear and other marine debris by Florida manatees, 1993 to 2012. *Endangered Species Research*, *32*, 415–427.

Riascos, J. M., Valencia, N., Peña, E. J., & Cantera, J. R. (2019). Inhabiting the technosphere: The encroachment of anthropogenic marine litter in Neotropical mangrove forests and its use as habitat by macrobenthic biota. *Marine Pollution Bulletin*, *142*, 559–568.

Rice, N., Hirama, S., & Witherington, B. (2021). High frequency of micro- and meso-plastics ingestion in a sample of neonate sea turtles from a major rookery. *Marine Pollution Bulletin*, *167*, 112363.

Richardson, K., Hardesty, B. D., & Wilcox, C. (2019). Estimates of fishing gear loss rates at a global scale: A literature review and meta-analysis. *Fish and Fisheries*, *20*, 1218–1231.

Rios, L. M., Jones, P. R., Moore, C., & Narayan, U. V. (2010). Quantitation of persistent organic pollutants adsorbed on plastic debris from the Northern Pacific Gyre's "eastern garbage patch" Lorena. *Journal of Environmental Monitoring*, *12*.

Rios, L. M., Moore, C., & Jones, P. R. (2007). Persistent organic pollutants carried by synthetic polymers in the ocean environment. *Marine Pollution Bulletin*, *54*, 1230–1237.

Robinson, K. J., Meyer, C., Underhill, L. G., & Calder, J.-L. (2012). Bank cormorant chick entangled in fishing line dies after more than 88 hours. *Ornithol Observations*, *3*, 188–194.

Rochman, C. M., Hoh, E., Kurobe, T., & Teh, S. J. (2013). Ingested plastic transfers hazardous chemicals to fish and induces hepatic stress. *Scientific Reports*, *3*, 3263.

Rochman, C. M., Kurobe, T., Flores, I., & Teh, S. J. (2014). Early warning signs of endocrine disruption in adult fish from the ingestion of polyethylene with and without sorbed chemical pollutants from the marine environment. *Science of the Total Environment*, *493*, 656–661.

Rodríguez, A., Ramírez, F., Carrasco, M. N., & Chiaradia, A. (2018). Seabird plastic ingestion differs among collection methods: Examples from the short-tailed shearwater. *Environmental Pollution*, *243*, 1750–1757.

Rodríguez, A., Rodríguez, B., & Nazaret Carrasco, M. (2012). High prevalence of parental delivery of plastic debris in Cory's shearwaters (*Calonectris diomedea*). *Marine Pollution Bulletin, 64*, 2219–2223.

Roff, G., Doropoulos, C., Rogers, A., Bozec, Y. M., Krueck, N. C., Aurellado, E., Priest, M., Birrell, C., & Mumby, P. J. (2016). The Ecological Role of Sharks on Coral Reefs. *Trends in Ecology & Evolution, 31*, 395–407.

Roman, L., Bell, E., Wilcox, C., Hardesty, B. D., & Hindell, M. (2019). Ecological drivers of marine debris ingestion in Procellariiform Seabirds. *Scientific Reports*, 9(1), 1–8.

Roman, L., Butcher, R. G., Stewart, D., Hunter, S., Jolly, M., Kowalski, P., Hardesty, B. D., & Lenting, B. (2021). Plastic ingestion is an underestimated cause of death for southern hemisphere albatrosses. *Conservation Letters, 14*(3), e12785.

Roman, L., Lowenstine, L., Parsley, L. M., Wilcox, C., Hardesty, B. D., Gilardi, K., & Hindell, M. (2019b). Is plastic ingestion in birds as toxic as we think? Insights from a plastic feeding experiment. *Science of the Total Environment, 665*, 660–667.

Ryan, P. G. (1987). The incidence and characteristics of plastic particles ingested by seabirds. *Marine Environmental Research, 23*, 175–206.

Ryan, P. G. (2018). Entanglement of birds in plastics and other synthetic materials. *Marine Pollution Bulletin, 135*, 159–164.

Ryan, P. G. (2020). Using photographs to record plastic in seabird nests. *Marine Pollution Bulletin, 156*, 111262.

Ryan, P. G., De Bruyn, P. J. N., & Bester, M. N. (2016). Regional differences in plastic ingestion among Southern Ocean fur seals and albatrosses. *Marine Pollution Bulletin, 104*, 207–210.

Sampaio, C. L., Leite, L., Reis-Filho, J. A., Loiola, M., Miranda, R. J., de Nunes, J. D. A. C. C., & Macena, B. C. (2018). New insights into whale shark Rhincodon typus diet in Brazil: An observation of ram filter-feeding on crab larvae and analysis of stomach contents from the first stranding in Bahia state. *Environmental Biology of Fishes, 101*(8), 1285–1293.

Sandin, S. A., Smith, J. E., DeMartini, E. E., Dinsdale, E. A., Donner, S. D., Friedlander, A. M., Konotchick, T., Malay, M., Maragos, J. E., Obura, D., Pantos, O., Paulay, G., Richie, M., Rohwer, F., Schroeder, R. E., Walsh, S., Jackson, J. B. C., Knowlton, N., & Sala, E. (2008). Baselines and degradation of coral reefs in the Northern Line Islands. *PLoS One, 3*(2), e1548.

Santos, R. G., Andrades, R., Boldrini, M. A., & Martins, A. S. (2015). Debris ingestion by juvenile marine turtles: An underestimated problem. *Marine Pollution Bulletin, 93*, 37–43.

Savoca, D., Arculeo, M., Barreca, S., Buscemi, S., Caracappa, S., Gentile, A., Persichetti, M. F., & Pace, A. (2018). Chasing phthalates in tissues of marine turtles from the Mediterranean sea. *Marine Pollution Bulletin, 127*, 165–169.

Savoca, D., Arculeo, M., Vecchioni, L., Cambera, I., Visconti, G., Melfi, R., Arizza, V., Palumbo, A., Buscemi, S., & Pace, A. (2021). Can phthalates move into the eggs of the loggerhead sea turtle *Caretta caretta*? The case of the nests on the Linosa Island in the Mediterranean Sea. *Marine Pollution Bulletin*, *168*, 112395.

Savoca, M. S., Wohlfeil, M. E., Ebeler, S. E., & Nevitt, G. A. (2016). Marine plastic debris emits a keystone infochemical for olfactory foraging seabirds. *Science Advances*, *2*, e1600395.

Schofield, G., Esteban, N., Katselidis, K. A., & Hays, G. C. (2019). Drones for research on sea turtles and other marine vertebrates — A review. *Biological Conservation*, *238*, 108214.

Schuyler, Q., Hardesty, B. D., Wilcox, C., & Townsend, K. (2014). Global analysis of anthropogenic debris ingestion by sea turtles. *Conservation Biology*, *28*, 129–39.

Schuyler, Q. A., Wilcox, C., Townsend, K. A., Wedemeyer-Strombel, K. R., Balazs, G., van Sebille, E., & Hardesty, B. D. (2016). Risk analysis reveals global hotspots for marine debris ingestion by sea turtles. *Global Change Biology*, *22*, 567–576.

Seif, S., Provencher, J. F., Gomm, S. A., Daoust, P. Y., Mallory, M. L., & Smith, P. A. (2018). Plastic and non-Plastic debris ingestion in three gull species feeding in an urban landfill environment. *Archives of Environmental Contamination and Toxicology*, *74*, 349–360.

Senko, J., Nelms, S., Reavis, J., Witherington, B., Godley, B., & Wallace, B. (2020). Understanding individual and population-level effects of plastic pollution on marine megafauna. *Endangered Species Research*, *43*, 234–252.

Setälä, O., Fleming-Lehtinen, V., & Lehtiniemi, M. (2014). Ingestion and transfer of microplastics in the planktonic food web. *Environmental Pollution*, *185*, 77–83.

Short, F. T., Polidoro, B., Livingstone, S. R., Carpenter, K. E., Bandeira, S., Bujang, J. S., Calumpong, H. P., Carruthers, T. J. B., Coles, R. G., Dennison, W. C., Erftemeijer, P. L. A., Fortes, M. D., Freeman, A. S., Jagtap, T. G., Kamal, A. H. M., Kendrick, G. A., Judson Kenworthy, W., La Nafie, Y. A., Nasution, I. M., Orth, R. J., Prathep, A., Sanciangco, J. C., van Tussenbroek, B., Vergara, S. G., Waycott, M., & Zieman, J. C. (2011). Extinction risk assessment of the world's seagrass species. *Biological Conservation*, *144*, 1961–1971.

Sievers, M., Brown, C. J., Tulloch, V. J. D., Pearson, R. M., Haig, J. A., Turschwell, M. P., & Connolly, R. M. (2019). The role of vegetated coastal wetlands for marine megafauna conservation. *Trends in Ecology & Evolution*, *34*, 807–817.

Stamper, M. A., Spicer, C. W., Neiffer, D. L., Mathews, K. S., & Fleming, G. J. (2009). Morbidity in a juvenile green sea turtle (Chelonia mydas) due to ocean-borne plastic. *Journal of Zoo and Wildlife Medicine*, 196–198.

Stamper, M. A., Whitaker, B. R., & Schofield, T. D. (2006). Case study: Morbidity in a pygmy sperm whale *Kogia breviceps* due to ocean-bourne plastic. *Marine Mammal Science, 22*, 719–722.

Stelfox, M., Bulling, M., & Sweet, M. (2019). Untangling the origin of ghost gear within the Maldivian archipelago and its impact on olive ridley (*Lepidochelys olivacea*) populations. *Endangered Species Research, 40*, 309–320.

Stelfox, M., Burian, A., Shanker, K., Rees, A. F., Jean, C., Willson, M. S., Manik, N. A., & Sweet, M. (2020). Tracing the origin of olive ridley turtles entangled in ghost nets in the Maldives: A phylogeographic assessment of populations at risk. *Biological Conservation, 245*, 108499.

Stelfox, M., Hudgins, J., & Sweet, M. (2016). A review of ghost gear entanglement amongst marine mammals, reptiles and elasmobranchs. *Marine Pollution Bulletin, 111*, 6–17.

Tavares, D. C., Da Costa, L. L., Rangel, D. F., De Moura, J. F., Zalmon, I. R., & Siciliano, S. (2016). Nests of the brown booby (*Sula leucogaster*) as a potential indicator of tropical ocean pollution by marine debris. *Ecological Indicators, 70*, 10–14.

Tavares, D. C., Moura, J. F., & Merico, A. (2019). Anthropogenic debris accumulated in nests of seabirds in an uninhabited island in West Africa. *Biological Conservation, 236*, 586–592.

Tavares, D. C., Moura, J. F., Acevedo-Trejos, E., Crawford, R. J. M., Makhado, A., Lavers, J. L., Witteveen, M., Ryan, P. G., & Merico, A. (2020). Confidence intervals and sample size for estimating the prevalence of plastic debris in seabird nests. *Environmental Pollution, 263*, 114394.

Teuten, E. L., Saquing, J. M., Knappe, D. R. U., Barlaz, M. A., Jonsson, S., Björn, A., Rowland, S. J., Thompson, R. C., Galloway, T. S., Yamashita, R., Ochi, D., Watanuki, Y., Moore, C., Viet, P. H., Tana, T. S., Prudente, M., Boonyatumanond, R., Zakaria, M. P., Akkhavong, K., Ogata, Y., Hirai, H., Iwasa, S., Mizukawa, K., Hagino, Y., Imamura, A., Saha, M., & Takada, H. (2009). Transport and release of chemicals from plastics to the environment and to wildlife. *Philosophical Transactions of the Royal Society B, 364*, 2027–2045.

Thompson, D. L., Ovenden, T. S., Pennycott, T., & Nager, R. G. (2020). The prevalence and source of plastic incorporated into nests of five seabird species on a small offshore island. *Marine Pollution Bulletin, 154*, 111076.

Triessnig, P. (2012). Beach condition and marine debris: New hurdles for sea turtle hatchling survival. Chelonian *Conservation Biology, 11*, 68–77.

Tulloch, A. I. T., Gordon, A., Runge, C. A., & Rhodes, J. R. (2019). Integrating spatially realistic infrastructure impacts into conservation planning to inform strategic environmental assessment. *Conservation Letters, 12*, 1–11.

Unger, B., Bravo, E. L., Deaville, R., Gröne, A., Ijsseldijk, L. L., Leopold, M. F., Siebert, U., Spitz, J., Wohlsein, P., & Herr, H. (2016). Large amounts of marine debris found in sperm whales stranded along the North Sea coast in early 2016. *Marine Pollution Bulletin, 112*, 134–141.

Unger, B., Herr, H., Benke, H., Böhmert, M., Burkhardt-Holm, P., Dähne, M., Hillmann, M., Wolff-Schmidt, K., Wohlsein, P., & Siebert, U. (2017). Marine debris in harbour porpoises and seals from German waters. *Marine Environmental Research, 130*, 77–84.

van Bijsterveldt, C. E. J., van Wesenbeeck, B. K., Ramadhani, S., Raven, O. V., van Gool, F. E., Pribadi, R., & Bouma, T. J. (2021). Does plastic waste kill mangroves? A field experiment to assess the impact of macro plastics on mangrove growth, stress response and survival. *Science of the Total Environment, 756*, 143826.

van Franeker, J. A., Blaize, C., Danielsen, J., Fairclough, K., Gollan, J., Guse, N., Hansen, P. L., Heubeck, M., Jensen, J. K., Le Guillou, G., Olsen, B., Olsen, K. O., Pedersen, J., Stienen, E. W. M., & Turner, D. M. (2011). Monitoring plastic ingestion by the northern fulmar *Fulmarus glacialis* in the North Sea. *Environmental Pollution, 159*, 2609–2615.

van Franeker, J. A., & Law, K. L. (2015). Seabirds, gyres and global trends in plastic pollution. *Environmental Pollution, 203*, 89–96.

van Sebille, E., Wilcox, C., Lebreton, L., Maximenko, N., Hardesty, B. D., van Franeker, J. A., Eriksen, M., Siegel, D., Galgani, F., & Law, K. L. (2015). A global inventory of small floating plastic debris. *Environmental Research Letters, 10*, 124006.

Vegter, A., Barletta, M., Beck, C., Borrero, J., Burton, H., Campbell, M., Costa, M., Eriksen, M., Eriksson, C., Estrades, A., Gilardi, K., Hardesty, B., Ivar do Sul, J., Lavers, J., Lazar, B., Lebreton, L., Nichols, W., Ribic, C., Ryan, P., Schuyler, Q., Smith, S., Takada, H., Townsend, K., Wabnitz, C., Wilcox, C., Young, L., & Hamann, M. (2014). Global research priorities to mitigate plastic pollution impacts on marine wildlife. *Endangered Species Research, 25*, 225–247.

Verlis, K. M., Campbell, M. L., & Wilson, S. P. (2014). Marine debris is selected as nesting material by the brown booby (*Sula leucogaster*) within the Swain Reefs, Great Barrier Reef, Australia. *Marine Pollution Bulletin, 87*, 180–190.

Votier, S. C., Archibald, K., Morgan, G., & Morgan, L. (2011). The use of plastic debris as nesting material by a colonial seabird and associated entanglement mortality. *Marine Pollution Bulletin, 62*, 168–172.

Wabnitz, C. C. C., Balazs, G., Beavers, S., Bjorndal, K. A., Bolten, A. B., Christensen, V., Hargrove, S., & Pauly, D. (2010). Ecosystem structure and processes at Kaloko Honoko-hau, focusing on the role of herbivores, including the green sea turtle *Chelonia mydas*, in reef resilience. *Marine Ecology Progress Series, 420*, 27–44.

Watts, A. J. R., Lewis, C., Goodhead, R. M., Beckett, S. J., Moger, J., Tyler, C. R., & Galloway, T. S. (2014). Uptake and retention of microplastics by the shore crab *Carcinus maenas*. *Environmental Science & Technology, 48*, 8823–8830.

Wegner, N. C. & Cartamil, D. P. (2012). Effects of prolonged entanglement in discarded fishing gear with substantive biofouling on the health and behavior of an adult shortfin mako shark, *Isurus oxyrinchus*. *Marine Pollution Bulletin*, *64*, 391–394.

Wilcox, C., Hardesty, B. D., Sharples, R., Griffin, D. A., Lawson, T. J., & Gunn, R. (2013). Ghostnet impacts on globally threatened turtles, a spatial risk analysis for northern Australia. *Conservation Letters*, *6*, 247–254.

Wilcox, C., van Sebille, E., & Hardesty, B. D. (2015). Threat of plastic pollution to seabirds is global, pervasive, and increasing. *Proceedings of the National Academy of Sciences of the United States of America*, *112*, 11899–11904.

Witteveen, M., Brown, M., & Ryan, P. G. (2017). Anthropogenic debris in the nests of kelp gulls in South Africa. *Marine Pollution Bulletin*, *114*, 699–704.

Woods J. S., Rødder, G., & Verones, F. (2019). An effect factor approach for quantifying the entanglement impact on marine species of macroplastic debris within life cycle impact assessment. *Ecological Indicators*, *99*, 61–66.

Yaghmour, F. & Al, A. S. (2020). Frequency and composition of anthropogenic debris in the nests of sooty gulls *Larus (Adelarus) hemprichii* Bruch, 1853 from Sir Bu Na'ir Island, United Arab Emirates. *Marine Pollution Bulletin*, *150*, 110715.

Yntema, A. C. L. & Mrosovsky, N. (1980). Sexual differentiation in hatchling loggerheads (*Caretta caretta*) incubated at different controlled temperatures. *Herpetologica*, *36*, 33–36.

Yoshikawa, T., & Asoh, K. (2004). Entanglement of monofilament fishing lines and coral death. *Biological Conservation*, *117*, 557–560.

Young, L. C., Vanderlip, C., Duffy, D. C., Afanasyev, V., & Shaffer, S. A. (2009). Bringing home the trash: Do colony-based differences in foraging distribution lead to increased plastic ingestion in laysan albatrosses? *PLoS One*, *4*, 11–13.

Zantis, L. J., Bosker, T., Lawler, F., Nelms, S. E., O'Rorke, R., Constantine, R., Sewell, M., & Carroll, E. L. (2022). Assessing microplastic exposure of large marine filter-feeders. *Science of The Total Environment*, *818*, 151815.

Zantis, L. J., Carroll, E. L., Nelms, S. E., & Bosker, T. (2021) Marine mammals and microplastics: A systematic review and call for standardisation. *Environmental Pollution*, *269*, 116142.

https://doi.org/10.1142/9789811259111_0006

Chapter 5

Microplastics in the Marine Environment — An Introduction

Thijs Bosker[*,†,‡] and Laura J. Zantis[†]

[*]*Leiden University College, Leiden University, P.O. Box 13228, 2501 EE The Hague, Netherlands*

[†]*Institute of Environmental Sciences, University of Leiden, P.O. Box 9518, 2300 RA Leiden, The Netherlands*

[‡]*t.bosker@luc.leidenuniv.nl*

1. Introduction

Microplastics are a hot topic at the moment. News articles in leading news outlets, such as the *Guardian*, the *NY Times* and the *BBC*, are common and highlight both the scale of the issue as well as the public interest on the topic. And for a good reason, microplastic levels in the ocean surface have been estimated at 15–51 trillion particles (van Sebille *et al.*, 2015), which is more than 500 times the number of stars in our galaxy. A recent study estimates that levels in ocean sediments are even higher, with a recent assessment coming to 14 Mt of microplastics buried in our ocean floors (Barrett *et al.*, 2020). Microplastics are found everywhere in our oceans: near urban centres (Zhao *et al.*, 2015) and also in the most remote places on our planet (Munari *et al.*, 2017). In addition, uptake has been demonstrated in a range of organisms from zooplankton (Cole *et al.*, 2013) to whales (Zantis *et al.*, 2021). Overall, microplastics enter the marine environment in large quantities through various sources, and it has been shown they are omnipresent worldwide. In addition, some studies have found they may be harmful to organisms, but the long-term effects of microplastic pollution on organisms and ecosystems remain unclear.

The goal of this chapter is to give a broad introduction to the topic of microplastic pollution in our marine environment, providing a foundation for the next, more in-depth chapters on specific topics. First, we will give a description of microplastics, including common physicochemical properties. We will then examine observed levels in the environment, followed by a brief discussion on uptake and effects. Finally, we will discuss the Trojan horse theory: a frequently mentioned concern regarding the role of microplastics as a vector of other pollutants, such as persistent organic pollutants and metals, into organisms. Finally, and even though there has been significant progress on the topic of microplastics, there are still a lot of areas that require further research, which will also be highlighted where applicable.

We will use some of our own research to demonstrate these concepts, for example, our work on microplastics in marine mammals and also levels of microplastics on beaches in Europe and the Caribbean.

2. What Is a Microplastic?

2.1 *Defining plastic particles*

Let's start with the most basic question: what is a microplastic? Although this seems a straightforward question, there is still ongoing debate on size limits (Hartmann *et al.*, 2019). A review of microplastics on beaches, for example, showed a range of different definitions, from <1 mm to 15 mm (Besley *et al.*, 2017). The most common size classifications are as follows (Hartmann *et al.*, 2019; Hurley *et al.*, 2020):

- Macroplastics: size range of >10 or >25 mm
- Mesoplastics: size range of 1 or 5 mm to 10 or 25 mm
- Microplastics: size range of 0.1 or 1 μm to 1 or 5 mm
- Nanoplastics: size range <0.1 μm or 1 μm.

The above list clearly indicates the ongoing *uncertainty* about the boundaries of plastic particles, which needs urgent resolving for clear and consistent reporting in the literature. For the remainder of this chapter, we will use a size range between 0.1 μm and 5 mm for microplastics and <0.1 μm for nanoplastics. Importantly, we are using the *largest* dimension of the plastic particle for classification (Hartmann *et al.*, 2019). Therefore, a piece of fibre (the most ubiquitous of shapes, as will be highlighted later on) of 20 μm in diameter but 7 mm in length will be classified as a meso-plastic, while a fibre of 20 μm in diameter and 5 mm in length will be considered a microplastic.

2.2 *Types and sources of microplastics*

Figure 1 provides a schematic diagram of the flow of microplastics into the marine environment. There are two types of microplastics distinguished in the literature: primary and secondary microplastics. Primary microplastics are manufactured to be small in origin, such as microbeads or nurdles, and are produced primarily for commercial purposes (Au *et al.*, 2017). This type of microplastic is commonly used in a variety of products and processes, including cosmetic products, pharmaceuticals

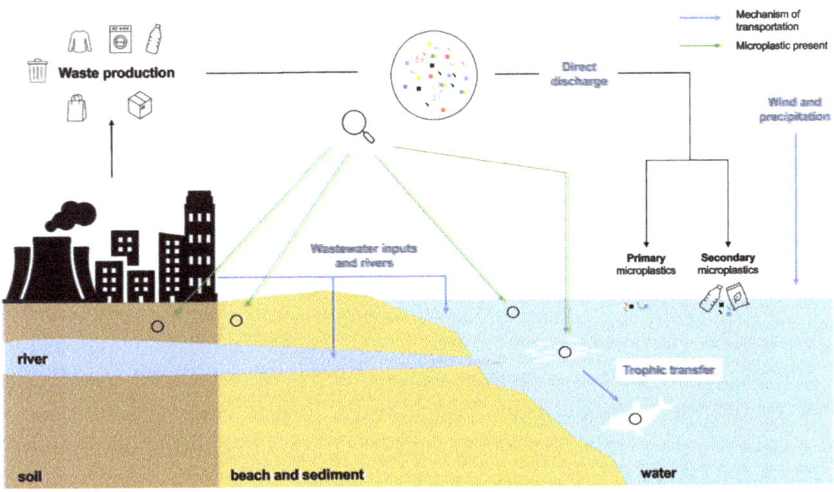

Figure 1. Summary figure highlighting the main sources, types and mechanisms of transportation of microplastics to the marine environment. Two main types, primary and secondary, of microplastics land in the sea, in sediments and in the water column through direct discharge of waste into the ocean (e.g., wastewater), transportation through natural factors (e.g., wind and waves) or through trophic transfer between organisms.

(e.g., production and packaging of drugs) and also in engineering and industrial applications (Auta *et al.*, 2018). Importantly, customers are often unaware that products they use on a daily basis contain microplastics (Chang, 2015; Kentin, 2018). Examples of these are a variety of cosmetic products, including facial exfoliating cleansers (Chang, 2015).

Several interest groups have tried to raise awareness about these primary microplastics, with the campaign "Beat the microbead" (www.beatthemicrobead.org) as a good example. As a result, efforts have been undertaken to ban these microbeads. For example, within the European Union (EU), there is a push to reduce the number of produced and used primary microplastics. The European Chemicals Agency (ECHA), for example, aims to tighten regulation of primary microplastics in consumer, commercial and industrial supply chains around the world in order to reduce the amount of primary microplastics in the EU (ECHA, 2020). Successful legislations have also been passed in the United States and the United Kingdom banning microplastics in rinse-off consumer products, such as microbeads in body scrubs (Rochman *et al.*, 2015). A similar ban was introduced by the EU, but other cosmetic products such as mascaras or glitters remain unregulated (Guerranti *et al.*, 2019). Microplastics from these and many other products, which are used domestically and/or in the industry, therefore, still end up in wastewaters. Although some of them are removed in wastewater treatment plants, a significant proportion is discharged and ends up in aquatic environments (Tanaka & Takada, 2016).

Secondary microplastics, on the other hand, are microplastics that have been broken down from large plastic items into smaller pieces. This type of fragmentation may be due to either mechanical (e.g., abrasion), chemical (e.g., hydrolysis and thermal) or biological (e.g., bacteria and fungi) degradation over time, both on land and in the sea (Cole *et al.*, 2011). For example, a plastic bag that ends up in the sea may break down into smaller particles over time due to UV degradation and wave action. However, this process is poorly understood, and significant uncertainty remains about the timescales of breakdown, the physical characteristics of the breakdown products and whether hazardous chemicals are released (Jahnke *et al.*, 2017).

In field studies, secondary microplastics are much more commonly found compared to primary microplastics. In most studies, the vast majority of microplastics detected are secondary plastics, of which fibres are very common (see Section 2.3). However, in contrast to primary microplastics, front-of-pipe solutions, such as bans, are not possible for secondary microplastics, and there have been limited regulatory actions to reduce their influx in the environment.

2.3 *Physicochemical characteristics of microplastics*

In this section, we will discuss the physical and chemical properties of microplastics which are commonly found in the environment. Due to the continuous weathering of plastics in the marine environment, microplastics come in a large range of different shapes, sizes, colours and chemical properties. In the following sections, we will describe the most common characteristics on a range of parameters.

2.3.1 *Microplastic shape*

Microplastics are present in the marine environment in different shapes due to fragmentation and degradation. Common shapes that are detected in the marine environment are fibres, fragments, films and pellets/beads (Figure 2). Plastic fibres, also known as microfibres, are among the most common types of microplastics found in the marine environment, both in deep sea and beach sediment (e.g., Blašković *et al.*, 2017; Graca *et al.*, 2017; Zantis *et al.*, 2021). For instance, Lots *et al.* (2017) found that over 90% of the microplastics found on European beaches are fibrous. Furthermore, the high fibre content in organisms most likely reflects the concentration of microplastics in the marine environment. For example,

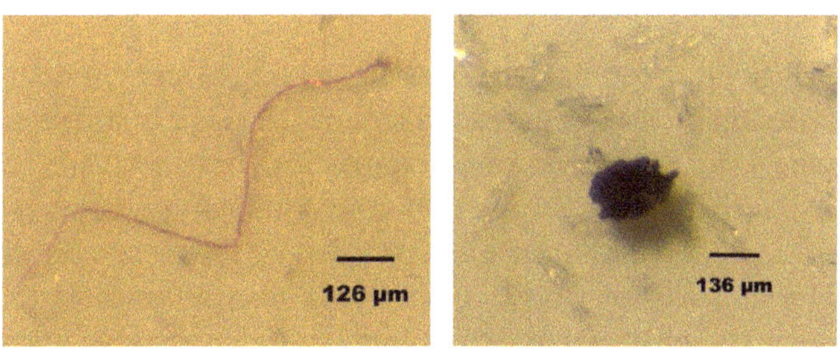

(a) (b)

Figure 2. Examples of two types of microplastic shapes: fibre (a) and fragment (b). Both the microplastics were extracted from the scat of Bryde's whales (*Balaenoptera brydei*), residing in the Hauraki Gulf of New Zealand. (a) Red fibre corresponding to cotton and (b) blue fragment identified as polyethylene.

a recent study on microplastics in the scat of whales showed that more than 99% out of 672 microplastics were fibres, with only 4 identified as fragments or films. Some examples of microplastics identified in whale scat are shown in Figure 2 (Zantis *et al.*, 2022).

A significant source of microfibres has been linked to clothing, and Napper and Thompson (2016) estimated that from a standard washing load, up to 700,000 fibres can be released. To this end, several incentives have been developed to remove these particles during washing, such as the XFiltra filter, which traps fibres before they are released into wastewaters, or the Guppyfriend bag, which is used to limit the shedding of clothing during washing (Napper *et al.*, 2020). When fibres are released during washing, the majority will end up in sewage treatment plants. Here, a significant portion can be captured: a study in Vancouver (BC, Canada) found that a single sewage treatment plant received 1.76 ± 0.31 trillion microplastics annually. Of these, the vast majority settled in sewage sludge, and only 1.7% were estimated to be released into the receiving environment (Gies *et al.*, 2018). Still, this would mean a total of 30 billion particles annually from a single wastewater treatment plant (Gies *et al.*, 2018).

Another important source of microfibres is the fishing industry, with microfibres being released from mariculture operations and fishing nets (Xue *et al.*, 2020). As these are released directly into the environment, removing them is more difficult compared to fibres originating from clothing (which can be captured in sewage treatment plants).

2.3.2 *Microplastic size*

It is very difficult to estimate the median or average sizes of microplastics in the environment. As discussed in Section 2.1, there is still some discussion on the size dimension of microplastics, but most are commonly defined as particles from 0.1 μm to 5 mm. Although particles as small as 0.1 μm are considered as microplastics, many of these are very difficult to detect. The majority of studies quantifying microplastics in the marine environment, therefore, restrict their analysis to larger-sized microplastics for two important reasons. First, the mesh size used during the collection of microplastics is often quite large, for example, when using mantra trawls a mesh size of >300 μm is often used. For this reason, several studies have highlighted that smaller-sized microplastic particles (<300 μm) are undersampled in the marine environment due to inappropriate and inadequate mesh sizes of sampling nets (Lindeque *et al.*, 2020; van Sebille *et al.*, 2015). In beach sediments, there are also major differences. For example, a large citizen science-based project in Chile used a mesh size of 1 mm when sieving the sand on beaches, thereby excluding particles <1 mm (Hidalgo-Ruz & Thiel, 2013).

However, even if the mesh size is very small, there is still an issue of identification. An important first step in most studies when quantifying microplastics is the use of a microscope. However, due to their small size, the identification under the microscope of microplastics is tricky, and most studies only include particles >150 μm or >300 μm in their study. This means that many of the smaller particles are often not included in the analysis. There are other techniques which allow the characterisation of particles <150 μm, such as Raman and Fourier transform infrared (FTIR), which rely less on human vision and can analyse particles >1 μm (Xu *et al.*, 2020). However, such techniques are not always available in laboratories and sample preparation can also be time-consuming. A recent study by Brandon *et al.* (2020) used epifluorescence microscopy to identify microplastics as small as 5 μm in seawater samples as well as in marine filter feeder organisms. This method allowed the authors to distinguish between plastic and non-plastic particles and to measure how many microplastics were present. Compared to other studies assessing the amount of microplastics in the marine environment, the particle concentration of their study was estimated to be 5–7 orders of magnitude higher than previous estimates, confirming that these smaller microplastics are likely dominant in the marine environment and that studies may be missing a large proportion of particles present.

2.3.3 *Microplastic colour*

Studies have described a variety of colours and shades of microplastic that were found in organisms, sediments and the water column. In studies investigating microplastics in both sediment and seawater, the most common colours were white or related (Hidalgo-Ruz *et al.*, 2012). On the contrary, in studies focussing on marine mammals, blue and black fibres were often the most common (Zantis *et al.*, 2021). There is no sure explanation of why this could be. However, the different results of the studies may be explained by the methodology or subjectivity of the observer. The colour determination of a microplastic can depend on both the perception of a colour and the experience of the observer (Martí *et al.*, 2020; Zantis *et al.*, 2021). In addition, opaque colours are more easily overlooked than particles with conspicuous colours, introducing a potential bias (Hidalgo-Ruz *et al.*, 2012). Furthermore, the background colour of filters or light, used, for example, in microscopic analysis, can also change the perception of the colour. For these reasons, the colour characterisation of microplastics should be interpreted with caution.

Importantly, due to forms of degradation and weathering (see Section 2.2), the colours of plastics fade and change their absorption and light reflection over time (Brandon *et al.*, 2020; Weinstein *et al.*, 2016). For example, clear plastics become yellow or brown, while coloured plastics can fade to white or clear. Martí *et al.* (2020) suggested using both the size and colour of debris as a proxy for exposure time in the environment, but this will require more in-depth research on the topic.

2.3.4 *Polymer types*

There are a variety of techniques and methods for identifying potential microplastic particles from water, sediment and other sample matrixes. The simplest method used is the visual identification of microplastics. However, only using basic microscopy or even the naked eye can result in the misidentification of organic particles as microplastics. In fact, the error rate has been estimated at up to 70% (Hidalgo-Ruz *et al.*, 2012). Therefore, results from studies using only visual identification should be treated with caution due to potentially higher error rates in the identification process (Lusher *et al.*, 2020a). For this reason, an increasing number of studies, especially those published more recently, are using more advanced techniques to reduce misidentification by subjecting potential microplastics to further analysis and identification (Dekiff *et al.*, 2014). More precise methods for determining polymer types range from thermal analysis to spectroscopy (Hidalgo-Ruz *et al.*, 2012; Shim *et al.*, 2017). These analyses are important to clarify whether a particle is a microplastic and to provide information on the type of plastic as well as its potential source and origin.

The most common methods for further analysis to characterise the polymer found are (micro)-FTIR analysis, Raman Spectroscopy and Phazir. FTIR identification, for example, can be a costly procedure, and often only a subset of suspected particles is analysed. Furthermore, when using these methods, it is important to have clear QA/QC protocols in place to minimise misclassification (Kühn *et al.*, 2020). The identification of the polymer type is an important key piece of information in order to determine the possible sources and properties of a potential microplastic particle.

As more and more studies are conducting further analyses on potential microplastics, we now have a growing database on the most common types of polymers detected in the environment: these are polyethylene terephthalate (PET or polyester), polypropylene (PP), polystyrene (PS), high-density polyethylene (HDPE), polyvinyl chloride (PVC), low-density polyethylene (LDPE) and polycarbonate (PC) (de Haan *et al.*, 2019; Lindeque *et al.*, 2020). Ultimately, these data can potentially be used to identify the main sources of microplastics in the environment.

3. Levels in the Environment

Based on their physicochemical properties, microplastics are distributed in the marine environment. Some will stay floating in the water column, while others will settle in the sediment or are taken up by marine organisms. In the following, we briefly discuss some of the main patterns of distribution observed in the environment, which is by no means meant to be an exhaustive review of the literature.

3.1 *Levels in the water column*

As highlighted in the introduction, the amount of microplastics floating on the ocean surface has been estimated at a staggering 15–51 trillion particles (van Sebille *et al.*, 2015), while a recent study estimated 12–21 Mt of microplastic in the top 200 m of the water column (Pabortsava & Lampitt, 2020). Importantly, they have been detected everywhere, even in the most remote places on our planet, including the deep sea (Barrett *et al.*, 2020). A multitude of studies have investigated microplastic concentration in different parts of the world. Lusher *et al.* (2014) estimated the density of microplastics in the northeast Atlantic Ocean at 2.46 items m^{-3}. A lower density of 1.15 ($\pm$ 1.45) items m^{-3} was identified in a later study that sampled the Atlantic several times from Germany to South Africa (Kanhai *et al.*, 2017). In the Arctic Ocean, a range of 38–234 items m^{-3} in the ice cores of this region was reported (Obbard *et al.*, 2014) suggesting a potential accumulation in ice.

These results show a wide range in concentrations of microplastics. As already mentioned, in many studies, only a proportion of larger microplastics were analysed, and the abundance of microplastics in the marine environment is underestimated (Brandon *et al.*, 2020). This can have significant impacts on the overall microplastic load in the water column, as highlighted by studies from Desforges *et al.* (2014) and Brandon *et al.* (2020). In these studies, two different sampling methods were used, one of which measured particles above 62.5 μm (Desforges *et al.*, 2014) and the other over 5 μm (Brandon *et al.*, 2020). The first study observed microplastics ranging in size from 62 μm to 5,000 μm and found an average level of 279 particles m^{-3} (Desforges *et al.*, 2014). In contrast, Brandon *et al.* (2020) used a more advanced quantification technique to detect particles >5 μm, and estimated average levels of ~4 million microplastic m^{-3} in the open ocean and 15 million microplastic m^{-3} in coastal waters (Brandon *et al.*, 2020). This is in line with a previous prediction that microplastic levels increase significantly with smaller sizes (Lenz *et al.*, 2016).

In addition, the distribution of microplastic around the globe varies geographically with locations and depends on environmental factors, such as ocean currents, tides or cyclones (Pirsaheb *et al.*, 2020). Sebille *et al.* (2015) found higher levels of microplastic loads in ocean gyres, with a hotspot in the North Atlantic subtropical gyre. Another potential hotspot might be the Arctic Ocean, where relatively high levels were found, which was influenced by the thermohaline circulation (Cózar *et al.*, 2017). Other studies have found a positive correlation between microplastics and urban density, such as population density and urban development (Dris *et al.*, 2018; Rochman *et al.*, 2013; Zhao *et al.*, 2015).

3.2 *Levels on beaches and in sediment*

A popular approach to study levels of microplastics in marine environments is investigating levels at beaches. Beaches are easily accessible and accumulate plastics from both the marine environment as well as adjacent lands. Weathering is particularly evident on beaches, where temperature and oxygen concentrations are higher than in the water (Andrady, 2011; GESAMP, 2015).

The most common types of microplastics found on beaches are fibres (Besley *et al.*, 2017; Bosker *et al.*, 2018; Lots *et al.*, 2017), with only limited fragments and beads detected. A large-scale investigation of microplastics on European beaches has found levels of microplastics ranging from 72 to 1,512 microplastics per kg of dry sediment, with high variability within sampling locations (Lots *et al.* 2017, Box 1). There are likely small-scale factors which contribute to the distribution and accumulation of microplastics on beaches, including the proximity to river mouths (Claessens *et al.*, 2011; Rech *et al.*, 2014) and densely populated zones (Mani *et al.*, 2015).

Tourism has also been linked to higher levels of microplastic contamination. On Mexican beaches along the Pacific coast with a high density of tourists, a significantly higher amount of microplastics were found (Retama *et al.*, 2016). In another study on beaches of the Southern Baltic Sea, tourism and urbanisation were identified as the most important factors contributing to high microplastic concentrations (Graca *et al.*, 2017).

Marine sediment is likely another major sink of microplastics, but they are harder to study compared to beaches, resulting in fewer studies. A recent study estimates that levels at 14 Mt of microplastics are buried

Box 1. Using citizen science to study plastic pollution.

Public awareness of plastic pollution allows for outreach and citizen science opportunities. Citizen science can be a powerful tool in research and simultaneously serve as a valuable outreach approach to inform and involve the public (Bonney *et al.*, 2009; Silvertown, 2009). For example, a citizen science project on beach microplastics was conducted in Chile, in which school children across the nation participated in collecting and extracting samples (Hidalgo-Ruz & Thiel, 2013). The majority of experimental steps were conducted by the participating children, including sieving and collection of the samples. In another example (Figure 3), the public was asked to provide sand samples from beaches across Europe for analyses for microplastics, allowing for a large-scale investigation of microplastic pollution across Europe using a standardised approach (Besley *et al.*, 2017; Lots *et al.*, 2017).

Citizen science is becoming increasingly popular when investigating (micro)plastic pollution, as it saves time and costs for researchers. A recent publication on citizen science projects on plastics in aquatic environments provides best practices and challenges, including a framework with

(*Continued*)

(*Continued*)

recommendations, which can be used by anyone who wants to develop or improve citizen science projects (Rambonnet *et al.*, 2019).

Figure 3. Volunteer participating in citizen science project by collecting sand samples from a beach in Norway.

in our ocean floors (Barrett *et al.*, 2020). Furthermore, a study on the level of microplastics at different sediment depths found plastics at all depths but higher levels in top layers (Willis *et al.*, 2017). Within sediments collected in harbours, very high microplastic levels were detected, with total microplastic concentrations of up to 390 particles kg^{-1} dry sediment, which was up to 50 times higher than the maximum concentration in other areas (Claessens *et al.*, 2011).

4. Assessment of Risk

In the previous sections, we have discussed levels of microplastics in the marine environment. Their ubiquitous presence has raised significant concern about the potential impacts on marine organisms. To assess the risk of microplastics for marine organisms, it is important to address two questions: first, are animals exposed to microplastics (i.e., uptake by animals), and second, does exposure cause adverse effects?

4.1 *Uptake in organisms*

There is overwhelming evidence that marine organisms are systematically exposed to microplastics, from zooplankton to top marine predators. For example, research in the South China Sea found microplastic present in a wide range of zooplankton, with levels increasing with trophic level (*Sun et al.*, 2017). A recent review study on microplastics in fish found that approximately 60% of all fish sampled contained microplastics in their gastrointestinal tract or stomach, with carnivorous fish having higher levels compared to omnivorous fish, with concentrations ranging up to 56 microplastics/fish (Sequeira *et al.*, 2020). In our recent study on microplastics in whale scat, we estimated the exposure levels of whales to be 2,934,986 microplastics (95% CI: 640,334–7,176,479) per day (Zantis *et al.*, 2022). From our results, we found four orders of magnitude difference between the predominant trophic transfer and the lesser environmental exposure. Therefore, we concluded that trophic transfer is the most important mechanism by which large filter-feeders are exposed to microplastic pollution in the marine environment (Zantis *et al.*, 2022).

4.2 *Impact on organisms*

Even though there is broad scientific consensus that marine organisms are exposed to microplastics, the potential adverse effects on organisms remain an important open question. To study adverse effects, most scientists rely on laboratory studies, in which model organisms are exposed to model microplastic particles (most often primary microplastics), often at relatively high and unrealistic exposure levels (Cunningham & Sigwart, 2019; Lenz *et al.*, 2016).

In addition, most studies use primary microplastics during exposure, as these are easy to obtain and relatively cheap. For example, in a review of 77 papers which conducted laboratory experiments on the impact of microplastics on organisms, only five papers (6.5%) compared degraded (secondary) to non-degraded plastics (Waldman & Rillig, 2020). However, as indicated in previous sections, secondary microplastics are the dominant type of microplastics observed in marine environments, and effects between primary and secondary microplastics can differ (Jaikumar *et al.*, 2019).

Another important consideration when studying impacts is the type of effect that is investigated: lethal or sub-lethal effects. If sub-lethal effects are studied, a wide range of endpoints can be selected, from molecular-level changes to direct impacts on fitness. Currently, there is an ongoing debate on the overall risk of microplastics. For example, a review on impact studies between 2016 and 2018 estimated the HC_5 (the hazard concentration at which plastic particles adversely affect 5% of species) to be 71.6 particles L^{-1} (with a 95% confidence interval between 3.45 and 1991 particles L^{-1}) (Skåre *et al.*, 2019). In addition, there are concerns about future levels of microplastics: a working group of experts concluded that the current risk of microplastics seems to be very rare, but, if emission remains constant in the future, widespread ecological risk may be present within a century (Koelmans *et al.*, 2019).

One concern which is often raised in the literature as well as on public websites is that microplastics can be a vehicle for the transfer of contaminants into organisms, which is often referred to as the Trojan Horse hypothesis (Burns & Boxall, 2018). Of specific concern are the adsorption and concentration of traditional persistent organic pollutants (POPs) (Andrady, 2011; Cole *et al.*, 2011) and metals (Holmes *et al.*, 2014) on the surface area of microplastics.

However, this hypothesis is widely debated (Burns & Boxall, 2018; Koelmans *et al.*, 2016), and there is no consensus in the scientific community that microplastics are a major source of transfer of organic contaminants into organisms as a result of adsorption (Burns & Boxall, 2018; Lohmann, 2017). For example, a laboratory study on lugworms (*Arenicola marina*) exposed to polystyrene microplastics and PCBs showed an increased accumulation of PCBs when simultaneously exposed to microplastics and PCB contaminated sediment (Besseling *et al.*, 2013). In contrast, a field study on the northern fulmar (*Fulmarus glacialis*) in Norwegian waters found negligible impacts of microplastics ingested by fulmars on the measured concentration of persistent organic pollutants in the tissue (Herzke *et al.*, 2016). The significant uncertainty about the Trojan Horse hypothesis means these statements have to be interpreted with significant caution, and more research is needed to better understand if and under which specific conditions the hypothesis holds.

Another area of concern is additives and plasticisers which are added to plastics to enhance their flexibility and durability. These could also bioaccumulate within organisms or be transferred from organism to organism. In addition, they could cause adverse effects on organisms, including impairment of development, reproduction and survival of organisms. For instance, mussels exposed to plasticisers had lower larval survival and growth rates, and reduced egg fertilisation (Capolupo *et al.*, 2020).

5. Challenges and Uncertainties

Although a significant and exponentially growing body of literature is being produced in microplastics, working with microplastics has proven hugely challenging. As mentioned previously, extracting and identifying small microplastic particles remain difficult. Modern and advanced techniques are being used, and studies have shown that very small microplastic particles can be more effectively detected. Various methods and protocols, ranging from the sample collection to the interpretation of results, are used to assess microplastic abundance in the marine environment and biota. This diversity in methods and protocols makes it difficult to compare study results and highlights the need for standardised methods, protocols and guidelines.

5.1 *Standardisation and harmonisation between studies*

Many protocols are in place to quantify microplastics and assess their abundance. However, it is precisely these methodological differences that make it problematic to compare the results between the studies. For example, the findings of studies focusing on microplastic in marine mammals were difficult to compare due to the diversity of techniques and the different materials used (Zantis *et al.*, 2021). When looking for microplastics taken up by marine mammals, one focuses on either the stomach, the intestines, both or sometimes even scat samples. However, different parts of the stomach and scat may contain different concentrations of microplastics, so the detection of microplastics will vary depending on the sample source where microplastic is detected. Furthermore, a difference in mesh and filter size changes the observed results. For instance, when a smaller mesh size is used, more microplastic particles are detected than when a larger mesh size is used. Studies investigating the occurrence of microplastics in sediment, beaches and the marine environment also encountered the same problem due to inconsistent sampling, extraction and interpretation methods (Liu *et al.*, 2020; Besley *et al.*, 2017; Shim & Thompson, 2015).

Another issue is the lack of transparency, which is important for replicating a study. In a review of microplastic studies on European beaches, Besley *et al.* (2017) identified that the density separation and the number of replicates contributed to the main differences between studies. Several extractions are required to recover microplastics from the sand. Besley *et al.* (2017) suggested that at least three extractions are needed to recover all the microplastics, but most studies did not report the number of extractions. As this information was not mentioned in most of the studies, the results of each study should be interpreted carefully as there is uncertainty whether all the microplastics were captured. Overall, there were many calls and suggestions for standardisation. Many studies have, therefore, provided guidelines for better comparison between studies and transparency of methods (e.g., Lusher *et al.*, 2020b; Zantis *et al.*, 2021).

5.2 *Issues with contamination prevention*

Over the last years, we have learned that microplastics are a widespread contaminant across the globe and in every medium. This means that during studies on microplastics, there is a significant risk of environmental contamination of samples, for example, from the air. This contamination of samples during collection, preparation and analysis may have a decisive impact on the results of a study. Therefore, measures to limit and account for contamination are necessary in order to obtain accurate estimates of microplastics and to reduce the risk of contamination (Hidalgo-Ruz *et al.*, 2012).

For example, in a recent review, comparing studies focusing on microplastics in marine mammals, a wide range of protocols for avoiding contamination, ranging from absent to extensive, were identified (Zantis *et al.*, 2021). These differences in contamination protocols between studies make it difficult to compare microplastic findings between studies. An important finding was that in some studies, no or only limited measures were in place to prevent external contamination. It is, therefore, doubtful that the suspected microplastics found in these studies are all actually microplastics from the samples taken.

On the other hand, several studies were fully aware of the risk of contamination. Some examples of methods used to limit microplastic contamination include use of clean equipment, wear of non-synthetic clothes and working under appropriate conditions that minimise environmental contamination in the laboratory (e.g., positive pressure laminar flow hood). In addition, recent publications contain more detailed and sophisticated protocols for contamination prevention than those published 10–15 years ago (Zantis *et al.*, 2021), suggesting increased awareness of the risk of contamination among scientists (Hidalgo-Ruz *et al.*, 2012; Löder & Gerdts, 2015; Norén, 2007). Overall, therefore, a standardised protocol to limit and address potential sources of contamination is essential and necessary in order to allow comparability, reproducibility and transparency.

6. Conclusion

In this chapter, we presented a brief introduction to microplastics. Although microplastics are widely studied, there is still not a clear definition of their size, which needs to be urgently resolved. We highlight that especially secondary microplastics are found everywhere in the marine environment: in water and sediment, on beaches as well as in organisms. Microplastics have a wide range of shapes and sizes, but fibres are the most common type of microplastic found in the marine environment.

Due to their small size, there are still technical difficulties in assessing overall environmental levels, as smaller microplastics are notoriously difficult to quantify in environmental samples. Therefore, reported levels in the environment are often a significant underestimation of actual concentrations in the environment. Although there is still a significant amount of variation and uncertainty about environmental levels, there is consensus that they are widespread and are found in every part of the marine systems. Levels in sediments seem to be especially high, and uptake is observed in a range of organisms, from zooplankton to whales.

There are several key areas which require attention in microplastic research. For example, the standardisation of research methods remains key in order to compare results across studies. In addition, the impacts of microplastics on marine organisms remain an important frontier in research, as there is no significant clarity about possible long-term impacts. Some studies have reported adverse impacts at environmentally relevant levels, while many other studies only observe effects at much higher exposure concentrations. Finally, there is still debate on whether microplastic can serve as a vector of additive or associated chemical pollutants into organisms.

References

Andrady, A. L., (2011). Microplastics in the marine environment. *Marine Pollution Bulletin, 62*, 1596–1605.

Andrady, A. L. (2017). The plastic in microplastics: A review. *Marine Pollution Bulletin, 119*(1), 12–22.

Au, S. Y., Lee, C. M., Weinstein, J. E., van den Hurk, P., & Klaine, S. J. (2017). Trophic transfer of microplastics in aquatic ecosystems: Identifying critical research needs. *Integrated Environmental Assessment and Management, 13*(3), 505–509.

Auta, H. S., Emenike, C. U., Jayanthi, B., & Fauziah, S. H. (2018). Growth kinetics and biodeterioration of polypropylene microplastics by Bacillus sp. and Rhodococcus sp. isolated from mangrove sediment. *Marine Pollution Bulletin, 127*, 15–21.

Barrett, J., Chase, Z., Zhang, J., Holl, M. M. B., Willis, K., Williams, A., Hardesty, B. D., & Wilcox, C. (2020). Microplastic pollution in deep-Sea sediments from the Great Australian Bight. *Frontiers in Marine Science, 7*, 808.

Besley, A., Vijver, M. G., Behrens, P., & Bosker, T. (2017). A standardized method for sampling and extraction methods for quantifying microplastics in beach sand. *Marine Pollution Bulletin, 114*, 77–83.

Besseling, E., Wegner, A., Foekema, E. M., van den Heuvel-Greve, M. J., & Koelmans, A. A. (2013). Effects of microplastic on fitness and PCB bioaccumulation by the lugworm *Arenicola marina* (L.). *Environmental Science & Technology, 47*, 593–600.

Blašković, A., Fastelli, P., Čižmek, H., Guerranti, C., & Renzi, M. (2017). Plastic litter in sediments from the Croatian marine protected area of the natural park of TelašČica bay (Adriatic Sea). *Marine Pollution Bulletin, 114*(1), 583–586.

Bonney, R., Cooper, C. B., Dickinson, J., Kelling, S., Phillips, T., Rosenberg, K. V., & Shirk, J. (2009). Citizen science: A developing tool for expanding science knowledge and scientific literacy. *BioScience, 59*(11), 977–984.

Bosker, T., Guaita, L., & Behrens, P. (2018). Microplastic pollution on Caribbean beaches in the Lesser Antilles. *Marine Pollution Bulletin, 133*, 442–447.

Brandon, J. A., Freibott, A., & Sala, L. M. (2020). Patterns of suspended and salp-ingested microplastic debris in the North Pacific investigated with epifluorescence microscopy. *Limnology and oceanography* letters, *5*, 46–53.

Burns, E. E. & Boxall, A. B. (2018). Microplastics in the aquatic environment: Evidence for or against adverse impacts and major knowledge gaps. *Environmental Toxicology and Chemistry, 37*, 2776–2796.

Capolupo, M., Sørensen, L., Jayasena, K. D. R., Booth, A. M., & Fabbri, E. (2020). Chemical composition and ecotoxicity of plastic and car tire rubber leachates to aquatic organisms. *Water Research, 169*, 115270.

Chang, M. (2015). Reducing microplastics from facial exfoliating cleansers in wastewater through treatment versus consumer product decisions. *Marine Pollution Bulletin, 101*, 330–333.

Claessens, M., De Meester, S., Van Landuyt, L., De Clerck, K., & Janssen, C. R. (2011). Occurrence and distribution of microplastics in marine sediments along the Belgian coast. *Marine Pollution Bulletin, 62*, 2199–2204.

Cole, M., Lindeque, P., Halsband, C., & Galloway, T. S. (2011). Microplastics as contaminants in the marine environment: A review. *Marine Pollution Bulletin, 62*, 2588–2597.

Cole, M., Lindeque, P., Fileman, E., Halsband, C., Goodhead, R., Moger, J., & Galloway, T. S. (2013). Microplastic ingestion by zooplankton. *Environmental Science & Technology, 47*, 6646–6655.

Cózar, A., Martí, E., Duarte, C. M., García-de-Lomas, J., van Sebille, E., Ballatore, T. J., Eguíluz, V. M., González-Gordillo, J. I., Pedrotti, M. L., & Echevarría, F. (2017). The Arctic Ocean as a dead end for floating plastics in the North Atlantic branch of the thermohaline circulation. *Science Advances, 3*, e1600582.

Cunningham, E. M. & Sigwart, J. D. (2019). Environmentally accurate microplastic levels and their absence from exposure studies. *Integrative and Comparative Biology, 59*, 1485–1496.

de Haan, W. P., Sanchez-Vidal, A., & Canals, M. (2019). Floating microplastics and aggregate formation in the Western Mediterranean Sea. *Marine Pollution Bulletin, 140*, 523–535.

Dekiff, J. H., Remy, D., Klasmeier, J., & Fries, E. (2014). Occurrence and spatial distribution of microplastics in sediments from Norderney. *Environmental Pollution, 186*, 248–256.
Desforges, J. P. W., Galbraith, M., & Ross, P. S. (2015). Ingestion of Microplastics by Zooplankton in the Northeast Pacific Ocean. *Archives of Environmental Contamination and Toxicology, 69*(3), 320–330.
Desforges, J. P., Galbraith, M., Dangerfield, N., & Ross, P. S. (2014). Widespread distribution of microplastics in subsurface seawater in the NE Pacific Ocean. *Marine Pollution Bulletin, 79*(1–2), 94–99.
Dris, R., Gasperi, J., & Tassin, B. (2018). Sources and fate of microplastics in urban areas: A focus on Paris megacity. *Handbook of Environmental Chemistry, 58*, 69–83.
ECHA (2020). Background document to the opinion on the Annex XV restriction report on intentionally added microplastic.
Eriksen, M., Mason, S., Wilson, S., Box, C., Zellers, A., Edwards, W., Farley, H., & Amato, S. (2013). Microplastic pollution in the surface waters of the Laurentian Great Lakes. *Marine Pollution Bulletin, 77*(1–2), 177–182.
Germanov, E. S., Marshall, A. D., Hendrawan, I. G., Admiraal, R., Rohner, C. A., Argeswara, J., Wulandari, R., Himawan, M. R., & Loneragan, N. R. (2019). Microplastics on the menu: Plastics pollute Indonesian manta ray and whale shark feeding grounds. *Frontiers in Marine Science, 6*, 679.
GESAMP (2015). Sources, fate and effects of microplastics in the marine environment (Part 1). In: Kershaw, P. J. (Ed.). IMO/FAO/UNESCO-IOC/UNIDO/WMO/IAEA/UN/UNEP/UNDP Joint Group of Experts on the Scientific Aspects of Marine Environmental Protection, No. 90, p. 96.
Gies, E.A., LeNoble, J. L., Noël, M., Etemadifar, A., Bishay, F., Hall, E. R., & Ross, P.S. (2018). Retention of microplastics in a major secondary wastewater treatment plant in Vancouver, Canada. *Marine Pollution Bulletin, 133*, 553–561.
Graca, B., Szewc, K., Zakrzewska, D., Dołęga, A., & Szczerbowska-Boruchowska, M. (2017). Sources and fate of microplastics in marine and beach sediments of the Southern Baltic Sea — A preliminary study. *Environmental Science and Pollution Research, 24*, 7650–7661.
Guerranti, C., Martellini, T., Perra, G., Scopetani, C., & Cincinelli, A. (2019). Microplastics in cosmetics: Environmental issues and needs for global bans. *Environmental Toxicology and Pharmacology, 68*, 75–79.
Hartmann, N. B., Hüffer, T., Thompson, R. C., Hassellöv, M., Verschoor, A., Daugaard, A. E., Rist, S., Karlsson, T., Brennholt, N., Cole, M., Herrling, M. P., Hess, M. C., Ivleva, N. P., Lusher, A. L., & Wagner, M. (2019). Are we speaking the same language? Recommendations for a definition and categorization framework for plastic debris. *Environmental Science & Technology, 53*, 1039–1047.

Herzke, D., Anker-Nilssen, T., Nøst, T. H., Götsch, A., Christensen-Dalsgaard, S., Langset, M., Fangel, K., & Koelmans, A. A. (2016). Negligible impact of ingested microplastics on tissue concentrations of persistent organic pollutants in northern fulmars off coastal Norway. *Environmental Science & Technology*, *50*, 1924–1933.

Hidalgo-Ruz, V., Gutow, L., Thompson, R. & Thiel, M. (2012). Microplastics in the marine environment: A review of the methods used for identification and quantification. *Environmental Science & Technology*, *46*, 3060–3075.

Hidalgo-Ruz, V. & Thiel, M. (2013). Distribution and abundance of small plastic debris on beaches in the SE Pacific (Chile): A study supported by a citizen science project. *Marine Environmental Research*, *87–88*, 12–18.

Holmes, L.A., Turner, A., & Thompson, R. C. (2014). Interactions between trace metals and plastic production pellets under estuarine conditions. *Marine Chemistry*, *167*, 25–32.

Hurley, R., Horton, A., Lusher, A., & Nizzetto, L. (2020). Plastic waste in the terrestrial environment. In *Plastic Waste and Recycling*, pp. 163–193. Academic Press London, San Diego, Cambridge, Oxford.

Jahnke, A., Arp, H. P. H., Escher, B. I., Gewert, B., Gorokhova, E., Kühnel, D., Ogonowski, M., Potthoff, A., Rummel, C., Schmitt-Jansen, M., Toorman, E., & MacLeod, M. (2017). Reducing uncertainty and confronting ignorance about the possible impacts of weathering plastic in the marine environment. *Environmental Science & Technology Letters*, *4*, 85–90.

Jaikumar, G., Brun, N. R., Vijver, M. G., & Bosker, T. (2019). Reproductive toxicity of primary and secondary microplastics to three cladocerans during chronic exposure. *Environmental Pollution*, *249*, 638–646.

Kanhai, L. D. K., Officer, R., Lyashevska, O., Thompson, R. C., & O'Connor, I. (2017). Microplastic abundance, distribution and composition along a latitudinal gradient in the Atlantic Ocean. *Marine Pollution Bulletin*, *115*(1–2), 307–314.

Kentin, E. (2018). Banning microplastics in cosmetic products in Europe: Legal challenges, *Proceedings of the International Conference on Microplastic Pollution in the Mediterranean Sea*, pp. 245–250.

Koelmans, A. A., Bakir, A., Burton, G. A., & Janssen, C. R. (2016). Microplastic as a vector for chemicals in the aquatic environment: Critical review and model-supported reinterpretation of empirical studies. *Environmental Science & Technology*, *50*, 3315–3326.

Koelmans, B., Pahl, S., Backhaus, T., Bessa, F., van Calster, G., Contzen, N., Cronin, R., Galloway, T., Hart, A., & Henderson, L. (2019). A scientific perspective on microplastics in nature and society, SAPEA.

Kühn, S., van Oyen, A., Bravo Rebolledo, E. L., Ask, A. V., & van Franeker, J. A. (2020). Polymer types ingested by northern fulmars (*Fulmarus glacialis*) and southern hemisphere relatives. *Environmental Science and Pollution Research*, *28*(2), 1643–1655.

Li, W. C., Tse, H. F., & Fok, L. (2016). Plastic waste in the marine environment: A review of sources, occurrence and effects. *Science of the Total Environment, 566–567*, 333–349.

Lindeque, P. K., Cole, M., Coppock, R. L., Lewis, C. N., Miller, R. Z., Watts, A. J. R., Wilson-McNeal, A., Wright, S. L., & Galloway, T. S. (2020). Are we underestimating microplastic abundance in the marine environment? A comparison of microplastic capture with nets of different mesh-size. *Environmental Pollution, 265*, 114721.

Liu, M., Lu, S., Chen, Y., Cao, C., Bigalke, M., & He, D. (2020). Analytical methods for microplastics in environments: Current advances and challenges. *Handbook of Environmental Chemistry, 95*, 3–24.

Lenz, R., Enders, K., & Nielsen, T. G. (2016). Microplastic exposure studies should be environmentally realistic. *Proceedings of the National Academy of Sciences, 113*, E4121–4122.

Löder, M. G. J. & Gerdts, G. (2015). Methodology used for the detection and identification of microplastics — A critical appraisal. *Marine Anthropogenic Litter*, 201–227.

Lohmann, R. (2017). Microplastics are not important for the cycling and bioaccumulation of organic pollutants in the oceans — But should microplastics be considered POPs themselves? *Integrated Environmental Assessment and Management, 13*, 460–465.

Lots, F. A., Behrens, P., Vijver, M. G., Horton, A. A., & Bosker, T. (2017). A large-scale investigation of microplastic contamination: Abundance and characteristics of microplastics in European beach sediment. Marine Pollution Bulletin, 123(1–2), 219–226.

Lusher, A. L., Bråte, I. L. N., Munno, K., Hurley, R. R., & Welden, N. A. (2020b). Is it or isn't it: The importance of visual classification in microplastic characterization. *Applied Spectroscopy, 74*(9), 1139–1153.

Lusher, A. L., Burke, A., O'Connor, I., & Officer, R. (2014). Microplastic pollution in the Northeast Atlantic Ocean: Validated and opportunistic sampling. *Marine Pollution Bulletin, 88*(1–2), 325–333.

Lusher, A. L., Munno, K., Hermabessiere, L., & Carr, S. (2020a). Isolation and extraction of microplastics from environmental samples: An evaluation of practical approaches and recommendations for further harmonization. *Applied Spectroscopy, 74*(9), 1049–1065.

Mani, T., Hauk, A., Walter, U., & Burkhardt-Holm, P. (2016). Microplastics profile along the Rhine River. *Scientific Reports, 5*, 17988.

Martí, E., Martin, C., Galli, M., Echevarría, F., Duarte, C. M., & Cózar, A. (2020). The colors of the ocean plastics. *Environmental Science and Technology, 54*(11), 6594–6601.

Munari, C., Infantini, V., Scoponi, M., Rastelli, E., Corinaldesi, C., & Mistri, M. (2017). Microplastics in the sediments of Terra Nova Bay (Ross Sea, Antarctica). *Marine Pollution Bulletin, 122*, 161–165.

Napper, I. E., Barrett, A. C., & Thompson, R. C. (2020). The efficiency of devices intended to reduce microfibre release during clothes washing. *Science of The Total Environment*, *738*, 140412.

Nelms, S. E., Galloway, T. S., Godley, B. J., Jarvis, D. S., & Lindeque, P. K. (2018). Investigating microplastic trophic transfer in marine top predators. *Environmental Pollution*, *238*, 999–1007.

Norén, F. (2007). Small plastic particles in Coastal Swedish waters. Retrieved from https://www.n-research.se/pdf/Small%20plastic%20particles%20in%20Swedish%20West%20Coast%20Waters.pdf.

Obbard, R. W., Sadri, S., Wong, Y. Q., Khitun, A. A., Baker, I., & Thompson, R. C. (2014). Global warming releases microplastic legacy frozen in Arctic Sea ice. *Earth's Future*, *2*(6), 315–320.

Pabortsava, K. & Lampitt, R. S. (2020). High concentrations of plastic hidden beneath the surface of the Atlantic Ocean. *Nature Communications*, *11*, 4073.

Pirsaheb, M., Hossini, H., & Makhdoumi, P. (2020). Review of microplastic occurrence and toxicological effects in marine environment: Experimental evidence of inflammation. *Process Safety and Environmental Protection*, *142*, 1–14.

Rambonnet, L., Vink, S. C., Land-Zandstra, A. M., & Bosker, T. (2019). Making citizen science count: Best practices and challenges of citizen science projects on plastics in aquatic environments. *Marine Pollution Bulletin*, *145*, 271–277.

Rech, S., Macaya-Caquilpán, V., Pantoja, J. F., Rivadeneira, M. M., Jofre Madariaga, D., & Thiel M. (2014). Rivers as a source of marine litter — A study from the SE Pacific. *Marine Pollution Bulletin*, *82*(1–2), 66–75.

Retama, I., Jonathan, M., Shruti, V., Velumani, S., Sarkar, S., Roy, P. D., & Rodríguez-Espinosa, P. (2016). Microplastics in tourist beaches of Huatulco Bay, Pacific coast of southern Mexico. *Marine Pollution Bulletin*, *113*, 530–535.

Rochman, C. M., Browne, M. A., Halpern, B. S., Hentschel, B. T., Hoh, E., Karapanagioti, H. K., Rios-Mendoza, L. M., Takada, H., Teh, S., & Thompson, R. C. (2013). Policy: Classify plastic waste as hazardous. *Nature*, *494*(7436), 169–170.

Rochman, C. M., Kross, S. M., Armstrong, J. B., Bogan, M. T., Darling, E. S., Green, S. J., Smyth, A. R., & Veríssimo, D. (2015). Scientific Evidence Supports a Ban on Microbeads. *Environmental Science and Technology*, *49*(18), 10759–10761.

Saley, A. M., Smart, A. C., Bezerra, M. F., Burnham, T. L. U., Capece, L. R., Lima, L. F. O., Carsh, A. C., Williams, S. L., & Morgan, S. G. (2019). Microplastic accumulation and biomagnification in a coastal marine reserve situated in a sparsely populated area. *Marine Pollution Bulletin*, *146*, 54–59.

Sequeira, I. F., Prata, J. C., da Costa, J. P., Duarte, A. C., & Rocha-Santos, T. (2020). Worldwide contamination of fish with microplastics: A brief global overview. *Marine Pollution Bulletin*, *160*, 111681.

Shahul Hamid, F., Bhatti, M. S., Anuar, N., Anuar, N., Mohan, P., & Periathamby, A. (2018). Worldwide distribution and abundance of microplastic: How dire is the situation? *Waste Management and Research*, *36*(10), 873–897.

Shim, W. J., Hong, S. H., & Eo, S. E. (2017). Identification methods in microplastic analysis: A review. *Analytical Methods*, *9*(9), 1384–1391.

Shim, W. J. & Thompson, R. C. (2015). Microplastics in the Ocean. *Archives of Environmental Contamination and Toxicology*, *69*(3), 265–268.

Silvertown, J. (2009). A new dawn for citizen science. *Trends in ecology & evolution, 24*(9), 467–471.

Skåre, J. U., Alexander, J., Haave, M., Jakubowicz, I., Knutsen, H. K., Lusher, A., Ogonowski, M., Rakkestad, K. E., Skaar, I., & Sverdrup, L.E. (2019). Microplastics; occurrence, levels and implications for environment and human health related to food. Scientific opinion of the Scientific Steering Committee of the Norwegian Scientific Committee for Food and Environment. Norwegian Scientific Committee for Food and Environment (VKM) Report.

Stolte, A., Forster, S., Gerdts, G., & Schubert, H. (2015). Microplastic concentrations in beach sediments along the German Baltic coast. *Marine Pollution Bulletin*, *99*(1–2), 216–229.

Sun, X., Li, Q., Zhu, M., Liang, J., Zheng, S., & Zhao, Y. (2017). Ingestion of microplastics by natural zooplankton groups in the northern South China Sea. *Marine Pollution Bulletin*, *115*, 217–224.

Tanaka, K. & Takada, H. (2016). Microplastic fragments and microbeads in digestive tracts of planktivorous fish from urban coastal waters. *Scientific Reports*, *6*(1), 1–8.

van Sebille, E., Wilcox, C., Lebreton, L., Maximenko, N., Hardesty, B. D., van Franeker, J. A., Eriksen, M., Siegel, D., Galgani, F., & Law, K. L. (2015). A global inventory of small floating plastic debris. *Environmental Research Letters*, *10*, 124006.

Waldman, W.R. & Rillig, M. C. (2020). Microplastic research should embrace the complexity of secondary particles. *Environmental Science & Technology*, *54*, 7751–7753.

Weinstein, J. E., Crocker, B. K., & Gray, A. D. (2016). From macroplastic to microplastic: Degradation of high-density polyethylene, polypropylene, and polystyrene in a salt marsh habitat. *Environmental Toxicology and Chemistry*, *35*(7), 1632–1640.

Willis, K. A., Eriksen, R., Wilcox, C., & Hardesty, B. D. (2017). Microplastic distribution at different sediment depths in an urban estuary. *Frontiers in Marine Science*, *4*(419), 1–8.

Wright, S. L., Thompson, R. C., & Galloway, T. S. (2013). The physical impacts of microplastics on marine organisms: A review. *Environmental pollution*, *178*, 483–492.

Xu, G., Cheng, H., Jones, R., Feng, Y., Gong, K., Li, K., Fang, X., Tahir, M. A., Valev, V. K., & Zhang, L. (2020). Surface-enhanced Raman spectroscopy facilitates the detection of microplastics <1 μm in the environment. *Environmental Science & Technology*, *54*(24), 15594–15603

Xue, B., Zhang, L., Li, R., Wang, Y., Guo, J., Yu, K., & Wang, S. (2020). Underestimated microplastic pollution derived from fishery activities and "hidden" in deep sediment. *Environmental Science & Technology*, *54*, 2210–2217.

Zantis, L. J., Bosker, T., Lawler, F., Nelms, S. E., O'Rorke, R., Constantine, R., Sewell, M., & Carroll, E. L. (2022). Assessing microplastic exposure of large marine filter-feeders. *Science of The Total Environment*, *818*, 151815.

Zantis, L. J., Carroll, E. L., Nelms, S. E., & Bosker, T. (2021). Marine mammals and microplastics: A systematic review and call for standardisation. *Environmental Pollution*, *269*, 1–2.

Zhao, S., Zhu, L., & Li, D. (2015). Microplastic in three urban estuaries, China. *Environmental Pollution*, *206*, 597–604.

https://doi.org/10.1142/9789811259111_0007

Chapter 6

Interactions and Effects of Microplastics on Marine Invertebrates

Xiaoxia Sun[*,†,‡,§,¶], Shan Zheng[*,†], Qingjie Li[*,†] and Kangning Zhang[*,§]

[*]*Jiaozhou Bay National Marine Ecosystem Research Station, Institute of Oceanology, Chinese Academy of Sciences, Qingdao 266071, China*

[†]*Laboratory for Marine Ecology and Environmental Science, Pilot National Laboratory for Marine Science and Technology, Qingdao 266237, China*

[‡]*Center for Ocean Mega-Science, Institute of Oceanology, Chinese Academy of Sciences, Qingdao 266071, China*

[§]*University of Chinese Academy of Sciences, Beijing 100049, China*

[¶]*xsun@qdio.ac.cn*

1. Introduction

Microplastics are considered a group of emerging contaminants that have generated increasing concern in the scientific and general community. Microplastics are present throughout the water column, in sediments, and in the digestive systems and tissues of marine biota. Marine invertebrates comprise more than 89% of all extant marine animal species (Horton *et al.*, 2018) and support all marine ecosystems by supplying energy derived from primary production to fish, marine birds and mammals (Dales, 1972).

Specifically, zooplankton form a critical link in the ocean food web between phytoplankton and other marine species and are essential to all life in oceans (Turner, 2015).

Microplastics are ingested by a wide range of marine invertebrates and may be transferred from prey to predators through food webs (Desforges *et al.*, 2014; Lusher *et al.*, 2013). The ingestion of microplastics might cause false satiation, growth rate reduction and reproduction effects. The presence of additives and absorbed chemicals might cause inflammation, hepatic stress and growth decrease and may ultimately lead to death (Setala *et al.*, 2016). At present, many investigation results have been documented on the abundance and distribution of microplastics ingested by invertebrates in different areas (Desforges *et al.*, 2015; Frias *et al.*, 2014; Graham & Thompson, 2009; Murray & Cowie, 2011; Van Cauwenberghe & Janssen, 2014; Thompson *et al.*, 2004), while a growing number of experimental studies have demonstrated that microplastics adversely affect the feeding, growth, development and survival of invertebrate keystone species, including copepods (Cole *et al.*, 2015) and oysters (Sussarellu *et al.*, 2016). However, many questions remain unknown.

To better understand the potential impacts of microplastics on marine invertebrates and explore the risks microplastics pose to individual biota and functional groups at the ecosystem level, this chapter reviewed the knowledge on the interaction between microplastics and invertebrates, including: (i) the abundance and characteristics of microplastics in zooplankton, bivalves and other invertebrates, (ii) the biological effects and toxicity of microplastics among keystone species, and (iii) the ecological impacts of microplastics ingested by zooplankton, bivalves and other invertebrates. Future research prospects are proposed.

2. Abundance and Characteristics of Microplastics in Invertebrates

2.1. *Zooplankton*

2.1.1. *Abundance of microplastics in zooplankton community in natural environments*

Almost all marine zooplankton have been found to feed on microplastics. Microplastics have been observed in the natural zooplankton community in a variety of functional groups and species, such as copepods, chaetognaths, cladocerans and amphipods, in addition to salp and fish larvae (Moore *et al.*, 2001; Desforges *et al.*, 2015; Steer *et al.*, 2017). Currently, the abundance of microplastics in zooplankton is often reported as the unit of volume, and the number of microplastics per individual (MP/ind) has been widely used, i.e., the number of organisms ingested microplastics/the total number of organisms processed (Desforges *et al.*, 2015; Steer *et al.*, 2017). Because zooplankton individuals are small and often take their group as a whole, it is rare to use weight as a unit to characterise the ingestion number of microplastics.

According to survey results in natural sea areas (Table 1, Annex A), the abundance of microplastics ingested by natural zooplankton groups in different areas varies widely (Botterell *et al.*, 2019). Generally, the abundance of microplastics ingested by zooplankton in coastal areas is higher than that in oceans. Considering the same group of zooplankton, the abundance of microplastics in zooplankton in coastal areas is higher than that in open oceans. For instance, the abundance of microplastics ingested by copepods in the northeastern Pacific Ocean is 0.025 MP/ind (Desforges *et al.*, 2014), while that in the East China Sea reaches 0.13 MP/ind (Sun *et al.*, 2018b). On the one hand, the abundance of microplastics in the marine environment affects the abundance of microplastics in zooplankton. The abundance of microplastics in seawater varies with the sea area. The areas with higher microplastic abundance in zooplankton usually show higher microplastic abundance in seawaters. Selecting the most widely distributed and abundant copepods as an example, the abundance of microplastics in copepods is 0.05 MP/ind in the north South China Sea (Sun *et al.*, 2017), which is higher than that in the Bohai Sea, 0.028 MP/ind. And with manta trawl (330 μm mesh) sampling at the surface seawater, the abundance of microplastics in seawater in the South China Sea, which is 0.44 MP/m^3 (Qi *et al.*, 2020), is higher than that in the Bohai Sea, 0.33 MP/m^3 (Zhang *et al.*, 2017). On the other hand, in areas with low productivity or low zooplankton abundance, even if the microplastic abundance is high, the quantity of microplastics consumed by zooplankton may be low, such as oceanic gyres, and the likelihood is predicted to be relatively low due to the low primary productivity and lower organism abundance (Clark *et al.*, 2016). Therefore, the abundance of zooplankton ingesting microplastics is largely affected by the abundance of microplastics and zooplankton in the environment.

For the different zooplankton groups, an investigation verified that the number of microplastics ingested by zooplankton tends to follow omnivorous > carnivorous > herbivorous. For example, in the North Pacific

Ocean, the abundance of microplastics in *Euphausia pacifica* is 0.058 MP/ind, while that in *Neocalanus cristatus* is 0.026 MP/ind. In the northern South China Sea, the abundance of microplastics in copepods is only 0.05 MP/ind, while that in fish larvae reaches as high as 1.43 MP/ind.

2.1.2. *Characteristics of the microplastics ingested by zooplankton*

2.1.2.1. Shape

The shapes of the microplastics ingested by zooplankton mainly include fibres, fragments and pellets (Table 1, Annex A). Fibres constitute the highest proportion and account for 66% of the microplastics ingested by zooplankton in the western English Channel (Steer *et al.*, 2017) and 92–93% of the microplastics ingested by zooplankton in the Bohai Sea (Zheng *et al.*, 2020). One different example is the South China Sea, where fragments are the most common type of microplastics (56–75%) ingested by zooplankton (Amin *et al.*, 2020). It is believed that the shapes of microplastics in zooplankton are consistent with those in seawater, and fibres are the main microplastic types, in general.

2.1.2.2. Colour

A small number of studies have revealed that the colours of microplastics ingested by zooplankton mainly include blue, black, red and colourless (Gove *et al.*, 2019; Zheng *et al.*, 2020). Blue is the most common microplastic colour in seawater and sediments (Gago *et al.*, 2018) because blue plastic products are very common and are often applied in fishing nets, ropes and other fishing gears. The percentages of blue microplastics in zooplankton are significantly higher than those in seawater, indicating that the colour of microplastics may affect the ingestion choices of marine organisms (Boerger *et al.*, 2010; Wright *et al.*, 2013).

2.1.2.3. Size

The size of the microplastics consumed by zooplankton greatly varies from several microns to several thousand microns (Cole *et al.*, 2013; Cole & Galloway, 2015; Desforges *et al.*, 2015). This is influenced by the feeding features of organisms. The size of the ingested microplastics is likely constrained by the size of the mouthparts of the species. Zooplankton exhibit feeding selectivity to different sizes of microplastics. Desforges *et al.* (2015) found that the euphausiid *Euphausia pacifica* (with a length of approximately 22 mm) ingested particles that were of a larger size on average (816 μm) than those ingested by the copepod *Neocalanus cristatus* (with a length of approximately 8.5 mm) preferentially ingesting particles with a size of 556 μm. However, studies have reported that there exists no significant difference in the size of microplastics ingested by diverse zooplankton groups along the Chinese coast (Sun *et al.*, 2017, 2018a, 2018b; Zheng *et al.*, 2020).

2.1.2.4. Chemical composition

The distribution of microplastics in the ocean is related to their density. Plastics with increased concentrations near the shore have been mostly identified as polypropylene (PP), polyester, polyethylene (PE), polyamide (nylon), acrylic and polyvinyl alcohol (PVA) (Browne *et al.*, 2011; Claessens *et al.*, 2013; Sun *et al.*, 2018a). Polystyrene (PS), PE, PP, polyester and nylon are the main types of microplastics ingested by zooplankton according to Table 1 in Annex A. This result is consistent with the main components of microplastics in surface seawater. Low-density microplastics, such as PS, PE and PP, tend to float on the sea surface and easily attach to microorganisms and phytoplankton, which makes microplastics "tasty". These two reasons make them more likely to be ingested by zooplankton. In contrast, high-density plastics, such as polyvinyl chloride (PVC), readily sink and become bioavailable to benthic suspension and deposit feeders (Wright *et al.*, 2013). Thus, the chemical composition of microplastics is an important characteristic impacting the abundance of microplastics in zooplankton.

2.2. *Bivalves*

2.2.1. *Microplastics in natural bivalves*

Bivalves are broadly distributed in the marine environment with extensive filter-feeding activity and limited movement capability; therefore, they are considered to be key bioindicators of microplastic pollution in nature. As important food resources, bivalves are of great economic value. Therefore, it is of great significance to study the microplastic pollution of bivalves. Microplastic ingestion by bivalves was first reported in 2008 (Browne *et al.*, 2008). At present, studies of microplastics in bivalves have mostly occurred in certain countries in Asia and subsequently in Europe (Figure 1). Mussels and oysters are the most frequently studied species, most of which are commercial bivalves (Figure 1).

Bivalves are contaminated by microplastics through ingestion or adherence. Not all microplastics accumulate in the body after ingestion, and organisms release them through biological processes. Bivalves are able to sort captured material, ingest nutritious particles and reject unwanted particles as pseudofaeces (Ward and Shumway, 2004). Over 40% of the ingested microplastics are either rejected in pseudofaeces or egested in faeces by oysters (Zhao *et al.*, 2018). After selective ingestion, microplastics accumulate, and their accumulation in different organs is observed. After exposure to fluorescent polypropylene (with a diameter of 53–63 μm) for 21 days at a concentration of 50 mg/L, accumulation of microplastics was found in the gill, stomach, stylus sac, secondary duct and intestine of mussels (Park *et al.*, 2017). Wang *et al.* (2021) conducted exposure experiments on mussels and found that the gill is not a target tissue for steady particle accumulation in contrast to the digestive tract. Nevertheless, microplastics have also been encountered in organs not involved in the process of ingestion. Adherence is a novel way for animals to take up microplastics beyond ingestion (Kolandhasamy *et al.*, 2018).

A review of current studies reveals the occurrence of microplastics in bivalves retrieved from natural waters or purchased from the market (Table 2, Annex A). At present, there are two commonly considered units

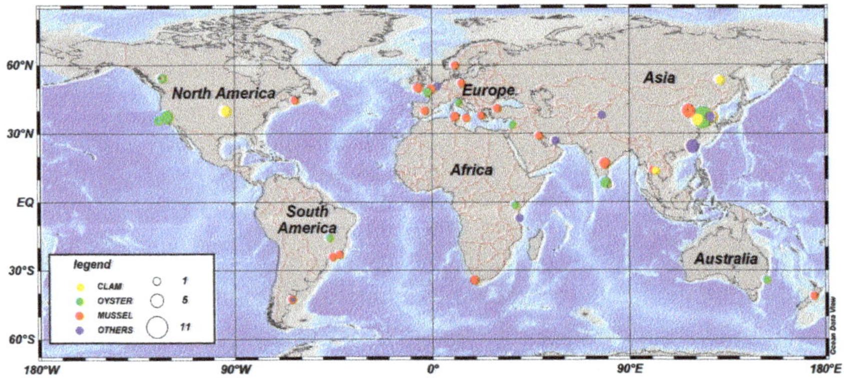

Figure 1. Microplastics reported in bivalves in the world. Dots of different colours represent different species, and the size of the dots represents the frequency of studies.

to express the abundance of microplastics: MP/ind and MP/g. Research has mainly focused on the microplastic content in the whole soft tissues of bivalves due to their wide consumption by people. Van Cauwenberghe and Janssen (2014) reported that mussels (*Mytilus edulis*) and oysters (*Crassostrea gigas*) contained 0.36 ± 0.07 and 0.47 ± 0.16 MP/g, respectively, on average. Nam Ngoc *et al.* (2018) found levels of 0.61 ± 0.56 and 2.1 ± 1.7 MP/ind, respectively. Li *et al.* (2015) investigated microplastic contamination in nine shellfish species obtained from a Chinese fishery market, determining that the number of total microplastics ranged from 2.1 to 10.5 MP/g. Microplastic contamination levels of 3.9–6.9 MP/ind in bivalves in Iran (Naji *et al.*, 2018) and 0.97 ± 0.74 MP/ind in bivalves in South Korea (Cho *et al.*, 2019) were reported. Compared to the earlier studies, the abundance of microplastics detected in bivalves was extremely high in only a few areas. The microplastic level in cultured oysters (*Crassostrea gigas*) in Sanggou Bay, China, reached 41.0 ± 15.5 MP/ind (Wang *et al.*, 2019). Adult oysters (*Crassostrea virginica*) exhibited an average value of 16.5 MP/ind in Mosquito Lagoon, which is a shallow, microtidal estuary along the east coast of central Florida (Waite *et al.*, 2018).

The factors influencing microplastic ingestion by bivalves have been investigated to a limited extent. Certain factors have been preliminarily studied and discussed, including biological and environmental factors. The trophic level and habitat (i.e., benthic, demersal and benthopelagic) of organisms do not obviously affect the ingestion of microplastics (Bour *et al.*, 2018; Walkinshaw *et al.*, 2020). Particle ingestion was positively correlated with time in the digestive tract of bivalves (Wang *et al.*, 2021). In addition to the organism itself, the environment in which it grows also affected the results. Microplastic quantities in Mosquito Lagoon were higher than those in other studies mentioned earlier due to the extensive urbanisation of the surrounding area (Waite *et al.*, 2018). However, there was no evidence that sampling seasons caused differences in microplastic ingestion results (Nam Ngoc *et al.*, 2018).

2.2.2. *Characteristics of the microplastics ingested by bivalves*

2.2.2.1. Shape

Fibres represent the predominant shape of microplastics ingested by bivalves (Table 2, Annex A). Other shapes include fragments, particles and films. Fibres comprise the highest proportion and account for more than 90% of the microplastics ingested by shellfish in Canada (Covernton *et al.*, 2019) and China (Wang *et al.*, 2019). Mussels retrieved from the Bizerte Lagoon in northern Tunisia were also highly polluted with fibres (approximately 82.3%) (Wakkaf *et al.*, 2020). Sixty-seven percent of the microplastics detected in eastern oysters (*Crassostrea virginica*) collected from a Florida estuary were fibres (Waite *et al.*, 2018). Fragments have also been reported as the dominant shape in other studies (Cho *et al.*, 2019; Gedik & Eryaşar, 2020; Cho *et al.*, 2021).

2.2.2.2. Colour

There is no consistent conclusion on the colour composition of microplastics in bivalves (Table 2, Annex A). The majority of microplastics found in mussels (*Perna perna*) sampled in the Santos estuary were white (Santana *et al.*, 2016). Li *et al.* (2018) determined that the proportion of microplastics with a light colour in wild oysters (*Saccostrea cucullata*) was 83%, which was significantly higher than that of microplastics with a dark colour. However, half of the detected microplastics were grey in blue mussels (*Mytilus edulis*) and Pacific oysters (*Crassostrea gigas*) sampled along the French Atlantic coast (Nam Ngoc *et al.*, 2018).

2.2.2.3. Size

Most microplastics detected in bivalves were smaller than 1 mm. Among all the microplastics ingested by wild oysters (*Saccostrea cucullata*) grown in the Pearl River Estuary in South China, 83.9% were smaller than 100 μm (Li *et al.*, 2018). Half of the microplastics detected in French bivalves ranged from 50–100 μm (Nam Ngoc *et al.*, 2018). Particles smaller than 300 μm dominated (more than 60%) in oysters, mussels and clams along the Korean coast (Cho *et al.*, 2021; Cho *et al.*, 2019). Different tissues of the same organism exhibit a distinct selectivity to particles. Differences in the size of microplastics were found between pseudofaeces, faeces and digestive gland/gut tissues, suggesting size-dependent particle ingestion (Zhao *et al.*, 2018). The size range of 100–200 μm was the most abundant size range among all the microplastics detected in the digestive system of commercial shellfish retrieved from China (Ding *et al.*, 2020). It was observed that with increasing size, microplastics were more likely to be biologically released (Ward *et al.*, 2019).

2.2.2.4. Chemical composition

According to the existing results (Table 2, Annex A), the common microplastics detected in shellfish include PP, PS, PE, poly(ethylene terephthalate) (PET) and nylon. The composition of microplastics in shellfish is consistent with production data worldwide. PE, PET and PP are significant polymer types in shellfish (Nam Ngoc *et al.*, 2018; Teng *et al.*, 2019; Cho *et al.*, 2019). The total amount of PP, PE and PET constituted 80% of the total microplastics in Mediterranean mussels (*Mytilus galloprovincialis*) sampled along the Turkish coast (Gedik & Eryaşar, 2020). The predominant components detected were also related to product consumption in the study region. The proportion of polystyrene in Korean samples was higher than that in other regions due to the wide use of polystyrene products in Korea (Cho *et al.*, 2021). According to research results, the main influencing factors on the characteristics of microplastics in bivalves have not yet been agreed upon. The influencing factors that have been studied include biological and environmental factors. The characteristics and typology of ingested polymers may be notably affected by the feeding mode of organisms (i.e., deposit feeder, filter feeder and predator). Environmental factors include the degree of urbanisation and weather. The microplastic level in bivalves was higher in urbanised areas with a wider diversity of polymer types than in non-urbanised areas (Cho *et al.*, 2021). Wang *et al.* (2019) found that typhoons caused changes in the size composition of the main microplastics in oysters.

2.3. *Other invertebrates*

Microplastic feeding in other invertebrates has also been observed. Reports on crustaceans (such as crabs, shrimp, lobsters and barnacles), gastropods and other mollusks (such as sea squirts and squid) have been published (Bour *et al.*, 2018; Murray & Cowie, 2011; Goldstein & Goodwin, 2013; Curren *et al.*, 2020; Villagran *et al.*, 2020; Bonello *et al.*, 2018). The accumulation of microplastics occurred at rates ranging from 0.2–0.6 MP/g in invertebrates (oysters, barnacles and periwinkles) along the Thailand coastline (Thushari *et al.*, 2017).

Among the other invertebrate organisms, crabs have been more widely researched. Waite *et al.* (2018) compared the occurrence of microplastics in oysters and crabs (*Panopeus herbstii*) and found that crabs contained more microplastic pieces (297.74 MP/g) than oysters (3.84 MP/g). A level of 3.2×10^{-3} MP/g was detected in the edible tissue of crabs (*Portunus pelagicus*) (Daniel *et al.*, 2021).

Curren *et al.* (2020) and Devriese *et al.* (2015) demonstrated that microplastic contamination was evident in commonly consumed shrimp (0.68 ± 0.55 MP/g or 1.23 ± 0.99 MP/ind), highlighting the role of shrimp in the trophic transfer and accumulation processes of microplastics in seafood. Levels of 1.153 ± 0.278 MP/g, 1.582 ± 0.448 MP/g and 0.678 ± 0.044 MP/g were measured in ophiuroids (*Ophiomusium lymani*), brittle stars (*Hymenaster pellucidus*) and gastropods (*Colus jeffreysianus*), respectively, in the deep sea (Courtene-Jones *et al.*, 2017).

Different marine taxa vary in their ability to feed on microplastics. A significantly higher abundance of suspected microplastics was found in gastropods than in crabs and bivalves in Hong Kong, China (Xu *et al.*, 2020). Daniel *et al.* (2021) first detected microplastics in the edible tissues of a squid species (*Uroteuthis (Photololigo) duvaucelii*) in India, and squids attained the highest microplastic abundance with an average value of 7.7×10^{-3} MP/g of edible tissue, higher than shrimp and crab levels. Five times more microplastics were detected in the sea squirt (*Ascidia* spp.) (0.62 MP/g) than in bivalve species sampled from the Ligurian Sea (Bonello *et al.*, 2018). As the current research is very limited, the conclusions obtained earlier have not been completely confirmed, and larger samples are thus needed to increase the robustness of this analysis.

Microplastic pollution in many important marine invertebrate species has not yet been reported. Many of the existing reports are qualitative reports. Therefore, more quantitative studies are needed to further enhance the understanding of microplastic pollution.

3. Biological Effects and Toxicity of Microplastics on Invertebrates

3.1. *Zooplankton*

The abundance and characteristics of microplastics ingested by zooplankton in the ocean have been studied, and the threats of zooplankton feeding on microplastics cannot be ignored (Figure 2). The toxicological study of

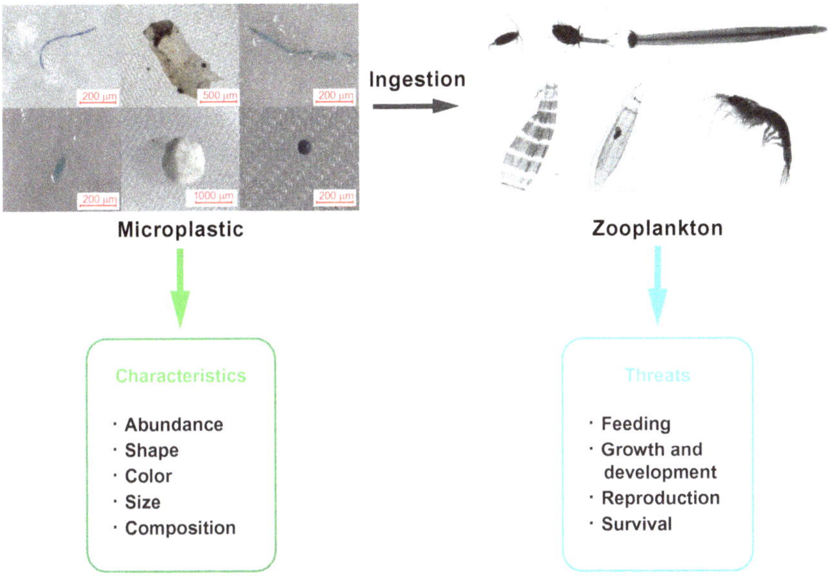

Figure 2. Characteristics and threats of MPs to zooplankton in the marine environment.

microplastics ingested by zooplankton has mainly occurred at the individual level, focusing on the feeding rate, reproduction, growth and longevity of the experimental subjects (Botterell *et al.*, 2019). Controlled experiments have been conducted to explore the ecotoxicological effects of microplastics on marine organisms. The obtained results demonstrated that the ecological effects of microplastics of different particle sizes, concentrations and chemical compositions varied.

3.1.1. *Feeding*

Zooplankton can get food by filtering water. They may mistakenly ingest microplastics when filtering water. Moreover, zooplankton may obtain microplastics from their prey with microplastics inside or outside their bodies. The ingestion of microplastics occupies the gastrointestinal space of organisms, resulting in false satiety, thus hindering the feeding of organisms, reducing the rate of ingestion, and even blocking or damaging gastrointestinal organs (Cole *et al.*, 2013). Copepods were exposed to microplastics in a series of studies. The results indicated that the feeding rate of copepods in the experimental group was significantly lower than that of copepods in the control group (Cole *et al.*, 2015). At a microplastic concentration of 1,000 MP/mL, the feeding rate of oyster larvae was significantly lower than that at a microplastic concentration of 1 MP/mL (Cole & Galloway, 2015). It was also found that the ingestion of microplastics affected zooplankton food choices: copepods exposed to 20 μm microplastics consumed the smallest available algae, and significant changes in the size range of the consumed algae were detected. Donaghay and Small (1979) found that if 20 μm microplastic particles were added to the environment, the copepod *Acartia clausi* preferentially fed on algae with a particle size of 14 μm. The consumption of smaller prey resulted in a significant reduction in the carbon biomass consumed, causing a carbon loss of 9.1 ± 3.7 mg per day in copepods. In addition to the microplastic concentration, the effect of the microplastic size on the feeding rate of zooplankton also differs. Compared to the control group, the filtration and feeding rates of microplastics (0.5 μm and 10 μm groups) of *Pseudomdiptomus annandalei* were highly reduced, while the feeding rate remained unchanged in the 2 μm microplastic group (Cheng *et al.*, 2020). The reason for the decrease in the copepod feeding rate may be the interruption of their swimming behaviour, thus reducing the risk of filtering microplastics (Van Colen *et al.*, 2020). In summary, microplastics not only alter the food selection of zooplankton but also reduce the energy intake, which may further affect the life span, growth and reproduction of zooplankton.

3.1.2. *Growth and development*

The energy obtained by zooplankton, except for respiration and excretion, is employed for individual growth and development (Widdows & Johnson, 1988). The occurrence of microplastics leads to a decrease in the predation behaviour and efficiency of zooplankton, which causes a lack of energy and affects their growth and development. The effects of microplastics on the growth and development of zooplankton are mainly reflected in the early life history stage. For instance, Lee *et al.* (2013) demonstrated that the consumption of microplastics reduced the consumption of algal prey, thus increasing the length of the nauplius stage of the copepod *Tigriopus japonicus*. Kaposi *et al.* (2014) determined that, at a concentration of 300 MP/mL, the average body width and postoral arms of *Tripneustes gratilla* larvae reached 94.8 and 400 μm, respectively, which were 111.4 and 480 μm smaller, respectively, than those of the control groups. There was no significant change in the *Crepidula onyx* larvae growth rate, survival rate or settling rate at environmentally relevant concentrations, but at a higher concentration, slower growth (control groups: 0.246 ± 0.0177 μm; trial groups: 0.191 ± 0.0168 μm) and faster settlement (control groups: 1.67 ± 0.64 ind/d; trial groups: 3.13 ± 0.49 ind/d) occurred among *Crepidula onyx* larvae (Lo & Chan, 2018). The effect of microplastics on marine zooplankton only occurs when the microplastic concentration and particle size reach a certain level. For example, PE with a particle size ranging from 27–32 μm and a concentration of 29 MP/mL imposed no notable effect on the growth rate of Antarctic krill (*Euphausia superba*) (Dawson *et al.*, 2018). In addition, all the life history endpoints of the measured barnacle *Amphibalanus amphitrite* were not significantly affected by all microplastic sizes at the tested concentrations (1, 10, 100 and 1,000 MP/mL) (Yu & Chan, 2020). When the concentration of microplastics in the environment is high, microplastic feeding exerts a negative impact on the growth and development of zooplankton because of the sensitivity of the early life stages to both internal and external microplastic exposure.

3.1.3. *Reproduction*

The zooplankton individual fecundity reflects the development trend of the whole population. Reproduction is a highly energy-consuming process, and it has been shown that diet influences egg production of the copepod *Acartia tonsa* (Teixeira *et al.*, 2010). Lee *et al.* (2013) reported a significant decrease in fecundity across two generations of the copepod *Tigriopus japonicas* exposed to multiple polystyrene microbead concentrations. They also found that many egg sacs failed to develop. Microplastics imposed no effect on the reproduction of the common marine copepod *Acartia tonsa* at concentrations lower than 0.01 MP/mL even though this concentration is 100 times higher than the environmentally relevant concentrations (Koski *et al.*, 2021). Cole *et al.* (2015) demonstrated that after long-term exposure to high concentrations (75 MP/mL) of microplastics, the hatching success rate and egg size of the copepod *Calanus helgolandicus* decreased, which was attributed to the food deficiency caused by the ingestion of microplastics. However, microplastics have been verified to promote the reproduction of zooplankton to a certain extent. Microplastics significantly promoted the reproduction of *Daphnia magna* ($F_{3,48} = 3.561$; $P = 0.021$) but did not affect the reproductive cycle resulting in a notable increase in the average number of offspring individuals when exposed to the highest test concentration over the control group (De Felice *et al.*, 2019). In addition, the oviposition rate of copepods exposed to microplastics was eight times higher than that of the control group, which suggested that microplastics could cause stress-induced oviposition of Arctic copepods, but there was no notable difference in egg size and quality between the two groups (Rodriguez-Torres *et al.*, 2020). Therefore, the effect of microplastics on the fecundity of zooplankton is complicated. Whether microplastics impact the fecundity of zooplankton depends on many factors, including the concentration of microplastics, abundance of prey, zooplankton species and experiment duration.

3.1.4. *Survival*

It has been verified that microplastics in seawater lead to a decrease in zooplankton feeding, but most studies have shown that microplastics do not significantly cause zooplankton death even though the concentrations of microplastics in the experiment are thousands of times higher than those in the environment (Kaposi *et al.*, 2014; Cole *et al.*, 2015). The average survival rate of *Tripneustes gratilla* larvae at day five was the lowest among larvae exposed to 300 MP/mL (38%) although this rate was not significantly different from that of the control group (Kaposi *et al.*, 2014). Similarly, Besseling *et al.* (2013) and Ugolini *et al.* (2013) found no effect of microplastic ingestion on the survival of the adult polychaete *Arenicola marina* and amphipod *Talitrus saltator*. However, an extremely high concentration of microplastics could lead to significant death. Compared to the control group, copepod survival was significantly affected when exposed to 0.05 μm PS beads at 1.25 μg/mL ($p < 0.05$) (Lee *et al.*, 2013). The mortality rate increased when 1–4 μm polyethylene (0.01–10 mg/L) particles were employed to treat the ephyra stage of jellyfish *Aurelia* sp. for 24 hours (Costa *et al.*, 2020). Therefore, the occurrence of microplastics in natural seawater may not necessarily cause the death of zooplankton, but if the concentration of microplastics in the environment continues to rise, zooplankton may become seriously threatened.

3.2. *Bivalves*

Bivalves are vulnerable to the potential risk posed by marine pollutants. Due to the occupation of ecological niches and the ability to filter a large water volume, the effect of microplastics on bivalves deserves much attention. The risk of bivalve exposure to microplastics has been studied broadly at different levels from the cell to the ecosystem.

3.2.1. *DNA/cell*

Studies of ecotoxicological effects in bivalves have been conducted at the cellular and molecular levels. Microplastics cause DNA damage and immune responses in cells. DNA damage in clams (*Scrobicularia plana*) has been observed after 14 days of exposure to 20 μm PS particles at a concentration of 1 mg/L (Ribeiro *et al.*, 2017). Exposure to microplastics may also cause cyto/genotoxicity (Tallec *et al.*, 2018). Canesi *et al.* (2015) exposed mussel hemocytes to different concentrations of polystyrene nanoparticles (1, 5 and 50 μg/mL) with a size of approximately 50 nm. They found that the activity of lysozyme was enhanced and the phagocytic activity of the tested cells decreased in a dose-dependent manner. In addition, the PS nanoparticles stimulated an increase in extracellular ROS (reactive oxygen species) and NO (nitric oxide) production, with maximal effects at lower concentrations. Three sizes of hemocytes, with particular tasks and functions in mussels, were also studied by Sendra *et al.* (2020). Similarly, an increase in the release of apoptotic signals and lysosome tracking and a decrease in phagocytic activity were found. Of the three sub-populations, the large granular cells were the most affected. In the marine environment, most bivalve species base their reproduction on external fertilisation. Hence, gametes face many threats associated with exposure to microplastics. Effects on spermatozoa and oocytes under microplastic exposure have been recorded in the literature. Sussarellu *et al.* (2016) conducted an exposure experiment on oysters in the reproductive cycle, which revealed a reduction in the oocyte diameter and motility rate of adult sperm over the control group.

3.2.2. *Tissue*

The obtained results regarding the alteration of histopathological parameters when bivalves are exposed to microplastics are different. No changes in histopathological parameters occurred in mussels exposed to a mixture of PE and PP at an environmental concentration over 10 days or in clams exposed to PET over seven days (Revel *et al.*, 2019; Parolini *et al.*, 2020; Goncalves *et al.*, 2019). However, tissue damage was observed in oysters (*Crassostrea gigas*) exposed to a mixture of PE and PET over 21 days (Teng *et al.*, 2021). These two opposite results may be related to the exposure time of organisms in the microplastic environment.

3.2.3. *Individual*

The effects of gametogenesis, fertilisation, embryogenesis, larval development (trochophore and D-veliger) and metamorphosis into the adult stage on the resulting ecophysiological parameters during the life cycle of bivalves have been investigated. These parameters include the energy budget, immunotoxicity, antioxidant defence, reproduction and development status, clearance rate and other activities.

Microplastics have been demonstrated to affect the reproduction of bivalves and the development of early embryos and larvae. The number of oocytes decreased when Pacific oysters were exposed to PS (Sussarellu *et al.*, 2016). High-density polyethylene (HDPE) caused larvae to exhibit abnormal swimming behaviour and subsequently affected survival (Bringer *et al.*, 2021). The formation and length of shells were altered (Sussarellu *et al.*, 2016; Luan *et al.*, 2019). Exposure to microplastics caused a reduction in energy intake and an alteration in the energy distribution of organisms (Sussarellu *et al.*, 2016; Bour *et al.*, 2018; Teng *et al.*, 2021; Xu *et al.*, 2017). Less energy was allocated to growth than before (Detree & Gallardo-Escarate, 2018). The impact of microplastics on byssus production has also been reported. Microplastics could cause a loss of byssus production, which makes mussel adhesion weak and easily swept away by waves (Rist *et al.*, 2016; Webb *et al.*, 2020).

The clearance rate is the volume of water filtered by filter-feeding organisms during the filtration of suspended particles in water per unit time (unit: L/h). The clearance rate quickly and sensitively reacts to pollutants and has been applied in marine environmental monitoring. Exposure to microplastics could reduce the clearance rate of bivalves (Wegner *et al.*, 2012; Rist *et al.*, 2016; Xu *et al.*, 2017). A decrease in the clearance rate of bivalves could reduce the energy transfer, benthic-pelagic coupling and water clarity of the ocean, thus decreasing the ecosystem service potential of the population as a whole (Christoforou *et al*., 2020; Harris & Carrington, 2020).

Oxidative stress of organisms could be triggered. Ribeiro *et al*. (2017) exposed clams to 1 mg/L (20 μm) PS for 14 days. The results revealed that microplastics affected antioxidant capacity and oxidative damage. Superoxide dismutase (SOD) and catalase (CAT) activities were significantly increased in the digestive gland of mussels (Revel *et al*., 2019). Oxidative stress reactions have also been reported in pearl oysters and scallops (Gardon *et al*., 2020).

The effects of exposure to microplastics on other activities were recorded. Pseudofaeces are produced, the faecal volume increases with exposure (Wegner *et al*., 2012), and prey consumption and absorption rates increase (Sussarellu *et al*., 2016) when microplastics are ingested by mussels and oysters. Decreased burrowing activity and neurotoxicity occurred in clams after exposure to PET for 40 days and 20 μm PS for 14 days, respectively (Hope *et al*., 2020; Ribeiro *et al*., 2017). The survival rate of mussels (*Perna viridis*) decreased as the concentration of exposed PVC increased (Rist *et al*., 2016). Inhibition of lipid metabolism was observed in oysters (*Crassostrea gigas*) after exposure to PE and PET for 21 days (Teng *et al*., 2021). However, in another study, after four weeks of exposure to PE microplastics in sediments, the survival, condition index and burrowing behaviour of the two bivalves remained unaffected (Bour *et al*., 2018). Hope *et al*. (2020) simulated an ecosystem containing microphytobenthos (MPB) and bivalves (*Macomona lilianal*) with PET microfibres in sediments. They found that the interaction between bivalves and MPB was enhanced, and the content of NOx in seawater increased, which affected the biogeochemical cycle of sediments.

The damage caused by microplastic exposure in organisms is irreversible, and the compensation mechanism (i.e., the upregulation of immune receptors and stress-related proteins) often insufficiently offsets the physiological stress caused by the initial exposure. Mussels establish stress memory upon microplastic exposure after repeated exposure to microplastics (Detree & Gallardo-Escarate, 2018).

PS microbeads and mussels are the most widely adopted experimental objects in ecotoxicological studies on microplastics. However, in reality, there are many components of microplastics in large quantities. Smaller microplastics (4–6 and 11–13 μm) yielded higher rates of malformations and developmental arrests than larger microplastics (20–25 μm)

(Bringer *et al.*, 2021). The risks upon microplastic exposure also varied with the exposure time and life cycle stage of the organism. Pearl oysters were found to achieve limited energy to cope with prolonged exposure (Gardon *et al.*, 2020). The toxic effect of microplastics on the life history of bivalves followed the order of hatching stage > metamorphosis > D-veliger larval stage, thus revealing stage-dependent toxic effects (Luan *et al.*, 2019).

4. Ecological Impacts of Microplastics Ingested by Invertebrates

Ingested microplastics could result in lethal or sub-lethal health impacts in a variety of marine organisms (Goldstein *et al.*, 2012; Wright *et al.*, 2013; Nelms *et al.*, 2018). Microplastics inside a given zooplankton can be transferred to the body of a predator through trophic transfer (Van Cauwenberghe & Janssen, 2014). The microplastic digestion ability of zooplankton is limited, and food experiences a certain retention time in the digestive tract and is then excreted with faeces. For example, microbeads remain in *Aurelia aurita* for six hours (Sucharitakul *et al.*, 2020; Rochman *et al.*, 2015). Laboratory experiments have also demonstrated the excretion process of microplastics by zooplankton: *Amphibalanus amphitrite* generally depurates microbeads within 200 min. Other species attain relatively longer gut retention times, generally ranging from 300 to 600 min to completely depurate microbeads (Yu *et al.*, 2021). Biological interactions are key to understanding the fate of microplastics in oceans (Clark *et al.*, 2016).

4.1. *Transport of microplastics from the surface to sediments*

The settlement of microplastics by marine organisms mainly occurs in the form of faeces, which generates marine snow and vertically settles in water. Cole *et al.* (2016) verified that faecal matter sinking is a mechanism by which floating plastics could be vertically transported away from surface waters. Faecal pellets are carriers of microplastics and promote the transfer of plastics to syntrophic biota. Therefore, faecal pellets may transport different shapes, sizes and compositions of microplastics from the surface water layer to the deep or sediments and increase their bioavailability to benthos. Regarding buoyant polymers, PE beads achieved a negative calculated sinking rate of −0.19 m/day as free particles but achieved a sinking rate of 818 m/day when incorporated into faecal pellets. Regarding polymers denser than seawater, polyamide (PA) fibres incorporated into faecal pellets sank at a rate of 855 m/day, an increase of 843 m/day over free PA fibres (Porter *et al.*, 2018). Moreover, microplastics alter the density and morphology of faecal pellets. The volume and sedimentation rate of faecal particles decreased with increasing microplastic concentration. The average settlement rate of normal faeces of *Harpacticus sp.* of the control group was 269 m/day, but it reached 31 m/day after 24 hours of exposure to 200 mg/L microplastic nylon 6 for 24 hours (Yu *et al.*, 2020). The sinking speed of faeces of the copepod *Calanus helgolandicus* containing low-density polyethylene was significantly lower than that of the control group, while the sinking speed of faeces containing high-density polyethylene increased (Coppock *et al.*, 2019). Faecal pellets affect the deposition and distribution of low-density microplastics and increase the recycling rate of microplastics in the different water layers. Therefore, the sinking of faecal pellets is an important mechanism in terms of the vertical migration of microplastics in seawater, with implications for nutrient and carbon transfer to the ocean floor.

4.2. *Transport of microplastics from the sediment surface to deeper layers*

Bioturbation includes all activities of benthos, such as burrowing, feeding, defecation and ventilation, which transport microplastics into sediments directly or indirectly (Kristensen *et al.*, 2012). Through sedimentation, such as natural and biological sedimentation, microplastics eventually reach the surface of seafloor sediments. These microplastic particles are ingested by benthos, especially burrowing organisms, or adhere to the body surface. When burrowing or burying activities occur, microplastics on the surface of sediments are carried into the interior of sediments (Karlson *et al.*, 2007). At present, research on microplastics in seafloor sediments has revealed that the depth of abundant microplastics ranges from 1 to 5 cm (Uddin *et al.*, 2021), and microplastics have even been encountered at a depth of 2 m in sediments (Turra *et al.*, 2014). Benthos carry microplastics from the surface of sediments to deep sediment layers through their own burial behaviour. Within three weeks, *Macoma balthica* transferred microplastics to a depth of 5.1 cm (Nakki *et al.*, 2017), while

1 mm polystyrene and 500 μm polyamide microspheres were embedded in sediments up to a depth of 20 cm by the lugworm *Arenicola marina* in 240 days (Gebhardt and Forster, 2018). The intensity and depth of bioturbation depend on local conditions and the characteristics of benthic communities (Van Colen *et al.*, 2021). Thus, bioturbation is very important to microplastic transportation from the sediment–water interface to deeper sediments.

5. Prospects

Generally, there are insufficient data available to formulate consistent conclusions regarding the interaction between microplastics and invertebrates. It is necessary to gain more insights into the impacts of microplastics at the community and ecosystem levels. A growing number of studies have emphasised the importance of irregular shapes, e.g., fibres, in natural ecosystems, calling for more future research efforts (Thompson *et al.*, 2009; Sun *et al.*, 2018a, 2018b, 2019). The chronic effects of low-level microplastic retention on invertebrates and microplastic risk assessment under different environmental conditions are imperative to achieve environmentally relevant risk assessment in further research. In addition, future research is needed on the persistent toxic chemicals adsorbed onto microplastics and ecotoxicology bioassays to assess the toxicological interactions between microplastics and other chemicals of high concern. The following points should be considered in follow-up research:

(1) Enhancement of the standardisation of field and laboratory protocols targeting microplastics and even nanoplastics, which include sampling collection, processing and characterisation of the isolated microplastics, in addition to data management, to ensure that the data collected and analysed in the different studies are comparable.
(2) Improvement of the geographic and taxonomic representations of the invertebrate species studied and examination of the biological exposure effect based on environmentally relevant microplastics to provide a greater understanding of the effects, trophic transfer and risk assessment.
(3) An increased emphasis on the description of microplastics as one of multiple environmental stressors is defined as a class of pollutants rather than a single pollutant to gain a better understanding of the

complexity of microplastics and even nanoplastics and their subsequent effects.

Acknowledgements

The authors are grateful for the financial support received from the National Natural Science Foundation of China (No. U2006206), the International Partnership Program of the Chinese Academy of Sciences (No. 133137KYSB20200002), the National Natural Science Foundation of China (No. 42006118), the International Partnership Program of the Chinese Academy of Sciences (No. 121311KYSB20190029) and the Strategic Priority Research Program of the Chinese Academy of Sciences (No. XDA19060204 and No. XDA23050303).

Annex A: Tables

Table 1. Abundance and characteristics of microplastics reported in zooplankton in different regions.

Region	Zooplankton	Microplastic size (μm)	Abundance (MP/ind)	Main type	Colour	Shapes	Reference
Northeast Pacific Ocean	*Neocalanus cristatus*	556 ± 149	0.026	PS	Black, red and blue	Fibre	Desforges *et al.* (2015)
	Euphausia pacifica	816 ± 108	0.058				
Northern South China Sea	Copepod	4–2399	0.05	Polyester	/	Fibres, particles, irregular shapes	Sun *et al.* (2017)
	Chaetognaths		0.15				
	Jellyfish		0.34				
	Shrimps		0.49				
	Fish larvae		1.20				
East China Sea	Brachyura larvae	/	0.18	Polymerised oxidised organic material, polyester, PP	/	Fibres, fragments, pellets	Sun *et al.* (2018b)
	Amphipoda spp.		0.66				
	Chaetognatha spp.		0.20				
	Cladocera spp.		0.18				
	Copepoda spp.		0.13				
	Euphausiacean spp.		0.20				
	Heteropoda spp.		0.30				
	Luciferidea spp.		0.31				
	Medusozoa spp.		0.21				
	Pteropoda spp.		0.34				

	Amphipoda spp.		0.09				
	Chaetognatha spp.		0.11				
	Euphausiacea spp.		0.07				
	Luciferidea spp.		0.10				
	Medusozoa spp.		0.24				
Yellow Sea	*Siphonophorae* spp.	154.62 ± 152.90	0.20	Organic oxidation polymers, poly-octenes, PP	/	Fibres, pellets and fragments	Sun *et al*. (2018a)
	Thaliacea spp.		0.07				
	Brachyura larvae		0.09				
	Fish larvae		1.12				
	Stomatopoda larvae		1.17				
West Coast of Hawaii Island	Fish larval	>1000	0.086	Polyester, nylon, PE, rayon, and artificial cellulose	Blue or translucent	Fibres	Gove *et al*. (2019)
Charleston Harbor	Zooplankton	/	0.024	/	/	Fragments or fibres	Payton *et al*. (2020)

(*Continued*)

Table 1. (*Continued*)

Region	Zooplankton	Microplastic size (μm)	Abundance (MP/ind)	Main type	Colour	Shapes	Reference
Southern South China Sea	Fish larvae Cyclopoid Shrimps Polychaete Calanoid Chaetognath	20–1680	0.14 0.007 0.01 0.01 0.13 0.003	/	/	Fibres and fragments	Amin *et al.* (2020)
Bohai Sea	Mantis shrimp Brachyura larvae Chaetognatha Copepoda Euphausia Fish larvae Medusae	1230 ± 1430	0.040 0.062 0.019 0.028 0.019 0.098 0.119	Cellophane, PET and polymerised and oxidised organic material	Blue, black, red, and transparent	Fibres and fragments	Zheng *et al.* (2020)
Chichester Harbour, United Kingdom	Temora longicornis	/	No MP	/	/	/	Outram *et al.* (2020)

Table 2. Abundance and characteristics of microplastics reported in bivalves in different regions.

Region	Source	Species	Microplastics abundance in bivalves (MP/ind)	Microplastics in bivalves (MP/g)	Composition	Shape	Colour	Size	Reference
		Mussel							
Germany	Market	Mussels (*Mytilus edulis*)	—	0.36 ± 0.07	—	—	—	—	Van Cauwenberghe and Janssen (2014)
France	Atlantic coast	Mussels (*M. edulis*)	0.61 ± 0.56	—	PP (47%), PE (38%)	Fragment (82%)	Grey (51%)	50–00 μm (52%)	Nam Ngoc *et al.* (2018)
Tunisia	Bizerte lagoon	Mussels (*M. galloprovincialis*)	0.4 ± 0.2	2.1 ± 1.0	—	—	—	—	Wakkaf *et al.* (2020)
Turkish		Mussels (*M. galloprovincialis*)	0.69	0.23	PET (32.9%), PP (28.4%), PE (19.4%)	Fragments (67.6%)	—	<500 μm (26.58%)	Gedik and Eryaşar (2020)
Turkish	Market	Mussels	0.6 ± 0.1	0.058	PE (35%), PP (15%)	Fibre (62.7%)	—	—	Gundogdu *et al.* (2020)
China	Along the coast	Mussels (*Mytilus edulis*)	2.93	0.62	—	Fibre (60.67%)		<1500 μm (81.89%)	Teng *et al.* (2019)
South Korea	Market	Mussels (*Mytilus edulis*)	—	0.12 ± 0.11	—	—	—	—	Cho *et al.* (2019)
		Oyster							
France	Atlantic coast	Oysters (*Crassostrea gigas*)	—	0.47 ± 0.16	—	—	—	—	Van Cauwenberghe and Janssen (2014)
France	Atlantic coast	Oysters (*C. gigas*)	2.1 ± 1.7	—	PE (47%), PP (25%)	Fragment (79%)	Grey (40%)	50–100 μm (53%)	Nam Ngoc *et al.* (2018)

(*Continued*)

Table 2. (*Continued*)

Region	Source	Species	Microplastics abundance in bivalves (MP/ind)	Microplastics in bivalves (MP/g)	Composition	Shape	Colour	Size	Reference
America	Mosquito Lagoon	Oysters (*C. virginica*)	16.5	—	—	—	—	—	Waite *et al.* (2018)
South Korea	Market	Oysters (*C. gigas*)	0.77 ± 0.74	0.07 ± 0.06	—	—	—	—	Cho *et al.* (2019)
China	Jiaozhou Bay	Oysters (*C. gigas*)	2.34	0.924	—	—	—	—	Zhang *et al.* (2021)
China	Pearl River Estuary	Oysters (*Saccostrea cucullata*)	—	—	—	Fibre (69.4%)	—	<100 μm (83.90%)	Li *et al.* (2018)
China	Sanggou Bay	Oysters (*C. gigas*)	41.0 ± 15.5	—	—	Fibre (99.2%)	Black (47.1%), transparent (17.9%)	<500 μm (65%)	Wang *et al.* (2019)
		Clam							
South Korea	Market	Clams (*Ruditapes philippinarum*)	1.15 ± 0.74	0.34 ± 0.31	—	—	—	—	Cho *et al.* (2019)
South Korea	Along the coast	Clams (*R. philippinarum*)	2.19 ± 1.20	0.43 ± 0.32	PP (40%), PE (20%)	Fragments (72%)	Colourless (85%)	<300 μm (83%)	Cho *et al.* (2021)
China	—	Clams (*R. philippinarum*)	—	3.50 ± 1.35	Cellophane (-), PET (-)	Fibre (56.13%)	Coloured (66.70%)	<1000 μm (55.20%)	Zhang *et al.* (2020)
China	Jiaozhou Bay	Clams (*R. philippinarum*)	2	1.509	—	—	—	—	Zhang *et al.* (2021)

South Korea	Market	Scallop (*Patinopecten yessoensis*)	—	0.08 ± 0.08	—	—	—	—	Cho *et al*. (2019)
		Mixture							
China	Market	Nine bivalves	4.3–57.2	2.1–10.5	—	Fibre (-)	—	<250 μm (33%–84%)	Li *et al*. (2015)
China	Jiaozhou Bay	Two bivalves	2.17	1.21	Cellophane (48.48%), PET (20.71%)	Fibre (92.97%)	Black (42.97%), blue (20.31%),	<500 μm (37.5%)	Zhang *et al*. (2021)
Iran	The Persian Gulf	Three bivalves	3.9–6.9	0.2–2.2	PE, PET, nylon	Fibre (58%)	—	10–25 μm (37–58%)	Naji *et al*. (2018)
South Korea	Along the coast	Two bivalves (Mussel and oyster)	1.21 ± 0.68	0.33 ± 0.23	PP (43%)	Fragments (69%)	Colourless (79%)	<300 μm (96%)	Cho *et al*. (2021)
South Korea	Market	Four bivalves	0.97 ± 0.74	0.15 ± 0.20	PEI, PP, PS, PET (>80%)	Fragments (76%)	—	<300 μm (65%)	Cho *et al*. (2019)

References

Amin, R. M., Sohaimi, E. S., Anuar, S. T., & Bachok, Z. (2020). Microplastic ingestion by zooplankton in Terengganu coastal waters, southern South China Sea. *Marine Pollution Bulletin, 150*, 110616. http://dx.doi.org/10.1016/j.marpolbul.2019.110616.

Besseling, E., Wegner, A., Foekema, E. M., van den Heuvel-Greve, M. J., & Koelmans, A. A. (2013). Effects of microplastic on fitness and PCB bioaccumulation by the lugworm *Arenicola marina* (L.). *Environmental Science & Technology, 47*, 593–600. http://dx.doi.org/10.1021/es302763x.

Boerger, C. M., Lattin, G. L., Moore, S. L., & Moore, C. J. (2010). Plastic ingestion by planktivorous fishes in the North Pacific Central Gyre. *Marine Pollution Bulletin, 60*, 2275–2278. http://dx.doi.org/10.1016/j.marpolbul.2010.08.007.

Bonello, G., Varrella, P., & Pane, L. (2018). First evaluation of microplastic content in benthic filter-feeders of the Gulf of La Spezia (Ligurian Sea). *Journal of Aquatic Food Product Technology, 27*, 284–291. http://dx.doi.org/10.1080/10498850.2018.1427820.

Botterell, Z. L. R., Beaumont, N., Dorrington, T., Steinke, M., Thompson, R. C., & Lindeque, P. K. (2019). Bioavailability and effects of microplastics on marine zooplankton: A review. *Environmental Pollution, 245*, 98–110. http://dx.doi.org/10.1016/j.envpol.2018.10.065.

Bour, A., Avio, C. G., Gorbi, S., Regoli, F., & Hylland, K. (2018). Presence of microplastics in benthic and epibenthic organisms: Influence of habitat, feeding mode and trophic level. *Environmental Pollution, 243*, 1217–1225. http://dx.doi.org/10.1016/j.envpol.2018.09.115.

Bringer, A., Cachot, J., Dubillot, E., Lalot, B., & Thomas, H. (2021). Evidence of deleterious effects of microplastics from aquaculture materials on pediveliger larva settlement and oyster spat growth of Pacific oyster, *Crassostrea gigas*. *Science of The Total Environment, 794*, 148708. http://dx.doi.org/10.1016/j.scitotenv.2021.148708.

Browne, M. A., Crump, P., Niven, S. J., Teuten, E., Tonkin, A., Galloway, T., & Thompson, R. (2011). Accumulation of microplastic on shorelines worldwide: Sources and sinks. *Environmental Science & Technology, 45*, 9175–9179. http://dx.doi.org/10.1021/es800249a.

Browne, M. A., Dissanayake, A., Galloway, T. S., Lowe, D. M., & Thompson, R. C. (2008). Ingested microscopic plastic translocates to the circulatory system of the mussel, *Mytilus edulis* (L.). *Environmental Science & Technology, 42*, 5026–5031. http://dx.doi.org/10.1021/es201811s.

Canesi, L., Ciacci, C., Bergami, E., Monopoli, M. P., Dawson, K. A., Papa, S., Canonico, B., & Corsi, I. (2015). Evidence for immunomodulation and apoptotic processes induced by cationic polystyrene nanoparticles in the hemocytes of the marine bivalve *Mytilus*. *Marine Environmental Research, 111*, 34–40. http://dx.doi.org/10.1016/j.marenvres.2015.06.008.

Cheng, Y., Wang, J., Yi, X., Li, L., Liu, X., & Ru, S. (2020). Low microalgae availability increases the ingestion rates and potential effects of microplastics on marine copepod *Pseudodiaptomus annandalei. Marine pollution bulletin, 152*, 110919. http://dx.doi.org/10.1016/j.marpolbul.2020.110919.

Cho, Y., Shim, W. J., Jang, M., Han, G. M., & Hong, S. H. (2019). Abundance and characteristics of microplastics in market bivalves from South Korea. *Environmental Pollution, 245*, 1107–1116. http://dx.doi.org/10.1016/j.envpol.2018.11.091.

Cho, Y., Shim, W. J., Jang, M., Han, G. M., & Hong, S. H. (2021). Nationwide monitoring of microplastics in bivalves from the coastal environment of Korea. *Environmental pollution (Barking, Essex: 1987), 270*, 116175–116175. http://dx.doi.org/10.1016/j.envpol.2020.116175.

Christoforou, E., Dominoni, D. M., Lindstrom, J., Stilo, G., & Spatharis, S. (2020). Effects of long-term exposure to microfibers on ecosystem services provided by coastal mussels. *Environmental Pollution, 266*, 115184. http://dx.doi.org/10.1016/j.envpol.2020.115184.

Claessens, M., Van Cauwenberghe, L., Vandegehuchte, M. B., & Janssen, C. R. (2013). New techniques for the detection of microplastics in sediments and field collected organisms. *Marine Pollution Bulletin, 70*, 227–233. http://dx.doi.org/10.1016/j.marpolbul.2013.03.009.

Clark, J. R., Cole, M., Lindeque, P. K., Fileman, E., Blackford, J., Lewis, C., Lenton, T. M., & Galloway, T. S. (2016). Marine microplastic debris: a targeted plan for understanding and quantifying interactions with marine life. *Frontiers in Ecology and the Environment, 14*, 317–324. http://dx.doi.org/10.1002/fee.1297.

Cole, M. & Galloway, T. S. (2015). Ingestion of Nanoplastics and Microplastics by Pacific Oyster Larvae. *Environmental Science & Technology, 49*, 14625–14632. http://dx.doi.org/10.1021/acs.est.5b04099.

Cole, M., Lindeque, P., Fileman, E., Halsband, C., & Galloway, T. S. (2015). The Impact of Polystyrene microplastics on feeding, function and fecundity in the marine copepod *Calanus helgolandicus. Environmental Science & Technology, 49*, 1130–1137. http://dx.doi.org/10.1021/es504525u.

Cole, M., Lindeque, P., Fileman, E., Halsband, C., Goodhead, R., Moger, J., & Galloway, T. S. (2013). Microplastic ingestion by zooplankton. *Environmental Science & Technology, 47*, 6646–6655. http://dx.doi.org/10.1021/es400663f.

Cole, M., Lindeque, P. K., Fileman, E., Clark, J., Lewis, C., Halsband, C., & Galloway, T. S. (2016). Microplastics alter the properties and sinking rates of zooplankton faecal pellets. *Environmental Science & Technology, 50*, 3239–3246. http://dx.doi.org/10.1021/acs.est.5b05905.

Coppock, R. L., Galloway, T. S., Cole, M., Fileman, E. S., Queiros, A. M., & Lindeque, P. K. (2019). Microplastics alter feeding selectivity and faecal density in the copepod, *Calanus helgolandicus. Science of the Total Environment, 687*, 780–789. http://dx.doi.org/10.1016/j.scitotenv.2019.06.009.

Costa, E., Gambardella, C., Piazza, V., Vassalli, M., Sbrana, F., Lavorano, S., Garaventa, F., & Faimali, M. (2020). Microplastics ingestion in the ephyra stage of *Aurelia* sp. triggers acute and behavioral responses. *Ecotoxicology and Environmental Safety*, *189*, 109983. http://dx.doi.org/10.1016/j.ecoenv.2019.109983.

Courtene-Jones, W., Quinn, B., Gary, S. F., Mogg, A. O. M., & Narayanaswamy, B. E. (2017). Microplastic pollution identified in deep-sea water and ingested by benthic invertebrates in the Rockall Trough, North Atlantic Ocean. *Environmental Pollution*, *231*, 271–280. http://dx.doi.org/10.1016/j.envpol.2017.08.026.

Covernton, G. A., Collicutt, B., Gurney-Smith, H. J., Pearce, C. M., Dower, J. F., Ross, P. S., & Dudas, S. E. (2019). Microplastics in bivalves and their habitat in relation to shellfish aquaculture proximity in coastal British Columbia, Canada. *Aquaculture Environment Interactions*, *11*, 357–374. http://dx.doi.org/10.3354/aei00316.

Curren, E., Leaw, C. P., Lim, P. T., & Leong, S. C. Y. (2020). Evidence of Marine Microplastics in commercially harvested seafood. *Frontiers in Bioengineering and Biotechnology*, *8*, 1–9. http://dx.doi.org/10.3389/fbioe.2020.562760.

Dales, R. P. (1972). Invertebrate Structure. *Nature*, *239*, 349. https://doi.org/10.1038/239349a0. http://dx.doi.org/10.1007/978-1-349-02495-7.

Daniel, D. B., Ashraf, P. M., Thomas, S. N., & Thomson, K. T. (2021). Microplastics in the edible tissues of shellfishes sold for human consumption. *Chemosphere*, *264*, 128554. http://dx.doi.org/10.1016/j.chemosphere.2020.128554.

Dawson, A., Huston, W., Kawaguchi, S., King, C., Cropp, R., Wild, S., Eisenmann, P., Townsend, K., & Nash, S. B. (2018). Uptake and depuration kinetics influence microplastic bioaccumulation and toxicity in Antarctic KriII (*Euphausia superba*). *Environmental Science & Technology*, *52*, 3195–3201. http://dx.doi.org/10.1021/acs.est.7b05759.

De Felice, B., Sabatini, V., Antenucci, S., Gattoni, G., Santo, N., Bacchetta, R., Ortenzi, M. A., & Parolini, M. (2019). Polystyrene microplastics ingestion induced behavioral effects to the cladoceran *Daphnia magna*. *Chemosphere*, *231*, 423–431. http://dx.doi.org/10.1016/j.chemosphere.2019.05.115.

Desforges, J.-P. W., Galbraith, M., Dangerfield, N., & Ross, P. S. (2014). Widespread distribution of microplastics in subsurface seawater in the NE Pacific Ocean. *Marine Pollution Bulletin*, *79*, 94–99. http://dx.doi.org/10.1007/s00244-015-0172-5.

Desforges, J.-P. W., Galbraith, M., & Ross, P. S. (2015). Ingestion of microplastics by zooplankton in the Northeast Pacific Ocean. *Archives of Environmental Contamination and Toxicology*, *69*, 320–330. http://dx.doi.org/10.1016/j.marpolbul.2013.12.035.

Detree, C. & Gallardo-Escarate, C. (2018). Single and repetitive microplastics exposures induce immune system modulation and homeostasis alteration in the edible mussel *Mytilus galloprovincialis*. *Fish & Shellfish Immunology*, *83*, 52–60. http://dx.doi.org/10.1016/j.fsi.2018.09.018.

Devriese, L. I., van Der Meulen, M. D., Maes, T., Bekaert, K., Paul-Pont, I., Frère, L., Robbens, J., & Vethaak, A. D. (2015). Microplastic contamination in brown shrimp (*Crangon crangon*, Linnaeus 1758) from coastal waters of the Southern North Sea and Channel area. *Marine Pollution Bulletin*, *98*, 179–187. http://dx.doi.org/10.1016/j.marpolbul.2015.06.051.

Ding, J., Li, J., Sun, C., Jiang, F., He, C., Zhang, M., Ju, P., & Ding, N. X. (2020). An examination of the occurrence and potential risks of microplastics across various shellfish. *Science of the Total Environment*, *739*, http://dx.doi.org/10.1016/j.scitotenv.2020.139887.

Donaghay, P. L. & Small, L. F. (1979). Food selection capabilities of the estuarine copepod *Acaritia clausi*. *Marine Biology*, *52*, 137–146. http://dx.doi.org/10.1007/bf00390421.

Frias, J. P. G. L., Otero, V., & Sobral, P. (2014). Evidence of microplastics in samples of zooplankton from Portuguese coastal waters. *Marine Environmental Research*, *95*, 89–95. http://dx.doi.org/10.1016/j.marenvres.2014.01.001.

Gago, J., Carretero, O., Filgueiras, A. V., & Vinas, L. (2018). Synthetic microfibers in the marine environment: A review on their occurrence in seawater and sediments. *Marine Pollution Bulletin*, *127*, 365–376. http://dx.doi.org/10.1016/j.marpolbul.2017.11.070.

Gardon, T., Morvan, L., Huvet, A., Quillien, V., Soyez, C., Le Moullac, G., & Le Luyer, J. (2020). Microplastics induce dose-specific transcriptomic disruptions in energy metabolism and immunity of the pearl oyster *Pinctada margaritifera*. *Environmental Pollution*, *266*, http://dx.doi.org/10.1016/j.envpol.2020.115180.

Gebhardt, C. & Forster, S. (2018). Size-selective feeding of *Arenicola marina* promotes long-term burial of microplastic particles in marine sediments. *Environmental Pollution*, *242*, 1777–1786. http://dx.doi.org/10.1016/j.envpol.2018.07.090.

Gedik, K. & Eryaşar, A. R. (2020). Microplastic pollution profile of Mediterranean mussels (*Mytilus galloprovincialis*) collected along the Turkish coasts. *Chemosphere*, *260*, 127570. http://dx.doi.org/10.1016/j.chemosphere.2020.127570.

Goldstein, M. C. & Goodwin, D. S. (2013). Gooseneck barnacles (*Lepas* spp.) ingest microplastic debris in the North Pacific Subtropical Gyre. *PeerJ*, *1*, e184. http://dx.doi.org/10.7717/peerj.184.

Goldstein, M. C., Rosenberg, M., & Cheng, L. (2012). Increased oceanic microplastic debris enhances oviposition in an endemic pelagic insect. *Biology Letters*, *8*, 817–820. http://dx.doi.org/10.1098/rsbl.2012.0298.

Goncalves, C., Martins, M., Sobral, P., Costa, P. M., & Costa, M. H. (2019). An assessment of the ability to ingest and excrete microplastics by filter-feeders:

A case study with the Mediterranean mussel. *Environmental Pollution, 245*, 600–606. http://dx.doi.org/10.1016/j.envpol.2018.11.038.

Gove, J. M., Whitney, J. L., McManus, M. A., Lecky, J., Carvalho, F. C., Lynch, J. M., Li, J., Neubauer, P., Smith, K. A., Phipps, J. E., Kobayashi, D. R., Balagso, K. B., Contreras, E. A., Manuel, M. E., Merrifield, M. A., Polovina, J. J., Asner, G. P., Maynard, J. A., & Williams, G. J. (2019). Prey-size plastics are invading larval fish nurseries. *Proceedings of the National Academy of Sciences of the United States of America, 116*, 24143–24149. http://dx.doi.org/10.1073/pnas.1907496116.

Graham, E. R. & Thompson, J. T. (2009). Deposit- and suspension-feeding sea cucumbers (*Echinodermata*) ingest plastic fragments. *Journal of Experimental Marine Biology and Ecology, 368*, 22–29. http://dx.doi.org/10.1016/j.jembe.2008.09.007.

Gundogdu, S., Cevik, C., & Atas, N. T. (2020). Stuffed with microplastics: Microplastic occurrence in traditional stuffed mussels sold in the Turkish market. *Food Bioscience, 37*, 100715. http://dx.doi.org/10.1016/j.fbio.2020.100715.

Harris, L. S. T. & Carrington, E. (2020). Impacts of microplastic vs. natural abiotic particles on the clearance rate of a marine mussel. *Limnology and Oceanography Letters, 5*, 66–73. http://dx.doi.org/10.1002/lol2.10120.

Hope, J. A., Hewitt, J., Pilditch, C. A., Savage, C., & Thrush, S. F. (2020). Effect of nutrient enrichment and turbidity on interactions between microphytobenthos and a key bivalve: Implications for higher trophic levels. *Frontiers in Marine Science, 7*, 1–17. http://dx.doi.org/10.3389/fmars.2020.00695.

Horton, T., Kroh, A., Ahyong, S., Bailly, N., Boury-Esnault, N., Brandão, S. N., *et al.* (2018). World register of marine species. Retrieved from https://www.marinespecies.org

Kaposi, K. L., Mos, B., Kelaher, B. P., & Dworjanyn, S. A. (2014). Ingestion of microplastic has limited impact on a marine larva. *Environmental Science & Technology, 48*, 1638–1645. http://dx.doi.org/10.1021/es404295e.

Karlson, K., Bonsdorff, E., & Rosenberg, R. (2007). The impact of benthic macrofauna for nutrient fluxes from Baltic Sea sediments. *Ambio, 36*, 161–167. http://dx.doi.org/10.1579/0044-7447(2007)36[161:Tiobmf]2.0.Co;2.

Kolandhasamy, P., Su, L., Li, J., Qu, X., Jabeen, K., & Shi, H. (2018). Adherence of microplastics to soft tissue of mussels: A novel way to uptake microplastics beyond ingestion. *Science of the Total Environment, 610*, 635–640. http://dx.doi.org/10.1016/j.scitotenv.2017.08.053.

Koski, M., Sondergaard, J., Christensen, A. M., & Nielsen, T. G. (2021). Effect of environmentally relevant concentrations of potentially toxic microplastic on coastal copepods. *Aquatic Toxicology, 230*, 105713. http://dx.doi.org/10.1016/j.aquatox.2020.105713.

Kristensen, E., Penha-Lopes, G., Delefosse, M., Valdemarsen, T., Quintana, C. O., & Banta, G. T. (2012). What is bioturbation? The need for a precise definition for fauna in aquatic sciences. *Marine Ecology Progress Series*, *446*, 285–302. http://dx.doi.org/10.3354/meps09506.

Lee, K.-W., Shim, W. J., Kwon, O. Y., & Kang, J.-H. (2013). Size-dependent effects of micro polystyrene particles in the marine copepod *Tigriopus japonicus*. *Environmental Science & Technology*, *47*, 11278–11283. http://dx.doi.org/10.1021/es401932b.

Li, H., Ma, L., Lin, L., Ni, Z, Xu, X., Shi, H., Yan, Y., Zheng, G., & Rittschof, D. (2018). Microplastics in oysters *Saccostrea cucullata* along the Pearl River Estuary, China. *Environmental Pollution*, *236*, 619–625. http://dx.doi.org/10.1016/j.envpol.2015.09.018.

Li, J., Yang, D., Li, L., Jabeen, K., & Shi, H. (2015). Microplastics in commercial bivalves from China. *Environmental Pollution*, *207*, 190–195. http://dx.doi.org/10.1016/j.envpol.2018.01.083.

Lo, H. K. A. & Chan, K. Y. K. (2018). Negative effects of microplastic exposure on growth and development of *Crepidula onyx*. *Environmental Pollution*, *233*, 588–595. http://dx.doi.org/10.1016/j.envpol.2017.10.095.

Luan, L., Wang, X., Zheng, H., Liu, L., Luo, X., & Li, F. (2019). Differential toxicity of functionalized polystyrene microplastics to clams (*Meretrix meretrix*) at three key development stages of life history. *Marine Pollution Bulletin*, *139*, 346–354. http://dx.doi.org/10.1016/j.marpolbul.2019.01.003.

Lusher, A. L., McHugh, M., & Thompson, R. C. (2013). Occurrence of microplastics in the gastrointestinal tract of pelagic and demersal fish from the English Channel. *Marine Pollution Bulletin*, *67*, 94–99. http://dx.doi.org/10.1016/j.marpolbul.2012.11.028.

Moore, C. J., Moore, S. L., Leecaster, M. K., & Weisberg, S. B. (2001). A comparison of plastic and plankton in the North Pacific central gyre. *Marine Pollution Bulletin*, *42*, 1297–1300. http://dx.doi.org/10.1016/s0025-326x(01)00114-x.

Murray, F. & Cowie, P. R. (2011). Plastic contamination in the decapod crustacean *Nephrops norvegicus* (Linnaeus, 1758). *Marine Pollution Bulletin*, *62*, 1207–1217. http://dx.doi.org/10.1016/j.marpolbul.2011.03.032.

Naji, A., Nuri, M., & Vethaak, A. D. (2018). Microplastics contamination in molluscs from the northern part of the Persian Gulf. *Environmental Pollution*, *235*, 113–120. http://dx.doi.org/10.1016/j.envpol.2017.12.046.

Nakki, P., Setala, O., & Lehtiniemi, M. (2017). Bioturbation transports secondary microplastics to deeper layers in soft marine sediments of the northern Baltic Sea. *Marine Pollution Bulletin*, *119*, 255–261. http://dx.doi.org/10.1016/j.marpolbul.2017.03.065.

Nam Ngoc, P., Poirier, L., Quoc Tuan, P., Lagarde, F., & Zalouk-Vergnoux, A. (2018). Factors influencing the microplastic contamination of bivalves from the French Atlantic coast: Location, season and/or mode of life? *Marine Pollution Bulletin*, *129*, 664–674. http://dx.doi.org/10.1016/j.marpolbul.2017.10.054.

Nelms, S. E., Galloway, T. S., Godley, B. J., Jarvis, D. S., & Lindeque, P. K. (2018). Investigating microplastic trophic transfer in marine top predators. *Environmental Pollution, 238*, 999–1007. http://dx.doi.org/10.1016/j.envpol.2018.02.016.

Outram, L., Hurley, J., Rott, A., Crooks, N., & Pernetta, A. P. (2020). No evidence of microplastic consumption by the copepod, *Temora longicornis* (Muller, 1785) in Chichester Harbour, United Kingdom. *Nauplius, 28*, http://dx.doi.org/10.1590/2358-2936e2020033.

Park, K., Park, J.-W., & Choi, J. S. (2017). Histological observation of micro/nano-plastic spheres in the mussel *Mytilus galloprovincialis*. *The Korean Journal of Malacology, 33*, 253–258. https://doi.org/10.9710/kjm.2017.33.4.253.

Parolini, M., De Felice, B., Gazzotti, S., Annunziata, L., Sugni, M., Bacchetta, R., & Ortenzi, M. A. (2020). Oxidative stress-related effects induced by micronized polyethylene terephthalate microparticles in the Manila clam. *Journal of Toxicology and Environmental Health-Part a-Current Issues, 83*, 168–179. http://dx.doi.org/10.1080/15287394.2020.1737852.

Payton, T. G., Beckingham, B. A., & Dustan, P. (2020). Microplastic exposure to zooplankton at tidal fronts in Charleston Harbor, SC USA. *Estuarine, Coastal and Shelf Science, 232*, http://dx.doi.org/10.1016/j.ecss.2019.106510.

Porter, A., Lyons, B. P., Galloway, T. S., & Lewis, C. (2018). Role of Marine Snows in Microplastic Fate and Bioavailability. *Environmental Science & Technology, 52*, 7111–7119. http://dx.doi.org/10.1021/acs.est.8b01000.

Qi, H., Fu, D., Wang, Z., Gao, M., & Peng, L. (2020). Microplastics occurrence and spatial distribution in seawater and sediment of Haikou Bay in the northern South China Sea. *Estuarine, Coastal and Shelf Science, 239*, http://dx.doi.org/10.1016/j.ecss.2020.106757.

Revel, M., Lagarde, F., Perrein-Ettajani, H., Bruneau, M., Akcha, F., Sussarellu, R., Rouxel, J., Costil, K., Decottignies, P., Cognie, B., & Châtel, A. (2019). Tissue-specific biomarker responses in the blue mussel Mytilus spp. exposed to a mixture of microplastics at environmentally relevant concentrations. *Frontiers in Environmental Science, 7*, 33. http://dx.doi.org/10.3389/fenvs.2019.00033.

Ribeiro, F., Garcia, A. R., Pereira, B. P., Fonseca, M., Mestre, N. C., Fonseca, T. G., Ilharco, L. M., & Bebianno, M. J. (2017). Microplastics effects in *Scrobicularia plana*. *Marine Pollution Bulletin, 122*, 379–391. http://dx.doi.org/10.1016/j.marpolbul.2017.06.078.

Rist, S. E., Assidqi, K., Zamani, N. P., Appel, D., Perschke, M., Huhn, M., & Lenz, M. (2016). Suspended micro-sized PVC particles impair the performance and decrease survival in the Asian green mussel *Perna viridis*. *Marine Pollution Bulletin, 111*, 213–220. http://dx.doi.org/10.1016/j.marpolbul.2016.07.006.

Rochman, C. M. (2015). The complex mixture, fate and toxicity of chemicals associated with plastic debris in the marine environment. *Marine Anthropogenic Litter*, 117–140. http://dx.doi.org/10.1007/978-3-319-16510-3.

Rodriguez-Torres, R., Almeda, R., Kristiansen, M., Rist, S., Winding, M. S., & Nielsen, T. G. (2020). Ingestion and impact of microplastics on arctic *Calanus* copepods. *Aquatic Toxicology*, *228*, http://dx.doi.org/10.1016/j.aquatox.2020.105631.

Santana, M. F. M., Ascer, L. G., Custodio, M. R., Moreira, F. T., & Turra, A. (2016). Microplastic contamination in natural mussel beds from a Brazilian urbanized coastal region: Rapid evaluation through bioassessment. *Marine Pollution Bulletin*, *106*, 183–189. http://dx.doi.org/10.1016/j.marpolbul.2016.02.074.

Sendra, M., Isabel Carrasco-Braganza, M., Maria Yeste, P., Vila, M., & Blasco, J. (2020). Immunotoxicity of polystyrene nanoplastics in different hemocyte subpopulations of *Mytilus galloprovincialis*. *Scientific Reports*, *10*. http://dx.doi.org/10.1038/s41598-020-65596-8.

Setala, O., Norkko, J., & Lehtiniemi, M. (2016). Feeding type affects microplastic ingestion in a coastal invertebrate community. *Marine Pollution Bulletin*, *102*, 95–101. http://dx.doi.org/10.1016/j.marpolbul.2015.11.053.

Steer, M., Cole, M., Thompson, R. C., & Lindeque, P. K. (2017). Microplastic ingestion in fish larvae in the western English Channel. *Environmental Pollution*, *226*, 250–259. http://dx.doi.org/10.1016/j.envpol.2017.03.062.

Sucharitakul, P., Pitt, K. A., & Welsh, D. T. (2020). Limited ingestion, rapid egestion and no detectable impacts of microbeads on the moon jellyfish, *Aurelia aurita*. *Marine Pollution Bulletin*, *156*. http://dx.doi.org/10.1016/j.marpolbul.2020.111208.

Sun, X., Li, Q., Zhu, M., Liang, J., Zheng, S., & Zhao, Y. (2017). Ingestion of microplastics by natural zooplankton groups in the northern South China Sea. *Marine Pollution Bulletin*, *115*, 217–224. http://dx.doi.org/10.1016/j.marpolbul.2016.12.004

Sun, X., Liang, J., Zhu, M., Zhao, Y., & Zhang, B. (2018a). Microplastics in seawater and zooplankton from the Yellow Sea. *Environmental Pollution*, *242*, 585–595. http://dx.doi.org/10.1016/j.envpol.2018.07.014.

Sun, X., Liu, T., Zhu, M., Liang, J., Zhao, Y., & Zhang, B. (2018b). Retention and characteristics of microplastics in natural zooplankton taxa from the East China Sea. *Science of the Total Environment*, *640*, 232–242. http://dx.doi.org/10.1016/j.scitotenv.2018.05.308.

Sun, Y., Xu, W., Gu, Q., Chen, Y., Zhou, Q., Zhang, L., Gu, L., Huang, Y., Lyu, K., & Yang, Z. (2019). Small-sized microplastics negatively affect rotifers: Changes in the key life-history traits and rotifer-phaeocystis population dynamics. *Environmental Science & Technology*, *53*, 9241–9251. http://dx.doi.org/10.1021/acs.est.9b02893.

Sussarellu, R., Suquet, M., Thomas, Y., Lambert, C., Fabioux, C., Pernet, M. E. J., Le Goic, N., Quillien, V., Mingant, C., Epelboin, Y., Corporeau, C., Guyomarch, J., Robbens, J., Paul-Pont, I., Soudant, P., & Huvet, A. (2016). Oyster reproduction is affected by exposure to polystyrene microplastics. *Proceedings of the National Academy of Sciences of the United States of America*, *113*, 2430–2435. http://dx.doi.org/10.1073/pnas.1519019113.

Tallec, K., Huvet, A., Di Poi, C., Gonzalez-Fernandez, C., Lambert, C., Petton, B., Le Goic, N., Berchel, M., Soudant, P., & Paul-Pont, I. (2018). Nanoplastics impaired oyster free living stages, gametes and embryos. *Environmental Pollution*, *242*, 1226–1235. http://dx.doi.org/10.1016/j.envpol.2018.08.020.

Teixeira, P. F., Kaminski, S. M., Avila, T. R., Cardozo, A. P., Bersano, J. G. F., & Bianchini, A. (2010). Diet influence on egg production of the copepod *Acartia tonsa* (Dana, 1896). *Anais da Academia Brasileira de Ciencias*, *82*, 333–339. http://dx.doi.org/10.1590/s0001-37652010000200009.

Teng, J., Wang, Q., Ran, W., Wu, D., Liu, Y., Sun, S., Liu, H., Cao, R., & Zhao, J. (2019). Microplastic in cultured oysters from different coastal areas of China. *Science of the Total Environment*, *653*, 1282–1292. http://dx.doi.org/10.1016/j.scitotenv.2018.11.057.

Teng, J., Zhao, J., Zhu, X., Shan, E., Zhang, C., Zhang, W., & Wang, Q. (2021). Toxic effects of exposure to microplastics with environmentally relevant shapes and concentrations: Accumulation, energy metabolism and tissue damage in oyster *Crassostrea gigas*. *Environmental Pollution (Barking, Essex: 1987)*, *269*, 116169. http://dx.doi.org/10.1016/j.envpol.2020.116169.

Thompson, R. C., Moore, C. J., vom Saal, F. S., & Swan, S. H. (2009). Plastics, the environment and human health: Current consensus and future trends. *Philosophical Transactions of the Royal Society B-Biological Sciences*, *364*, 2153–2166. http://dx.doi.org/10.1098/rstb.2009.0053.

Thompson, R. C., Olsen, Y., Mitchell, R. P., Davis, A., Rowland, S. J., John, A. W. G., McGonigle, D., & Russell, A. E. (2004). Lost at sea: Where is all the plastic? *Science*, *304*, 838–838. http://dx.doi.org/10.1126/science.1094559.

Thushari, G. G. N., Senevirathna, J. D. M., Yakupitiyage, A., & Chavanich, S. (2017). Effects of microplastics on sessile invertebrates in the eastern coast of Thailand: An approach to coastal zone conservation. *Marine Pollution Bulletin*, *124*, 349–355. http://dx.doi.org/10.1016/j.marpolbul.2017.06.010.

Turner, J. T. (2015). Zooplankton fecal pellets, marine snow, phytodetritus and the ocean's biological pump. *Progress in Oceanography*, *130*, 205–248. http://dx.doi.org/10.1016/j.pocean.2014.08.005.

Turra, A., Manzano, A. B., Dias, R. J. S., Mahiques, M. M., Barbosa, L., Balthazar-Silva, D., & Moreira, F. T. (2014). Three-dimensional distribution of plastic pellets in sandy beaches: Shifting paradigms. *Scientific Reports*, *4*(1), 1–7. http://dx.doi.org/10.1038/srep04435.

Uddin, S., Fowler, S. W., Uddin, M. F., Behbehani, M., & Naji, A. (2021). A review of microplastic distribution in sediment profiles. *Marine Pollution Bulletin, 163*, http://dx.doi.org/10.1016/j.marpolbul.2021.111973.

Ugolini, A., Ungherese, G., Ciofini, M., Lapucci, A., & Camaiti, M. (2013). Microplastic debris in sandhoppers. *Estuarine Coastal and Shelf Science, 129*, 19–22. http://dx.doi.org/10.1016/j.ecss.2013.05.026.

Van Cauwenberghe, L. & Janssen, C. R. (2014). Microplastics in bivalves cultured for human consumption. *Environmental Pollution, 193*, 65–70. http://dx.doi.org/10.1016/j.envpol.2014.06.010.

Van Colen, C., Moereels, L., Vanhove, B., Vrielinck, H., & Moens, T. (2021). The biological plastic pump: Evidence from a local case study using blue mussel and infaunal benthic communities. *Environmental Pollution, 274*, http://dx.doi.org/10.1016/j.envpol.2020.115825.

Van Colen, C., Vanhove, B., Diem, A., & Moens, T. (2020). Does microplastic ingestion by zooplankton affect predator-prey interactions? An experimental study on larviphagy. *Environmental Pollution, 256*, http://dx.doi.org/10.1016/j.envpol.2019.113479.

Villagran, D. M., Truchet, D. M., Buzzi, N. S., Forero Lopez, A. D., & Fernandez Severini, M. D. (2020). A baseline study of microplastics in the burrowing crab (*Neohelice granulata*) from a temperate southwestern Atlantic estuary. *Marine Pollution Bulletin, 150*, http://dx.doi.org/10.1016/j.marpolbul.2019.110686.

Waite, H. R., Donnelly, M. J., & Walters, L. J. (2018). Quantity and types of microplastics in the organic tissues of the eastern oyster *Crassostrea virginica* and Atlantic mud crab *Panopeus herbstii* from a Florida estuary. *Marine Pollution Bulletin, 129*, 179–185. http://dx.doi.org/10.1016/j.marpolbul.2018.02.026.

Wakkaf, T., El Zrelli, R., Kedzierski, M., Balti, R., Shaiek, M., Mansour, L., Tlig-Zouari, S., Bruzaud, S., & Rabaoui, L. (2020). Microplastics in edible mussels from a southern Mediterranean lagoon: Preliminary results on seawater-mussel transfer and implications for environmental protection and seafood safety. *Marine Pollution Bulletin, 158*, 111355. http://dx.doi.org/10.1016/j.marpolbul.2020.111355.

Walkinshaw, C., Lindeque, P. K., Thompson, R., Tolhurst, T., & Cole, M. (2020). Microplastics and seafood: lower trophic organisms at highest risk of contamination. *Ecotoxicology and Environmental Safety, 190*, 110066. http://dx.doi.org/10.1016/j.ecoenv.2019.110066.

Wang, J., Lu, L., Wang, M., Jiang, T., Liu, X., & Ru, S. (2019). Typhoons increase the abundance of microplastics in the marine environment and cultured organisms: A case study in Sanggou Bay, China. *Science of the Total Environment, 667*, 1–8. http://dx.doi.org/10.1016/j.scitotenv.2019.02.367.

Wang, S., Hu, M., Zheng, J., Huang, W., Shang, Y., Fang, J. K.-H., Shi, H., & Wang, Y. (2021). Ingestion of nano/micro plastic particles by the mussel *Mytilus coruscus* is size dependent. *Chemosphere*, *263*. http://dx.doi.org/10.1016/j.chemosphere.2020.127957.

Ward, J. E. & Shumway, S. E. (2004). Separating the grain from the chaff: Particle selection in suspension- and deposit-feeding bivalves. *Journal of Experimental Marine Biology and Ecology*, *300*, 83–130. http://dx.doi.org/10.1016/j.jembe.2004.03.002.

Ward, J. E., Zhao, S., Holohan, B. A., Mladinich, K. M., Griffin, T. W., Wozniak, J., & Shumway, S. E. (2019). Selective ingestion and egestion of plastic particles by the Blue Mussel (*Mytilus edulis*) and Eastern Oyster (*Crassostrea virginica*): Implications for using bivalves as bioindicators of microplastic pollution. *Environmental Science & Technology*, *53*, 8776–8784. http://dx.doi.org/10.1021/acs.est.9b02073.

Webb, S., Gaw, S., Marsden, I. D., & McRae, N. K. (2020). Biomarker responses in New Zealand green-lipped mussels *Perna canaliculus* exposed to microplastics and triclosan. *Ecotoxicology and Environmental Safety*, *201*, http://dx.doi.org/10.1016/j.ecoenv.2020.110871.

Wegner, A., Besseling, E., Foekema, E. M., Kamermans, P., & Koelmans, A. A. (2012). Effects of nanopolystyrene on the feeding behavior of the blue mussel (*Mytilus edulis* L.). *Environmental Toxicology and Chemistry*, *31*, 2490–2497. http://dx.doi.org/10.1002/etc.1984.

Widdows, J. & Johnson, D. (1988). Physiological energetics of *Mytilus edulis*: Scope for Growth. *Marine Ecology Progress Series*, *46*, 113–121. http://dx.doi.org/10.3354/meps046113.

Wright, S. L., Thompson, R. C., & Galloway, T. S. (2013). The physical impacts of microplastics on marine organisms: A review. *Environmental Pollution*, *178*, 483–492. http://dx.doi.org/10.1016/j.envpol.2013.02.031.

Xu, X. Y., Lee, W. T., Chan, A. K. Y., Lo, H. S., Shin, P. K. S., & Cheung, S. G. (2017). Microplastic ingestion reduces energy intake in the clam *Atactodea striata*. *Marine Pollution Bulletin*, *124*, 798–802. http://dx.doi.org/10.1016/j.marpolbul.2016.12.027.

Xu, X., Wong, C. Y., Tam, N. F. Y., Lo, H.-S., & Cheung, S.-G. (2020). Microplastics in invertebrates on soft shores in Hong Kong: Influence of habitat, taxa and feeding mode. *Science of the Total Environment*, *715*, http://dx.doi.org/10.1016/j.scitotenv.2020.136999.

Yu, J., Xu, R., Wei, Y., Tian, J., Wang, X., Lai, J., Chen, R., Yang, G., Zhang, Z., & Shao, Y. (2020). Effect of microplastics on feeding, excretion and reproduction of marine copepod. *Periodical of Ocean University of China*, *50*, 73–80. http://dx.doi.org/10.16441/j.cnki.hdxb.20190379.

Yu, S.-P. & Chan, B. K. K. (2020). Effects of polystyrene microplastics on larval development, settlement, and metamorphosis of the intertidal barnacle

Amphibalanus amphitrite. *Ecotoxicology and Environmental Safety*, *194*, http://dx.doi.org/10.1016/j.ecoenv.2020.110362.

Yu, S.-P., Nakaoka, M., & Chan, B. K. K. (2021). The gut retention time of microplastics in barnacle naupliar larvae from different climatic zones and marine habitats. *Environmental Pollution*, *268*, http://dx.doi.org/10.1016/j.envpol.2020.115865.

Zhang, K., Liang, J., Liu, T., Li, Q., Zhu, M., Zheng, S., & Sun, X. (2021). Abundance and characteristics of microplastics in shellfish from Jiaozhou Bay, China. *Journal of Oceanology and Limnology*, *40*(1), 163–172. http://dx.doi.org/10.1007/s00343-021-0465-7.

Zhang, W., Zhang, S., Wang, J., Wang, Y., Mu, J., Wang, P., Lin, X., & Ma, D. (2017). Microplastic pollution in the surface waters of the Bohai Sea, China. *Environmental Pollution*, *231*, 541–548. http://dx.doi.org/10.1016/j.envpol.2017.08.058.

Zhang, X., Yan, B., & Wang, X. (2020). Selection and optimization of a protocol for extraction of microplastics from *Mactra veneriformis*. *Science of the Total Environment*, *746*, http://dx.doi.org/10.1016/j.scitotenv.2020.141250.

Zhao, S., Ward, J. E., Danley, M., & Mincer, T. J. (2018). Field-based evidence for microplastic in marine aggregates and mussels: Implications for trophic transfer. *Environmental Science & Technology*, *52*, 11038–11048. http://dx.doi.org/10.1021/acs.est.8b03467.

Zheng, S., Zhao, Y., Liangwei, W., Liang, J., Liu, T., Zhu, M., Li, Q., & Sun, X. (2020). Characteristics of microplastics ingested by zooplankton from the Bohai Sea, China. *Science of the Total Environment*, *713*, http://dx.doi.org/10.1016/j.scitotenv.2019.136357.

https://doi.org/10.1142/9789811259111_0008

Chapter 7

Ecotoxicity of Micro- and Nano-Sized Plastics

Shima Ziajahromi* **and Frederic D. L. Leusch**

Australian Rivers Institute, School of Environment and Science, Griffith University, Gold Coast QLD 4222, Australia

**s.ziajahromi@griffith.edu.au*

1. Introduction

In the light of their versatility, durability, flexibility, lightweight and low costs, plastic materials have been widely used in industry and all areas of our day-to-day lives (Wagner *et al.*, 2014). However, the gloom of this gleam is persistence and accumulation in the environment. Microplastics (MPs), defined as plastic particles between 1 and 5,000 μm (Frias & Nash, 2019), are contaminants of emerging concern that have been found in a variety of shapes, polymer types, sizes, colours and concentrations throughout aquatic environments worldwide, from tropical areas to the poles, including both freshwater (Li *et al.*, 2020a) and marine ecosystems (Ivar do Sul & Costa, 2014).

To date, the majority of studies on MPs have focused on the marine environment (Akdogan & Guven, 2019). Freshwater systems are considered a source and critical pathway for pollutants to enter the marine environment (Lu *et al.*, 2021). It is estimated that 1.15–2.41 Mt of plastic debris is transported into the ocean via rivers annually (Lebreton *et al.*, 2017), and the concentration of MPs in the oceans is predicted to further increase in the future. It is projected that by 2060, the concentrations of pelagic MPs will increase by approximately fourfold from concentrations in 2016 (Isobe *et al.*, 2019). While the current overall estimate of risk of MPs to aquatic organisms is low, this increasing trend has raised concern about their potential future impacts on aquatic organisms (Adam *et al.*, 2019, Besseling *et al.*, 2018, Everaert *et al.*, 2020).

In aquatic environments, MPs undergo a process of fragmentation and degradation. Fragmentation leads to the breakdown of larger polymer chains to smaller polymeric fragments, while degradation breaks the bonds holding the polymers together leading to changes in the properties of those polymers (Mattsson *et al.*, 2018). MPs of various morphology, polymer types and sizes are present in aquatic environments, and these differences alter their degradation rate and speed. Physical, chemical and biological degradation processes can further break down MPs and produce nanoplastics, defined as particles between 1 and 1,000 nm (Mattsson *et al.*, 2018).

Compared to MPs, nanoplastics have greater surface area to volume ratios, which increases their ability to sorb toxic chemicals from the environment (Gangadoo *et al.*, 2020). They also exhibit different physical properties, and their smaller size enhances their ability to pass through biological barriers and accumulate in organs and tissues (Gangadoo *et al.*, 2020; Hu & Palic, 2020). Nanoplastics are reported to have weaker aggregation ability and higher stability in aqueous environment, which could increase their persistence and abundance, and consequently increase their toxicity (Mao *et al.*, 2020). Compared to larger MPs (1–5 mm), smaller microplastics and nanoplastics may induce more pronounced impacts on aquatic ecosystems due to their ability to disperse more easily throughout the water (Gao *et al.*, 2021; Isobe *et al.*, 2019).

Microplastics and nanoplastics can be ingested by a variety of aquatic biota (e.g., pelagic and benthic organisms and marine mammals) and thus enter the trophic food web. The amount and likelihood of ingestion are affected by the size of an animal's mouthparts, its feeding mode and food selectivity. Filter feeders, for example, are particularly susceptible as they can filter a large amount of water and thus concentrate MPs from the water. Ingestion of microplastics and nanoplastics can pose significant threat to individual organisms (including physical damage such as clogging, perforated gut, internal abrasion and gastric rupture), populations and wider ecosystems (Botterell *et al.*, 2019; Pinheiro *et al.*, 2020). It has been reported that over 250 taxa of marine fauna, including 32 marine invertebrate species, are impacted by MPs present in water (e.g., entanglement and ingestion) (Wojcik-Fudalewska *et al.*, 2016).

Microplastics and nanoplastics can also be vectors of adsorbed (i.e., attached to the surface) or absorbed/incorporated (i.e., contained within) toxic chemicals. These can be chemicals used as additives during the plastic production process (e.g., plasticisers and UV filters) or environmental contaminants such as persistent organic pollutants (POPs), trace metals, pharmaceutical and other micropollutants. These toxic chemicals can enter organisms exposed to contaminated MPs and cause lethal and sub-lethal adverse effects on survival, growth, respiration, breeding, development and metabolism, especially in the early life stages (Gandara *et al.*, 2016; Pinheiro *et al.*, 2020). The presence of MPs in the aquatic environment represents not only an ecological risk but also a potential hazard for food safety and, consequently, for human health through consumption of MP-contaminated seafoods (Corami *et al.*, 2020).

Since there is no consistent requirement for reporting MPs in aquatic organisms (Lu *et al.*, 2021), monitoring studies on ingestion of MPs by aquatic organisms to date often lack information about MP characteristics such as polymer type, size range (lower and upper size limits), colours, morphotypes and concentration reporting unit (e.g., number of particles *vs.* mass per litre). This can limit the comparability and availability of baseline data to assess the actual impacts and risks of MPs using environmentally realistic scenarios. For example, while 1 μm is an agreed-upon lower size limit for MPs, studies often only report MPs >100 micron in aquatic species, with no information on the smaller MPs. Likewise, while the number of *in vivo* and *in vitro* laboratory studies on the effects of nanoplastics on aquatic species is growing (Gangadoo *et al.*, 2020), there is lack of understanding of the actual concentrations and properties of nanoplastics in the aquatic organisms, thus making it difficult to put these toxicity studies in an environmental context.

This chapter will provide an overview of the toxicity of microplastics and nanoplastics to aquatic ecosystems and will discuss (i) effects associated with size, type, morphology and concentration, (ii) combined effects of MPs and associated chemicals, (iii) bioaccumulation and biomagnification through aquatic food webs, and (iv) potential risks to aquatic organisms under environmentally realistic conditions.

2. Toxic Effects of MPs to Aquatic Organisms

Currently, the uptake of MPs by a variety of aquatic organisms (both invertebrates and vertebrates) has been thoroughly documented, with MPs reported in the digestive tract of over 690 wild marine species

(Carbery *et al.*, 2018) where they may be translocated to organs and tissues. MPs can cause various adverse effects on reproduction, growth and energy budget, as well as inflammatory responses, histological changes, cytotoxicity, DNA damage, physical damage and mortality (Bajt, 2021). However, the mechanisms of toxicity of microplastics and nanoplastics to organisms are less understood. In addition, it is important to remember that MPs are not a homogeneous contaminant but instead come in a variety of sizes, shapes and chemical compositions that can affect their ecotoxicity to aquatic organisms.

2.1. *Dose-related effects*

Results of ecotoxicology studies on aquatic organisms are reported as the concentration of MPs (either by the mass or number of particles, and ideally both) causing adverse effects on exposed organism individuals/populations, usually either the lethal concentration (e.g., LC_{50} = Lethal concentration to 50% of the test population) or as an effective concentration in sub-lethal tests (e.g., EC_x = Effective concentration to induce an adverse effect in x% the test population) (Miloloža *et al.*, 2020). Evidence shows that toxic effects of microplastics and nanoplastics to exposed organisms are strongly dependent on the exposure concentrations, with a dose-response relationship usually reported for various adverse effects including mortality, growth, development, reproduction, cellular response (genotoxicity and immunotoxicity) and behavioural alteration.

Acute toxicity usually occurs at the low mg/L range, although of course varies depending on the type of organisms, size, shape type and concentration of MPs. For example, LC_{50} values of 2.2 mg/L (7.4×10^4 particles/L) and approximately 90 mg/L (4.6×10^7 particles/L) have been reported for zooplankton (*Ceriodaphnia dubia*) and crustacean (*Hyalella azteca*) exposed to 10–27 μm polyethylene (PE) beads (Au *et al.*, 2015; Ziajahromi *et al.*, 2017), while an LC_{50} of approximately 2 mg/L (7.1×10^4 particles/L) has been reported for *H. azteca* exposed to 20–75 μm polypropylene (PP) fibres (Au *et al.*, 2015).

Furthermore, exposure to smaller MPs produced an LC_{50} value of 16.9 mg/L for brine shrimps exposed to 5 μm polystyrene (PS) beads (Suman *et al.*, 2020), and 13.3 and 83 mg/L for marine rotifer and marine algae, respectively, exposed to 40 nm polymethylmethacrylate (PMMA) nanoplastics (Venancio *et al.*, 2019).

Reported effective concentrations (EC_{50}) for sub-lethal effects are also usually in the low mg/L to high μg/L range. For example, the EC_{50} for immobilisation in *D. magna* exposed to 1 μm PE MPs was calculated to be between 52 and 57.5 mg/L (Eltemsah & Bohn 2019; Rehse *et al.*, 2016). Ziajahromi *et al.* (2017) reported greater toxicity (lower EC_{50} values) for reproduction in *C. dubia* exposed to polyester fibres (EC_{50} = 429 μg/L) compared to 1–4 μm PE beads (EC_{50} = 958 μg/L).

In toxicology, the main dose-response models include the linear no-threshold (LNT) model, the threshold model and the hormetic model (Sun *et al.*, 2021) (Figure 1).

The threshold model is most commonly used to illustrate the adverse effects of microplastics and nanoplastics to aquatic organisms above a threshold dose ("point of departure") (Agathokleous *et al.*, 2021). Recently, Sun *et al.* (2021) evaluated 270 concentration-related effects of MPs to 3765 individual freshwater and marine organisms. The authors reported that while high doses almost always produced adverse effects, lower doses (<1 mg/L, usually more environmentally realistic doses) could sometimes produce hormetic responses, with positive responses for some endpoints such as oxidative stress, genotoxicity, immunotoxicity, neurotoxicity and behavioural alteration. This was observed for various MP types, shapes and sizes. However, it is hard to conclude about the

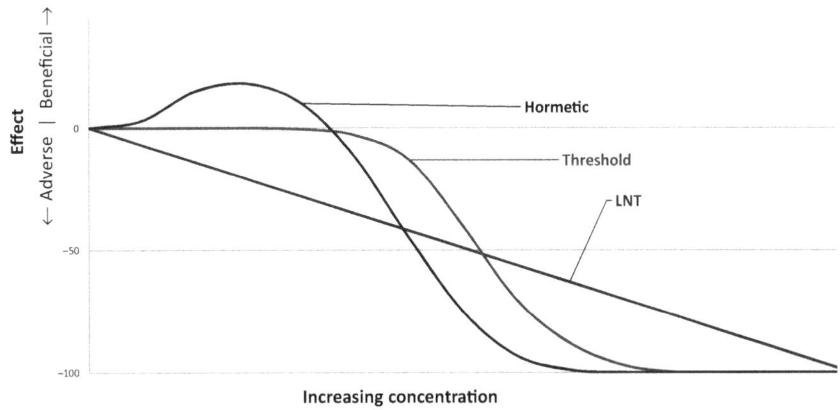

Figure 1. Main dose-response models in toxicology: Idealised linear no-threshold (LNT), threshold and hormetic dose-response curves.

ultimate benefits of the low-dose positive stimulation effects, as it appears that they can ultimately lead to negative outcomes due to compensatory mechanisms (Agathokleous *et al.*, 2021). For example, the swimming speed of pacific oysters increased after exposure to a low concentration of MPs (0.1 mg/L), but body length was decreased at the same low concentration, compared to the negative control (Bringer *et al.*, 2020). However, the potential compensatory mechanisms remain poorly understood. It has been suggested that the linear no-threshold dose-response model is probably meaningless for exposure to MPs due to the capability of organisms to recover (Agathokleous *et al.*, 2021). A full recovery after exposure to MPs and depuration has been reported in various aquatic organisms including bivalves, fishes, phytoplankton, mussels and zooplanktons (Detree & Gallardo-Escarate, 2018; Hamed *et al.*, 2019; Mao *et al.*, 2018; Oliveira *et al.*, 2018; Solomando *et al.*, 2020).

2.2. *Size-related effects*

Another important factor affecting the ecotoxicity of MPs is their size. Size can influence the amount of MPs ingested and their retention time in an organism's digestive tract, leading to different physiological effects. While larger MPs (300–5,000 μm) have been widely reported in the marine environment, this reporting bias is likely an artefact of the widespread use of plankton nets with a 300 μm mesh, and particles between 10 and 100 μm have been shown to be more abundant in marine waters (Enders *et al.*, 2015).

Wang *et al.* (2021a) reported that accumulation of MPs in mussel guts was inversely related to the size of the particles (i.e., smaller MP particles accumulated significantly more than larger ones). In general, smaller microplastics and nanoplastics have longer gut retention time possibly due to being readily trapped in the villi of digestive tract and, therefore, likely pose a greater toxicological risk (Gray & Weinstein 2017; Kinjo *et al.*, 2019; Roch *et al.*, 2021). While faster excretion time has been reported for smaller MP beads (1 μm) compared to larger beads (90 μm) in mussels (Kinjo *et al.*, 2019), the larger MPs were completely excreted while a small number of the smaller beads were retained for a significantly longer period of time. Similarly, a more effective egestion of 6 μm MPs compared to smaller nano-sized particles (0.5 μm) has been reported in a rotifer (Jeong *et al.*, 2016).

Size also has a dramatic impact on the ability of microplastics and nanoplastics to cross biological membranes. Particles larger than 20 μm are unlikely to be able to cross the lipid bilayer of cells, but particles <5 μm have been shown to cross this protective barrier (Ragusa *et al.*, 2021). Longer retention in an organism's digestive tract (or exposure via gills) greatly increases the risk of translocation of small micro- and nano-sized plastics to the circulatory system and ultimately tissue accumulation. It has been shown that the transport of smaller microplastics and nanoplastics to the digestive glands and epithelial cells can have more pronounced negative impacts at the cellular level in mussels (von Moos *et al.*, 2012). Abarghouei *et al.* (2021) reported higher concentrations of small MPs (0.25 μm) compared to 8 μm particles in fish liver tissue as a result of translocation from gills to this organ through the blood stream.

Size of MPs is an important factor causing hepatic toxicity and lipid metabolism disorder in fish, with smaller MPs (2 and 10 μm) causing greater hepatic injury and lipid accumulation compared to bigger MPs (200 μm) (Zhang *et al.*, 2021). To date, many studies have revealed a strong size-dependent toxicity of MPs to a variety of aquatic organisms with exposure to smaller MPs often inducing more pronounced adverse effects than larger particles. For example, greater effects associated with exposure to smaller sizes of MPs include reduced growth, reduced fecundity, activation of antioxidant-related enzymes in rotifer (Jeong *et al.* 2016), increased dysfunction of intestine immune cells (Gu *et al.*, 2020), significant change in gene expression in fishes (Abarghouei *et al.*, 2021), increased cellular internalisation in microalgae (Chen *et al.*, 2020), and reduced growth, reproduction and feeding in zooplankton (An *et al.*, 2021). While most studies suggest that smaller MPs are more toxic in the μm range, some laboratory studies at the even smaller microplastic to nanoplastic scale have shown that a smaller size does not necessarily mean higher toxicity. Wu *et al.* (2021) recently reported that 1 μm MPs were more toxic than 100 nm MPs in cyanobacteria (*Microcystis aeruginosa*). Zimmermann *et al.* (2020) likewise reported that smaller MPs (polyvinylchloride particles <20 μm) were less toxic to daphnia than larger particles (polyurethane and polylactic acid particles, 40 μm).

While it is clear that size influences toxicity, it is currently difficult to draw a firm general conclusion on whether smaller or larger MPs are most toxic, and there is a need to better understand the ecotoxicological effects associated with environmentally relevant sizes of MPs on aquatic organism health. In a review article, de Sa *et al.* (2018) reported a mismatch between the sizes of MPs reported in marine organisms (400–1,600 μm) and those used in laboratory studies (<50 μm).

2.3. *Shape/morphology-related effects*

MPs occur in different shapes, most commonly fibres, fragments and beads. Shape is one of the most important determinants of MP accumulation in an organism's gut, influencing both ingestion and egestion rate, and thus producing varying toxicological profiles. Currently, the majority of MPs found in the marine environment are fibres (>50%), followed by fragments, film and pellets (Harris, 2020). Fibres are likewise the most common MPs in freshwater environments (53.8%), also followed by fragments (30.2%) (Lu *et al.*, 2021).

Non-spherical MPs (fragments and fibres) have a longer intestinal passage time and larger aspect ratio, which gives them a higher potential to accumulate and embed in tissues (Qiao *et al.*, 2019) ultimately leading to more harmful effects. For example, MP fragments have higher gut accumulation in daphnia than beads (An *et al.*, 2021). Shape-dependent accumulation of MPs was also observed in zebrafish (Qiao *et al.*, 2019) and shrimp (Gray & Weinstein, 2017), with fibres and fragments constantly showing higher accumulation rate than beads. High surface area and increased accumulation rate of non-spherical MPs can lead to greater negative effects (such as physical damage) to the exposed organisms. Laboratory studies have shown that exposure to MP fibres and fragments is usually associated with more severe toxicity than spherical MPs, for example, severe intestinal epithelial damage in zebrafish exposed to fibres (Gray & Weinstein, 2017) and greater oxidative stress and mortality in zooplankton exposed to MP fragments (Na *et al.*, 2021). That said, some studies have reported no significant differences in accumulation and effects of spherical and non-spherical MPs in fish after short-term (acute) exposure (Choi *et al.*, 2018; Hamer *et al.*, 2014).

The shape of MPs can also influence toxicity beyond just ingestion-related effects. For example, while MP fibres were not ingested by zooplankton (*Ceriodaphnia dubia*), they were more toxic to reproduction and growth than beads (which were ingested) mainly due to physical damage and stress to the organism's body (e.g., entanglement) (Ziajahromi *et al.*, 2017).

Overall, it appears that irregular-shaped particles are more harmful than smooth (bead-shaped) particles, however, this can depend on the size and type of organisms, their feeding behaviour and the duration of exposure to MPs.

2.4. *Composition-related effects*

Finally, the chemical composition (polymer type) of MPs can influence their toxicity. Currently, PE, polyethylene terephthalate (PET), PP and polyamide (PA) are the dominant types of polymers detected in the aquatic environments, in both marine and freshwater ecosystems (Amelia *et al*., 2021; Lu *et al*., 2021). There are two aspects that could differentially affect the toxicity of polymers: (i) different polymers differ in the main chemical chain as well as the type and number of side chains that deviate from the main chain, which could potentially impact the toxicity of polymers to aquatic organisms (Hollerová *et al*., 2021), and (ii) various polymers can have different degradation mechanisms, which may influence the movement and migration of embedded chemical additives from the polymer (which can vary depending on its intended form and use) and ultimately influence the effects associated with the polymer type (Hahladakis *et al*., 2018). Zimmermann *et al*. (2020) showed that toxicity of MPs to zooplankton is polymer-specific, with polyvinyl chloride (PVC) MPs exhibiting higher toxicity to reproduction than polyurethane MPs of similar shape and size. Genotoxicity in fish embryos was also linked with the type of polymer, with PS consistently more genotoxic than PE (Jakubowska *et al*., 2020). It has been suggested that PS may be more toxic in laboratory experiments because of its higher density: more dense PS MPs settle in close proximity to the fish larvae (at the bottom of the experimental vessel) and developing organisms compared with lower-density MP such as polyethylene, leading to higher exposure and consequently greater toxic effects (Jakubowska *et al*., 2020). Thus, toxicity of MPs of different types can also be an artefact of the laboratory toxicity testing approach. Thus, while recent evidence shows that different polymers might have different toxic effects to the organisms, the mechanisms underlying these effects and their environmental relevance have yet to be elucidated.

3. Combined Effects of MPs and Associated Chemicals on Aquatic Biota

3.1. *MPs and sorbed contaminants*

In addition to effects caused by MPs themselves to aquatic organisms, their role as vector of pollutants (e.g., persistent organic pollutants, metals and micropollutants) has raised concerns about exacerbating the effects, given that the sorbed contaminants can be accumulated on the surface of plastics and transferred to the organisms when MPs are ingested. A large number of laboratory and field studies have now documented the capacity of MPs to sorb (absorb/adsorb) a wide range of organic pollutants (e.g., PAH, PCB, HCH, DDT and organochlorine pesticides), heavy metals (e.g., Cd, Co, Cr, Cu, Ni, Pb and Zn) as well as pharmaceuticals, as reviewed in Ziajahromi *et al.* (2016). The bioavailability, bioaccumulation and ultimate toxicity of sorbed contaminants vary depending on the type of contaminant, the type of polymer, and environmental and physiological conditions (e.g., pH and ionic strength) (Wang *et al.*, 2020b). The bioavailability of pollutants sorbed to MPs mainly depends on their tendency to desorb during gut passage. Under simulated gut conditions, Bakir *et al.* (2014) showed an increase in bioavailability of POPs under lower-gastric pH and gut surfactant of warm-blooded organisms, which may lead to an increase in the uptake and bioaccumulation of pollutants in the organism's tissue and organs. Likewise, a laboratory study demonstrated enhanced bioaccumulation of pharmaceuticals sorbed to PS MPs in fish tissues after waterborne exposure (Zhang *et al.*, 2019b). However, there is still debate about the enhanced bioavailability and bioaccumulation of contaminants sorbed to MPs. Devriese *et al.* (2017) demonstrated negligible bioaccumulation of polychlorinated biphenyls (PCB) after lobster exposure to PCB-spiked PE and PS. This laboratory study also did not find desorption of PCB from PE MPs after being ingested by lobster. Modelling studies suggest that uptake of MPs may not be a significant contributor to the transfer of sorbed chemicals to marine organisms via the gut compared to natural particles (Bakir *et al.*, 2016).

There is still uncertainty about potential mechanisms of desorption and accumulation of MP-sorbed chemicals inside the organism's body. It is also not entirely clear whether co-exposure to MPs and chemical contaminants is more or less toxic than the single exposure itself. Some studies have reported more pronounced toxic effects after co-exposure in organisms such as fish (He *et al.*, 2021; Luis *et al.*, 2015; Rainieri *et al.*, 2018; Zhang *et al.*, 2019b), mussel (Webb *et al.*, 2020) and zooplankton (Wang *et al.*, 2020c). Others have found no such effect in daphnia (Horton *et al.*, 2018) and sea urchin larvae (Beiras & Tato 2019). Yet other studies have instead reported antagonistic effects when organisms were exposed to MPs in combination with chemicals in algae (Yang *et al.*, 2020), copepods (Li *et al.*, 2020b) or midge larvae (Ziajahromi *et al.*,

2019). This antagonism was explained by the fact that sorbed chemicals were strongly bound to MPs, resulting in decreased bioavailability and toxicity of the chemicals to the organisms. Ziajahromi *et al.* (2019) noted that natural organic matter (NOM) present in environmental waters provides a large number of sorption sites which may, under environmentally realistic concentrations of MPs, sorb the majority of contaminants present in the water. It is important also to note here that many laboratory studies to date have tested MP concentration orders of magnitude higher than those found in the environment and types of MPs and contaminants not commonly found in the aquatic environment.

It is clear that the presence of MPs can impact the toxicity and fate of chemicals in exposed organisms. However, the significance of this effect may depend on the amount of chemicals bound to those particles, how this is influenced by environmental conditions (e.g., aging and formation of biofilm on plastics) (Liu *et al.*, 2019), and to what extent these chemicals can desorb and transfer to exposed organisms, which are impacted by the residence time of MPs in the organism's gastrointestinal tract. Li *et al.* (2020b) reported a link between the lower combined toxicity of MPs and chemicals and a short residence time of MPs in an organism's gastrointestinal tract. Research is needed to broaden our knowledge about biological effects of MPs and sorbed contaminants, especially under environmental realistic conditions: relevant concentrations of MPs and contaminants using real (not synthetic) water samples in the presence of natural organic and inorganic material.

3.2. *MPs and additives/plasticisers*

Plastics can also include a wide range of chemicals (plasticisers, antioxidants, fillers, colourants, reinforcements, solvents and catalysts) that are embedded in various polymers to improve plastic properties and enhance durability (Bridson *et al.*, 2021). Some additives such as phthalates, bisphenol A, brominated flame retardants and chlorinated paraffins are categorised as highly toxic to aquatic life (Groh *et al.*, 2019). Once plastics are released to the aquatic environment, some additives that are not chemically bound to the plastic polymer can leach and present a hazard to exposed aquatic organisms. It has been suggested that effects associated with additives could be greater than exposure to contaminants sorbed to plastic polymers from the environment (Kwon *et al.*, 2017). The lethal and sub-lethal toxicity of plastic leachates has been demonstrated on various organisms such as bacteria (Tetu *et al.*, 2019), zooplankton (Na *et al.*, 2021) and invertebrates (Capolupo *et al.*, 2020) with effects observed on survival, development and reproduction. Beiras *et al.* (2021) tested the effects of three MPs and their associated additives (PVC, PE and PA) on three different organisms (microalgae, crustaceans and echinoderms) and concluded that the toxicity of MPs was entirely due to leaching of additives and not the MPs themselves. While the hazard of plastic leachate is obvious, there is uncertainty regarding the rate of transfer of additives to organisms, bioavailability and bioaccumulation. While growing evidence from experimental and modelling studies reveals transfer and accumulation of additives from MPs to exposed aquatic organisms (Di Renzo *et al.*, 2021; Diepens & Koelmans 2018; Tanaka *et al.*, 2015), there is still debate as to whether this is a significant pathway to transfer additives to exposed organisms or whether it represents a comparatively small risk (Schrank *et al.*, 2019).

4. Trophic Transfer of Microplastics/Nanoplastics (Biomagnification)

Within organisms, bioaccumulation (increase over time) and biomagnification (increase over trophic level) are two important concepts applied in ecological risk assessments to determine the extent to which pollutants are transported within the food chain (Miller *et al.*, 2020). Evidence shows that MPs are ingested by a wide range of aquatic organisms from primary producers to secondary consumers and predators of higher trophic levels. The uptake of MPs by lower trophic organisms (e.g., zooplanktons, benthic invertebrates and small fishes) can negatively impact the quality of food supply to other aquatic organisms at upper trophic levels (e.g., larger fishes and mammals) (Amelia *et al.*, 2021) leading to MP accumulation and potentially transferring along the aquatic food chain. This pathway may also cause higher effects to marine ecosystems due to a corresponding accumulation of associated toxic substances at higher trophic levels (Wang *et al.*, 2021b). While concerns have been raised regarding the potential bioaccumulation and biomagnification of MPs, this may vary across different organisms and mainly depends on the organism's feeding type and their capacity to ingest, accumulate and egest MPs (depuration capacity) (Wang *et al.*, 2021b; Zhang *et al.*, 2019a). Recent studies have shown negligible biomagnification of MPs through food webs (Santana *et al.*, 2017; Walkinshaw *et al.*, 2020; Wang *et al.*, 2021b). While several laboratory studies have reported transfer of MPs from prey to predator (Farrell & Nelson, 2013; Santana *et al.*, 2017; Setala *et al.*, 2014; Zhang *et al.*, 2019a), Wang *et al.* (2021b) recently reported that direct exposure (via waterborne MPs) and indirect exposure (via feeding on contaminated food) both led to roughly similar transfer of small MPs (1–5 μm) to the exposed organisms. This highlights the difficulty of meaningfully evaluating the impact of the exposure pathways (e.g., water and prey) and assessing the actual risk of biomagnification.

Concerns about bioaccumulation and biomagnification have also been raised in the context of human health, as humans consume marine organisms. It has been estimated that humans may ingest up to 55,000 MP particles/year through consumption of seafood alone (Danopoulos *et al.*, 2020). A recent review has performed a broad-scale evaluation of biomagnification through assessing MP content of commercial marine organisms from different trophic levels including fish, shellfish, crustaceans and macroalgae (Walkinshaw *et al.*, 2020). The study concluded that, when measured by mass of organism tissue, the average MP content in organisms at lower trophic levels (e.g., shellfish) was four orders of magnitude higher than the MP content in higher trophic levels (e.g., piscivorous fish). Therefore, unlike other persistent toxic contaminants such as mercury (which have shown to biomagnify), MP concentrations in marine organisms did not correlate with trophic levels (Figure 2). This means that although laboratory and field studies have indicated potential bioaccumulation of MPs (over time) in various organisms at different trophic levels, there is currently no evidence that MPs biomagnify in higher trophic organisms.

5. Environmental Realism and Complexity of the Issue

One of the major (mostly unresolved) questions in the context of ecotoxicological impact of MPs is to what extent the reported laboratory effects compare with "the real world". There is currently a remarkable mismatch between laboratory and natural conditions, with many toxicity studies to date using concentrations that are orders of magnitude higher than environmentally relevant concentrations, or using polymer types, morphology

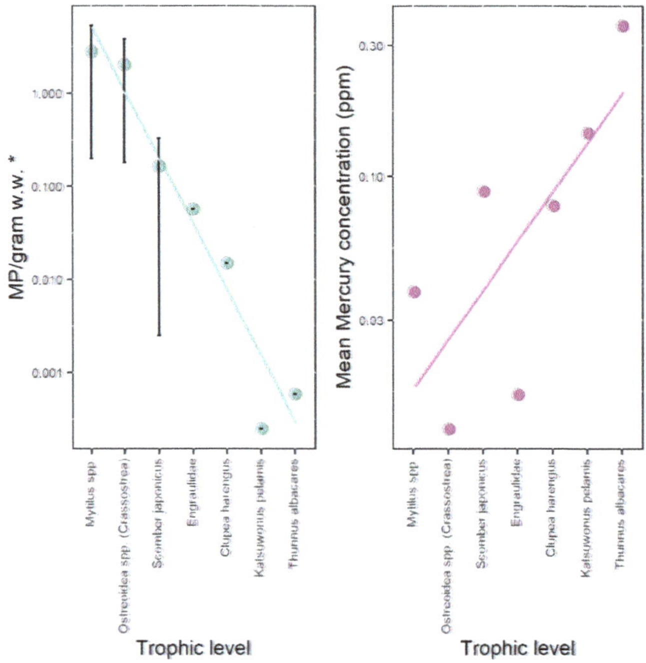

Figure 2. Left: concentration of MPs (number per gram wet weight of organisms). Right: concentration of mercury (ppm) in tissue of the marine organisms from different trophic levels.

Source: Walkinshaw *et al.* (2020).

and size ranges that are not environmentally relevant (Agathokleous *et al.*, 2021). Further obfuscating this issue is the fact that concentrations in laboratory exposures are often reported in mass units (e.g., mg/L), while concentrations in the environment are usually reported in counts (e.g., particles/L or particles/m^3). Fortunately, it is in relatively straightforward to convert from one concentration unit to the other (Leusch & Ziajahromi, 2021), and Figure 3 and Eqs. (1)–(4) provide a method to convert between the two and enable this comparison.

$$x\,(\text{beads/L}) = \frac{y\,(\text{mg/L}) \times 10^9\ (\text{unit conversion factor})}{(\pi/6) \times \text{density}\ (\text{g/cm}^3) \times [\text{diameter}\ (\mu\text{m})]^3} \qquad (1)$$

$$y\,(\text{mg/L}) = \frac{(\pi/6)\times \text{density (g/cm}^3) \times x\,(\text{beads/L}) \times [\text{diameter } (\mu\text{m})]^3}{10^9 \text{ (unit conversion factor)}} \tag{2}$$

$$x\,(\text{fibres / L}) = \frac{y\,(\text{mg/L}) \times 10^9 \text{ (unit conversion factor)}}{\pi \times [\text{radius } (\mu\text{m})]^2 \times \text{length } (\mu\text{m}) \times \text{density (g/cm}^3)} \tag{3}$$

$$y(\text{mg/L}) = \frac{\pi \times [\text{radius } (\mu\text{m})]^2 \times \text{length } (\mu\text{m}) \times \text{density (g/cm}^3) \times x\,(\text{fibres/L})}{10^9 \text{ (unit conversion factor)}} \tag{4}$$

Using a meta-analysis approach, Bucci *et al.* (2020) assessed the environmental relevance of ecotoxicology studies and showed that >80% of the studies (out of 139 lab and field studies on a variety of taxa) used concentrations and size ranges of MPs that are not detected in the natural environment. Pinheiro *et al.* (2020) showed that the type of MPs used for laboratory experiments is not reflective of those ingested by aquatic organisms, with PS and PP as the most commonly used polymers for laboratory assays while PA and polyester (PES) were the main polymer type detected in wild organisms.

Everaert *et al.* (2018) thoroughly assessed the toxicity data available in the literature for marine MPs and suggested that adverse effects (not considering potential effects of sorbed chemicals) are unlikely to occur at concentrations of MPs below 6,650 particles/m^3. Using an improved dataset, the same authors calculated a Predicted No Effect Concentration (PNEC) of 1.21×10^5 particles/m^3 (Everaert *et al.*, 2020) and suggested that this safe level may be exceeded in hotspot areas (such as coastal regions, harbours and narrow straits). There is still some debate about this issue, with some suggesting that given the ability of organisms to recover after even short periods of depuration, the current data do not appear to support the conclusion that current environmental concentrations of MPs cause negative impact on individual organisms (Agathokleous *et al.*, 2021; Burns & Boxall, 2018). However, MP toxicity research is a very active field, and as more studies are produced with more environmentally realistic conditions, more data will become available to provide better picture of risk (Adam *et al.*, 2019). Improvement of analytical techniques

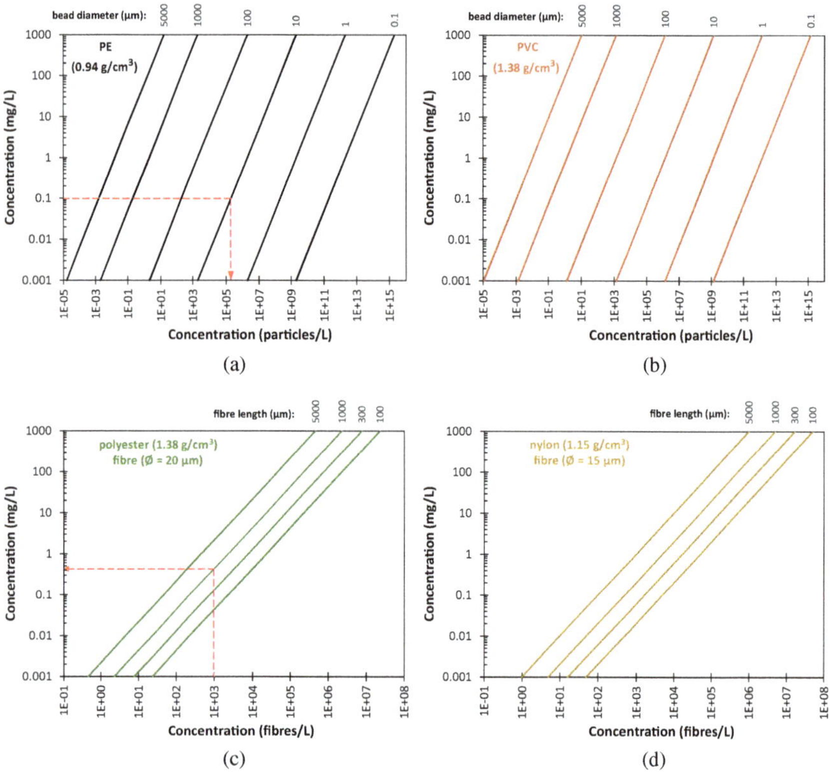

Figure 3. Conversion chart from mass to particle/L units for beads (top) and fibres (bottom) to facilitate comparison of laboratory studies with natural conditions. The concentrations of microplastic beads (spheres) can be converted using Eqs. (1) and (2), while the concentrations of microplastic fibres (cylinders) can be converted using Eqs. (3) and (4). As worked examples, 0.1 mg/L of 10 μm PE beads equals 2.0×10^5 particles/L, while 1,000 polyester fibres/L (300 μm length, 20 μm diameter) equals 0.4 mg/L.

Source: Adapted from Leusch and Ziajahromi (2021).

to detect smaller MPs will also improve our estimate of exposure and likely change our understanding of risk.

In the case of nanoplastics, the majority of studies (>90%) have only tested the effects of PS nanoparticles (usually in the form of beads) on aquatic organisms, with the remaining studies focused mainly on PMMA and polycarbonate (Barria *et al.*, 2020; Gangadoo *et al.*, 2020). Also, the currently used commercial PS nanoplastics are often functionalised with amino groups and carboxyl acids, which may influence their toxicity (Kokalj *et al.*, 2021). While the presence of nanoplastics in the aquatic environment has been predicted (and confirmed) (Cai *et al.*, 2021), there is a severe knowledge gap about the actual form and concentration of nanoplastics in the aquatic environments with only one study so far detecting nanoplastics (>90% composed of PVC and PET) in seawater samples (Ter Halle *et al.*, 2017). Our knowledge of quantitative concentrations, size ranges, complex forms and aggregation behaviour of nanoplastics in the natural environment is largely limited due to major methodological issues that are as yet unresolved (Cai *et al.*, 2021). A growing number of *in vitro* and *in vivo* toxicity studies have reported adverse effects of single-form nanoplastics on aquatic organisms (usually using very high concentrations, of the order of millions of particles/L), including the potential for nanoplastics to induce growth inhibition and behaviour alteration through oxidative stress and inflammatory responses (Hu & Palic 2020). However, the complexity of the issue in current research is often disregarded and, therefore, there is need to better assess the adverse effects associated with the nanoplastics existing in the environment.

A final concern regarding environmental realism of the current ecotoxicological research is that MPs used for toxicity assays do not generally mirror the properties and format of natural MPs. Many studies to date have used "virgin" MPs (i.e., new) for toxicity testing, however natural environmental processes such as UV irradiation, physical abrasion and biofilm formation affect the physical and chemical structure of "aged" MPs. Aged MPs may produce different toxic effects in exposed organisms as the ageing process may induce changes in the chemical structure of MPs (Wang *et al.*, 2020a). Martinez-Gomez *et al.* (2017) showed that aged MPs cause greater toxic effects on larval development of sea urchins than virgin MPs. Jovanovic *et al.* (2018) concluded that six different polymer types of virgin MPs at environmentally realistic concentrations did not cause any significant adverse effects on development of fish, while Guimaraes *et al.* (2021) showed that naturally aged MPs at environmentally relevant concentrations can cause neurotoxic and cytotoxic effects to fish. These studies highlight the importance and necessity for current research to use aged/weathered MPs for toxicity testing to provide a more realistic approach. This is easier said than done though, as separating aged plastics from environmental media requires significant sample handling (including acid digestion), and it is unclear whether artificial ageing adequately mimics natural processes.

6. Conclusions

Micro- and nano-sized plastics are ubiquitous in the aquatic environment and, because of increasing worldwide plastic use and their extreme persistence, will likely occur at ever-increasing concentrations into the future. It is clear that a vast number of aquatic species are exposed to MPs and this exposure may induce a variety of toxic effects in individuals, populations and ultimately ecosystems. Moreover, the ability of MPs to sorb chemical contaminants could exacerbate their toxicity. MPs in the environment occur in a variety of concentrations, shapes, types and size ranges, which are major determinants of their ultimate toxicological effects in exposed organisms. While MPs can bioaccumulate in organisms over time, they appear not likely to biomagnify through the food chain from lower trophic to higher trophic organisms. To date, the majority of studies have used unrealistic exposure conditions to assess the effects associated with microplastics and nanoplastics and there is not enough reliable evidence to show to what extent environmentally realistic microplastic and nanoplastic concentrations can cause toxic effects to various aquatic organisms. But with an improved understanding of the importance of environmental relevance in toxicity studies and more research teams lifting the lid on occurrence of small microplastics and nanoplastics in our environment, new data are coming in fast to improve our understanding of the risks that MPs pose to our aquatic ecosystems. This is important for two reasons: (1) MP concentrations are predicted to increase over the next decades, which will increase the threat they pose to aquatic biota, and (2) unlike macroplastics, microplastics cannot be scooped out without also removing microscopic planktonic organisms. We must deal with the threat of microplastic to our environment up-front, at the consumer and policy stage — this is unlikely to occur until we can more accurately estimate and demonstrate the risk that they pose to our ecosystem.

References

Abarghouei, S., Hedayati, A., Raeisi, M., Hadavand, B. S., Rezaei, H., & Abed-Elmdoust, A. (2021). Size-dependent effects of microplastic on uptake, immune system, related gene expression and histopathology of goldfish (Carassius auratus). *Chemosphere*, *276*, 129977.

Adam, V., Yang, T., & Nowack, B. (2019). Toward an ecotoxicological risk assessment of microplastics: Comparison of available hazard and exposure data in freshwaters. *Environmental Toxicology and Chemistry*, *38*(2), 436–447.

Agathokleous, E., Iavicoli, I., Barceló, D., & Calabrese, E. J. (2021). Micro/nano-plastics effects on organisms: A review focusing on 'dose'. *Journal of Hazardous Materials*, *417*, 126084.

Akdogan, Z. & Guven, B. (2019). Microplastics in the environment: A critical review of current understanding and identification of future research needs. *Environmental Pollution*, *254*(Pt A), 113011.

Amelia, T. S. M., Khalik, W. M. A. W. M., Ong, M. C., Shao, Y. T., Pan, H.-J., & Bhubalan, K. (2021). Marine microplastics as vectors of major ocean pollutants and its hazards to the marine ecosystem and humans. *Progress in Earth and Planetary Science*, *8*(1), 1–26.

An, D., Na, J., Song, J., & Jung, J. (2021). Size-dependent chronic toxicity of fragmented polyethylene microplastics to Daphnia magna. *Chemosphere*, *271*, 129591.

Au, S. Y., Bruce, T. F., Bridges, W. C., & Klaine, S. J. (2015). Responses of Hyalella azteca to acute and chronic microplastic exposures. *Environmental Toxicology and Chemistry*, *34*(11), 2564–2572.

Bajt, O. (2021). From plastics to microplastics and organisms. *FEBS Open Bio*, *11*(4), 954–966.

Bakir, A., O'Connor, I. A., Rowland, S. J., Hendriks, A. J., & Thompson, R. C. (2016). Relative importance of microplastics as a pathway for the transfer of hydrophobic organic chemicals to marine life. *Environmental Pollution*, *219*, 56–65.

Bakir, A., Rowland, S. J., & Thompson, R. C. (2014). Enhanced desorption of persistent organic pollutants from microplastics under simulated physiological conditions. *Environmental Pollution*, *185*, 16–23.

Barria, C., Brandts, I., Tort, L., Oliveira, M., & Teles, M. (2020). Effect of nano-plastics on fish health and performance: A review. *Marine Pollution Bulletin*, *151*, 110791.

Beiras, R. & Tato, T. (2019). Microplastics do not increase toxicity of a hydrophobic organic chemical to marine plankton. *Marine Pollution Bulletin*, *138*, 58–62.

Beiras, R., Verdejo, E., Campoy-Lopez, P., & Vidal-Linan, L. (2021). Aquatic toxicity of chemically defined microplastics can be explained by functional additives. *Journal of Hazardous Materials*, *406*, 124338.

Besseling, E., Redondo-Hasselerharm, P., Foekema, E. M., & Koelmans, A. A. (2018). Quantifying ecological risks of aquatic micro- and nanoplastic. *Critical Reviews in Environmental Science and Technology*, *49*(1), 32–80.

Botterell, Z. L. R., Beaumont, N., Dorrington, T., Steinke, M., Thompson, R. C., & Lindeque, P. K. (2019). Bioavailability and effects of microplastics on marine zooplankton: A review. *Environmental Pollution*, *245*, 98–110.

Bridson, J. H., Gaugler, E. C., Smith, D. A., Northcott, G. L., & Gaw, S. (2021). Leaching and extraction of additives from plastic pollution to inform environmental risk: A multidisciplinary review of analytical approaches. *Journal of Hazardous Materials*, *414*, 125571.
Bringer, A., Cachot, J., Prunier, G., Dubillot, E., Clerandeau, C., & Helene, T. (2020). Experimental ingestion of fluorescent microplastics by pacific oysters, Crassostrea gigas, and their effects on the behaviour and development at early stages. *Chemosphere*, *254*, 126793.
Bucci, K., Tulio, M., & Rochman, C. M. (2020). What is known and unknown about the effects of plastic pollution: A meta-analysis and systematic review. *Ecological Applications*, *30*(2), e02044.
Burns, E. E. & Boxall, A. B. A. (2018). Microplastics in the aquatic environment: Evidence for or against adverse impacts and major knowledge gaps. *Environmental Toxicology and Chemistry*, *37*(11), 2776–2796.
Cai, H., Xu, E. G., Du, F., Li, R., Liu, J., & Shi, H. (2021). Analysis of environmental nanoplastics: Progress and challenges. *Chemical Engineering Journal*, *410*, Article No. 128208, 1–12.
Capolupo, M., Sorensen, L., Jayasena, K. D. R., Booth, A. M., & Fabbri, E. (2020). Chemical composition and ecotoxicity of plastic and car tire rubber leachates to aquatic organisms. *Water Research*, *169*, 115270.
Carbery, M., O'Connor, W., & Palanisami, T. (2018). Trophic transfer of microplastics and mixed contaminants in the marine food web and implications for human health. *Environment International*, *115*, 400–409.
Chen, Y., Ling, Y., Li, X., Hu, J., Cao, C., & He, D. (2020). Size-dependent cellular internalization and effects of polystyrene microplastics in microalgae *P. helgolandica* var. *tsingtaoensis* and *S. quadricauda*. *Journal of Hazardous Materials*, *399*, 123092.
Choi, J. S., Jung, Y. J., Hong, N. H., Hong, S. H., & Park, J. W. (2018). Toxicological effects of irregularly shaped and spherical microplastics in a marine teleost, the sheepshead minnow (*Cyprinodon variegatus*). *Marine Pollution Bulletin*, *129*(1), 231–240.
Corami, F., Rosso, B., Roman, M., Picone, M., Gambaro, A., & Barbante, C. (2020). Evidence of small microplastics (<100 μm) ingestion by Pacific oysters (*Crassostrea gigas*): A novel method of extraction, purification, and analysis using Micro-FTIR. *Marine Pollution Bulletin*, *160*, 111606.
Danopoulos, E., Jenner, L. C., Twiddy, M., & Rotchell, J. M. (2020). Microplastic contamination of seafood intended for human consumption: A systematic review and meta-analysis. *Environmental Health Perspectives*, *128*(12), 126002.
de Sa, L. C., Oliveira, M., Ribeiro, F., Rocha, T. L., & Futter, M. N. (2018). Studies of the effects of microplastics on aquatic organisms: What do we

know and where should we focus our efforts in the future? *Science of the Total Environment*, *645*, 1029–1039.

Detree, C. & Gallardo-Escarate, C. (2018). Single and repetitive microplastics exposures induce immune system modulation and homeostasis alteration in the edible mussel *Mytilus galloprovincialis*. *Fish and Shellfish Immunology*, *83*, 52–60.

Devriese, L. I., De Witte, B., Vethaak, A. D., Hostens, K., & Leslie, H. A. (2017). Bioaccumulation of PCBs from microplastics in Norway lobster (Nephrops norvegicus): An experimental study. *Chemosphere*, *186*, 10–16.

Di Renzo, L., Mascilongo, G., Berti, M., Bogdanović, T., Listeš, E., Brkljača, M., Notarstefano, V., Gioacchini, G., Giorgini, E., Olivieri, V., Silvestri, C., Matiddi, M., D'Alterio, N., Ferri, N., & Di Giacinto, F. (2021). Potential impact of microplastics and additives on the health status of loggerhead turtles (*Caretta caretta*) stranded along the Central Adriatic Coast. *Water, Air, & Soil Pollution*, *232*(3).

Diepens, N. J. & Koelmans, A. A. (2018). Accumulation of plastic debris and associated contaminants in aquatic food webs. *Environmental Science & Technology*, *52*(15), 8510–8520.

Eltemsah, Y. S. & Bohn, T. (2019). Acute and chronic effects of polystyrene microplastics on juvenile and adult *Daphnia magna*. *Environmental Pollution*, *254*(Pt A), 112919.

Enders, K., Lenz, R., Stedmon, C. A., & Nielsen, T. G. (2015). Abundance, size and polymer composition of marine microplastics ≥10 mum in the Atlantic Ocean and their modelled vertical distribution. *Marine Pollution Bulletin*, *100*(1), 70–81.

Everaert, G., De Rijcke, M., Lonneville, B., Janssen, C. R., Backhaus, T., Mees, J., van Sebille, E., Koelmans, A. A., Catarino, A. I., & Vandegehuchte, M. B. (2020). Risks of floating microplastic in the global ocean. *Environmental Pollution*, *267*, 115499.

Everaert, G., Van Cauwenberghe, L., De Rijcke, M., Koelmans, A. A., Mees, J., Vandegehuchte, M., & Janssen, C. R. (2018). Risk assessment of microplastics in the ocean: Modelling approach and first conclusions. *Environmental Pollution*, *242*(Pt B), 1930–1938.

Farrell, P. & Nelson, K. (2013). Trophic level transfer of microplastic: *Mytilus edulis* (L.) to *Carcinus maenas* (L.). *Environmental Pollution*, *177*, 1–3.

Frias, J. & Nash, R. (2019). Microplastics: Finding a consensus on the definition. *Marine Pollution Bulletin*, *138*, 145–147.

Gandara, E. S. P. P., Nobre, C. R., Resaffe, P., Pereira, C. D. S., & Gusmao, F. (2016). Leachate from microplastics impairs larval development in brown mussels. *Water Research*, *106*, 364–370.

Gangadoo, S., Owen, S., Rajapaksha, P., Plaisted, K., Cheeseman, S., Haddara, H., Truong, V. K., Ngo, S. T., Vu, V. V., Cozzolino, D., Elbourne, A.,

Crawford, R., Latham, K., & Chapman, J. (2020). Nano-plastics and their analytical characterisation and fate in the marine environment: From source to sea. *Science of the Total Environment, 732*, 138792.

Gao, G., Zhao, X., Jin, P., Gao, K., & Beardall, J. (2021). Current understanding and challenges for aquatic primary producers in a world with rising micro- and nano-plastic levels. *Journal of Hazardous Materials, 406*, 124685.

Gray, A. D. & Weinstein, J. E. (2017). Size- and shape-dependent effects of microplastic particles on adult daggerblade grass shrimp (Palaemonetes pugio). *Environmental Toxicology and Chemistry, 36*(11), 3074–3080.

Groh, K. J., Backhaus, T., Carney-Almroth, B., Geueke, B., Inostroza, P. A., Lennquist, A., Leslie, H. A., Maffini, M., Slunge, D., Trasande, L., Warhurst, A. M., & Muncke, J. (2019). Overview of known plastic packaging-associated chemicals and their hazards. *Science of the Total Environment, 651*(Pt 2), 3253–3268.

Gu, W., Liu, S., Chen, L., Liu, Y., Gu, C., Ren, H. Q., & Wu, B. (2020). Single-cell RNA sequencing reveals size-dependent effects of polystyrene microplastics on immune and secretory cell populations from Zebrafish Intestines. *Environmental Science & Technology, 54*(6), 3417–3427.

Guimaraes, A. T. B., Charlie-Silva, I., & Malafaia, G. (2021). Toxic effects of naturally-aged microplastics on zebrafish juveniles: A more realistic approach to plastic pollution in freshwater ecosystems. *Journal of Hazardous Materials, 407*, 124833.

Hahladakis, J. N., Velis, C. A., Weber, R., Iacovidou, E., & Purnell, P. (2018). An overview of chemical additives present in plastics: Migration, release, fate and environmental impact during their use, disposal and recycling. *Journal of Hazardous Materials, 344*, 179–199.

Hamed, M., Soliman, H. A. M., Osman, A. G. M., & Sayed, A. E. H. (2019). Assessment the effect of exposure to microplastics in Nile Tilapia (*Oreochromis niloticus*) early juvenile: I. blood biomarkers. *Chemosphere, 228*, 345–350.

Hamer, J., Gutow, L., Kohler, A., & Saborowski, R. (2014). Fate of microplastics in the marine isopod Idotea emarginata. *Environmental Science & Technology, 48*(22), 13451–13458.

Harris, P. T. (2020). The fate of microplastic in marine sedimentary environments: A review and synthesis. *Marine Pollution Bulletin, 158*, 111398.

He, J., Yang, X., & Liu, H. (2021). Enhanced toxicity of triphenyl phosphate to zebrafish in the presence of micro- and nano-plastics. *Science of the Total Environment, 756*, 143986.

Hollerová, A., Hodkovicová, N., Blahová, J., Faldyna, M., Maršálek, P., & Svobodová, Z. (2021). Microplastics as a potential risk for aquatic environment organisms — A review. *Acta Veterinaria Brno, 90*(1), 99–107.

Horton, A. A., Vijver, M. G., Lahive, E., Spurgeon, D. J., Svendsen, C., Heutink, R., van Bodegom, P. M., & Baas, J. (2018). Acute toxicity of organic pesticides to *Daphnia magna* is unchanged by co-exposure to polystyrene microplastics. *Ecotoxicology and Environmental Safety, 166*, 26–34.

Hu, M. & Palic, D. (2020). Micro- and nano-plastics activation of oxidative and inflammatory adverse outcome pathways. *Redox Biology, 37*, 101620.

Isobe, A., Iwasaki, S., Uchida, K., & Tokai, T. (2019). Abundance of non-conservative microplastics in the upper ocean from 1957 to 2066. *Nature Communications, 10*(1), 417.

Ivar do Sul, J. A. & Costa, M. F. (2014). The present and future of microplastic pollution in the marine environment. *Environmental Pollution, 185*, 352–364.

Jakubowska, M., Bialowas, M., Stankeviciute, M., Chomiczewska, A., Pazusiene, J., Jonko-Sobus, K., Hallmann, A., & Urban-Malinga, B. (2020). Effects of chronic exposure to microplastics of different polymer types on early life stages of sea trout Salmo trutta. *Science of the Total Environment, 740*, 139922.

Jeong, C. B., Won, E. J., Kang, H. M., Lee, M. C., Hwang, D. S., Hwang, U. K., Zhou, B., Souissi, S., Lee, S. J., & Lee, J. S. (2016). Microplastic size-dependent toxicity, oxidative stress induction, and p-JNK and p-p38 activation in the Monogonont Rotifer (*Brachionus koreanus*). *Environmental Science & Technology, 50*(16), 8849–8857.

Jovanovic, B., Gokdag, K., Guven, O., Emre, Y., Whitley, E. M., & Kideys, A. E. (2018). Virgin microplastics are not causing imminent harm to fish after dietary exposure. *Marine Pollution Bulletin, 130*, 123–131.

Kinjo, A., Mizukawa, K., Takada, H., & Inoue, K. (2019). Size-dependent elimination of ingested microplastics in the Mediterranean mussel *Mytilus galloprovincialis*. *Marine Pollution Bulletin, 149*, 110512.

Kokalj, A. J., Hartmann, N. B., Drobne, D., Potthoff, A., & Kühnel, D. (2021). Quality of nanoplastics and microplastics ecotoxicity studies: Refining quality criteria for nanomaterial studies. *Journal of Hazardous Materials, 415*, 125751.

Kwon, J. H., Chang, S., Hong, S. H., & Shim, W. J. (2017). Microplastics as a vector of hydrophobic contaminants: Importance of hydrophobic additives. *Integrated Environmental Assessment and Management, 13*(3), 494–499.

Lebreton, L. C. M., van der Zwet, J., Damsteeg, J. W., Slat, B., Andrady, A., & Reisser, J. (2017). River plastic emissions to the world's oceans. *Nature Communications, 8*, 15611.

Leusch, F. D. L. & Ziajahromi, S. (2021). Converting mg/L to particle/L: Reconciling the occurrence and toxicity literature on microplastics. *Environmental Science & Technology, 55*(17), 11470–11472.

Li, C., Busquets, R., & Campos, L. C. (2020a). Assessment of microplastics in freshwater systems: A review. *Science of the Total Environment*, *707*, 135578.

Li, Z., Zhou, H., Liu, Y., Zhan, J., Li, W., Yang, K., & Yi, X. (2020b). Acute and chronic combined effect of polystyrene microplastics and dibutyl phthalate on the marine copepod *Tigriopus japonicus*. *Chemosphere*, *261*, 127711.

Liu, G., Zhu, Z., Yang, Y., Sun, Y., Yu, F., & Ma, J. (2019). Sorption behavior and mechanism of hydrophilic organic chemicals to virgin and aged microplastics in freshwater and seawater. *Environmental Pollution*, *246*, 26–33.

Lu, H. C., Ziajahromi, S., Neale, P. A., & Leusch, F. D. (2021). A systematic review of freshwater microplastics in water and sediments: recommendations for harmonisation to enhance future study comparisons. *Science of the Total Environment*, *781*, 146693.

Luis, L. G., Ferreira, P., Fonte, E., Oliveira, M., & Guilhermino, L. (2015). Does the presence of microplastics influence the acute toxicity of chromium(VI) to early juveniles of the common goby (Pomatoschistus microps)? A study with juveniles from two wild estuarine populations. *Aquatic Toxicology*, *164*, 163–174.

Mao, Y., Ai, H., Chen, Y., Zhang, Z., Zeng, P., Kang, L., Li, W., Gu, W., He, Q., & Li, H. (2018). Phytoplankton response to polystyrene microplastics: Perspective from an entire growth period. *Chemosphere*, *208*, 59–68.

Mao, Y., Li, H., Huangfu, X., Liu, Y., & He, Q. (2020). Nanoplastics display strong stability in aqueous environments: Insights from aggregation behaviour and theoretical calculations. *Environmental Pollution*, *258*, 113760.

Martinez-Gomez, C., Leon, V. M., Calles, S., Gomariz-Olcina, M., & Vethaak, A. D. (2017). The adverse effects of virgin microplastics on the fertilization and larval development of sea urchins. *Marine Environmental Research*, *130*, 69–76.

Mattsson, K., Jocic, S., Doverbrat, I., & Hansson, L.-A. (2018). Nanoplastics in the aquatic environment. In Zeng, E. Y. (Ed.). Amsterdam: Elsevier.

Miller, M. E., Hamann, M., & Kroon, F. J. (2020). Bioaccumulation and biomagnification of microplastics in marine organisms: A review and meta-analysis of current data. *PLoS One*, *15*(10), e0240792.

Miloloža, M., Kučić Grgić, D., Bolanča, T., Ukić, Š., Cvetnić, M., Ocelić Bulatović, V., Dionysiou, D. D., & Kušić, H. (2020). Ecotoxicological assessment of microplastics in freshwater sources — A review. *Water*, *13*(1).

Na, J., Song, J., Achar, J. C., & Jung, J. (2021). Synergistic effect of microplastic fragments and benzophenone-3 additives on lethal and sublethal Daphnia magna toxicity. *Journal of Hazardous Materials*, *402*, 123845.

Oliveira, P., Barboza, L. G. A., Branco, V., Figueiredo, N., Carvalho, C., & Guilhermino, L. (2018). Effects of microplastics and mercury in the freshwater bivalve *Corbicula fluminea* (Muller, 1774): Filtration rate, biochemical biomarkers and mercury bioconcentration. *Ecotoxicology and Environmental Safety*, *164*, 155–163.

Pinheiro, L. M., do Sul, J. A. I., & Costa, M. F. (2020). Uptake and ingestion are the main pathways for microplastics to enter marine benthos: A review. *Food Webs*, *24*, e00150.

Qiao, R., Deng, Y., Zhang, S., Wolosker, M. B., Zhu, Q., Ren, H., & Zhang, Y. (2019). Accumulation of different shapes of microplastics initiates intestinal injury and gut microbiota dysbiosis in the gut of zebrafish. *Chemosphere*, *236*, 124334.

Ragusa, A., Svelato, A., Santacroce, C., Catalano, P., Notarstefano, V., Carnevali, O., Papa, F., Rongioletti, M. C. A., Baiocco, F., Draghi, S., D'Amore, E., Rinaldo, D., Matta, M., & Giorgini, E. (2021). Plasticenta: First evidence of microplastics in human placenta. *Environment International*, *146*, 106274.

Rainieri, S., Conlledo, N., Larsen, B. K., Granby, K., & Barranco, A. (2018). Combined effects of microplastics and chemical contaminants on the organ toxicity of zebrafish (*Danio rerio*). *Environmental Research*, *162*, 135–143.

Rehse, S., Kloas, W., & Zarfl, C. (2016). Short-term exposure with high concentrations of pristine microplastic particles leads to immobilisation of Daphnia magna. *Chemosphere*, *153*, 91–99.

Roch, S., Ros, A. F. H., Friedrich, C., & Brinker, A. (2021). Microplastic evacuation in fish is particle size-dependent. *Freshwater Biology*, *66*(5), 926–935.

Santana, M. F. M., Moreira, F. T., & Turra, A. (2017). Trophic transference of microplastics under a low exposure scenario: Insights on the likelihood of particle cascading along marine food-webs. *Marine Pollution Bulletin*, *121*(1–2), 154–159.

Schrank, I., Trotter, B., Dummert, J., Scholz-Bottcher, B. M., Loder, M. G. J., & Laforsch, C. (2019). Effects of microplastic particles and leaching additive on the life history and morphology of *Daphnia magna*. *Environmental Pollution*, *255*(Pt 2), 113233.

Setala, O., Fleming-Lehtinen, V., & Lehtiniemi, M. (2014). Ingestion and transfer of microplastics in the planktonic food web. *Environmental Pollution*, *185*, 77–83.

Solomando, A., Capo, X., Alomar, C., Alvarez, E., Compa, M., Valencia, J. M., Pinya, S., Deudero, S., & Sureda, A. (2020). Long-term exposure to microplastics induces oxidative stress and a pro-inflammatory response in the gut of Sparus aurata Linnaeus, 1758. *Environmental Pollution*, *266*(Pt 1), 115295.

Suman, T. Y., Jia, P. P., Li, W. G., Junaid, M., Xin, G. Y., Wang, Y., & Pei, D. S. (2020). Acute and chronic effects of polystyrene microplastics on brine shrimp: First evidence highlighting the molecular mechanism through transcriptome analysis. *Journal of Hazardous Materials*, *400*, 123220.

Sun, T., Zhan, J., Li, F., Ji, C., & Wu, H. (2021). Effect of microplastics on aquatic biota: A hormetic perspective. *Environmental Pollution, 285*, 117206.

Tanaka, K., Takada, H., Yamashita, R., Mizukawa, K., Fukuwaka, M. A., & Watanuki, Y. (2015). Facilitated leaching of additive-derived PBDEs from plastic by seabirds' stomach oil and accumulation in tissues. *Environmental Science & Technology*, *49*(19), 11799–11807.

Ter Halle, A., Jeanneau, L., Martignac, M., Jarde, E., Pedrono, B., Brach, L., & Gigault, J. (2017). Nanoplastic in the North Atlantic Subtropical Gyre. *Environmental Science & Technology*, *51*(23), 13689–13697.

Tetu, S. G., Sarker, I., Schrameyer, V., Pickford, R., Elbourne, L. D. H., Moore, L. R., & Paulsen, I. T. (2019). Plastic leachates impair growth and oxygen production in Prochlorococcus, the ocean's most abundant photosynthetic bacteria. *Communications Biology*, *2*, 184.

Venancio, C., Ferreira, I., Martins, M. A., Soares, A., Lopes, I., & Oliveira, M. (2019). The effects of nanoplastics on marine plankton: A case study with polymethylmethacrylate. *Ecotoxicology and Environmental Safety, 184*, 109632.

von Moos, N., Burkhardt-Holm, P., & Kohler, A. (2012). Uptake and effects of microplastics on cells and tissue of the blue mussel *Mytilus edulis L.* after an experimental exposure. *Environmental Science & Technology*, *46*(20), 11327–11335.

Wagner, M., Scherer, C., Alvarez-Munoz, D., Brennholt, N., Bourrain, X., Buchinger, S., Fries, E., Grosbois, C., Klasmeier, J., Marti, T., Rodriguez-Mozaz, S., Urbatzka, R., Vethaak, A. D., Winther-Nielsen, M., & Reifferscheid, G. (2014). Microplastics in freshwater ecosystems: What we know and what we need to know. *Environmental Sciences Europe*, *26*(1), 12.

Walkinshaw, C., Lindeque, P. K., Thompson, R., Tolhurst, T., & Cole, M. (2020). Microplastics and seafood: Lower trophic organisms at highest risk of contamination. *Ecotoxicology and Environmental Safety*, *190*, 110066.

Wang, Q., Wangjin, X., Zhang, Y., Wang, N., Wang, Y., Meng, G., & Chen, Y. (2020a). The toxicity of virgin and UV-aged PVC microplastics on the growth of freshwater algae *Chlamydomonas reinhardtii*. *Science of the Total Environment*, *749*, 141603.

Wang, S., Hu, M., Zheng, J., Huang, W., Shang, Y., Kar-Hei Fang, J., Shi, H., & Wang, Y. (2021a). Ingestion of nano/micro plastic particles by the mussel Mytilus coruscus is size dependent. *Chemosphere*, *263*, 127957.

Wang, T., Hu, M., Xu, G., Shi, H., Leung, J. Y. S., & Wang, Y. (2021b). Microplastic accumulation via trophic transfer: Can a predatory crab counter the adverse effects of microplastics by body defence? *Science of the Total Environment*, *754*, 142099.

Wang, T., Wang, L., Chen, Q., Kalogerakis, N., Ji, R., & Ma, Y. (2020b). Interactions between microplastics and organic pollutants: Effects on toxicity, bioaccumulation, degradation, and transport. *Science of the Total Environment, 748*, 142427.

Wang, Z., Dong, H., Wang, Y., Ren, R., Qin, X., & Wang, S. (2020c). Effects of microplastics and their adsorption of cadmium as vectors on the cladoceran Moina monogolica Daday: Implications for plastic-ingesting organisms. *Journal of Hazardous Materials, 400*, 123239.

Webb, S., Gaw, S., Marsden, I. D., & McRae, N. K. (2020). Biomarker responses in New Zealand green-lipped mussels Perna canaliculus exposed to microplastics and triclosan. *Ecotoxicology and Environmental Safety, 201*, Article No. 110871, 1–91.

Wojcik-Fudalewska, D., Normant-Saremba, M., & Anastacio, P. (2016). Occurrence of plastic debris in the stomach of the invasive crab Eriocheir sinensis. *Marine Pollution Bulletin, 113*(1–2), 306–311.

Wu, D., Wang, T., Wang, J., Jiang, L., Yin, Y., & Guo, H. (2021). Size-dependent toxic effects of polystyrene microplastic exposure on *Microcystis aeruginosa* growth and microcystin production. *Science of the Total Environment, 761*, 143265.

Yang, W., Gao, X., Wu, Y., Wan, L., Tan, L., Yuan, S., Ding, H., & Zhang, W. (2020). The combined toxicity influence of microplastics and nonylphenol on microalgae *Chlorella pyrenoidosa. Ecotoxicology and Environmental Safety, 195*, 110484.

Zhang, F., Wang, X., Xu, J., Zhu, L., Peng, G., Xu, P., & Li, D. (2019a). Food-web transfer of microplastics between wild caught fish and crustaceans in East China Sea. *Marine Pollution Bulletin, 146*, 173–182.

Zhang, S., Ding, J., Razanajatovo, R. M., Jiang, H., Zou, H., & Zhu, W. (2019b). Interactive effects of polystyrene microplastics and roxithromycin on bioaccumulation and biochemical status in the freshwater fish red tilapia (*Oreochromis niloticus*). *Science of the Total Environment, 648*, 1431–1439.

Zhang, X., Wen, K., Ding, D., Liu, J., Lei, Z., Chen, X., Ye, G., Zhang, J., Shen, H., Yan, C., Dong, S., Huang, Q., & Lin, Y. (2021). Size-dependent adverse effects of microplastics on intestinal microbiota and metabolic homeostasis in the marine medaka (*Oryzias melastigma*). *Environment International, 151*, 106452.

Ziajahromi, S., Kumar, A., Neale, P. A., & Leusch, F. D. L. (2017). Impact of microplastic beads and fibers on waterflea (*Ceriodaphnia dubia*) survival, growth, and reproduction: Implications of single and mixture exposures. *Environmental Science & Technology, 51*(22), 13397–13406.

Ziajahromi, S., Kumar, A., Neale, P. A., & Leusch, F. D. L. (2019). Effects of polyethylene microplastics on the acute toxicity of a synthetic pyrethroid to midge larvae (*Chironomus tepperi*) in synthetic and river water. *Science of the Total Environment*, *671*, 971–975.

Ziajahromi, S., Neale, P. A., & Leusch, F. D. (2016). Wastewater treatment plant effluent as a source of microplastics: Review of the fate, chemical interactions and potential risks to aquatic organisms. *Water Science and Technology*, *74*(10), 2253–2269.

Zimmermann, L., Gottlich, S., Oehlmann, J., Wagner, M., & Volker, C. (2020). What are the drivers of microplastic toxicity? Comparing the toxicity of plastic chemicals and particles to *Daphnia magna*. *Environmental Pollution*, *267*, 115392.

https://doi.org/10.1142/9789811259111_0009

Chapter 8

Plastic Pollution in the Arctic Marine Environment

Hollie L. Ball* and Crispin Halsall

Lancaster Environment Centre (LEC), Lancaster University, Lancaster LA1 4YQ, UK

**h.ball@lancaster.ac.uk*

1. Introduction

Concern about pollution in the Arctic and growing environmental pressures across this region gave rise to the Arctic Environmental Protection Strategy in 1991, a circumpolar agreement focused on environmental monitoring, protection of the Arctic environment and later, environmental sustainability (see UNEP). A number of working groups with specific environmental remits were created that operate under the Arctic Council. One of these groups, the Arctic Monitoring Assessment Programme (AMAP) reports widely on pollutants like radioactive substances, mercury/heavy metals, and persistent organic pollutants, particularly with regard to human health and ecosystem exposure (https://www.amap.no/). A growing area of concern is marine litter and microplastics, and a programme to assess these contaminants and their impact on food webs is underway (https://litterandmicroplastics.amap.no/). This has included an extensive assessment of marine plastic litter coordinated through the Protection of the Arctic Marine Environment (PAME), another Arctic Council working group whose remit covers the protection and sustainable use of the marine environment.

Polar regions are increasingly receiving and being impacted by plastic waste (Horton & Barnes, 2020). Arctic beaches are visibly polluted with plastic debris, and the rates of this pollution are continuing to rise as Arctic populations grow and tourism increases (Polasek *et al.*, 2017). Tourism has been identified as an important source of litter in temperate coastal regions and it is suspected that it will soon have a large contribution to plastic pollution in the Arctic (Tekman *et al.*, 2017). Polasek *et al.* (2017) conducted a project to remove and document marine debris from five coastal National Park Service units in Alaska. The remote, relatively pristine nature of the Alaskan coastline was found to have a surprisingly high accumulation of anthropogenically derived debris. Approximately 80 km of coastline were cleaned, and over 10,000 kg of marine debris were collected. Out of the 28 beaches surveyed, hard plastics were found everywhere, and ropes/netting were found at almost all beaches. Overall, plastic contributed to 60% of the total weight of debris. The results suggest local geography and substrate of a beach likely has a strong effect on the amount and concentration of debris collected. There is a significant lack of information regarding the concentration, distribution and type of plastic debris on Arctic beaches, especially those in more remote higher-latitude locations (Bergmann *et al.*, 2017).

The issue of macroplastic pollution is one that we can visually acknowledge, however a growing concern comes from plastic particles that can be difficult to observe, e.g., microplastics ($\leq$5 mm in diameter) and nanoplastics ($\leq$1 μm in diameter). Currently, microplastics in the Arctic have been observed in sea ice (Obbard *et al.*, 2014; Bergmann *et al.*, 2016; Peeken *et al.*, 2018), snow (Bergmann *et al.*, 2019), surface and subsurface waters (Lusher *et al.*, 2015; Kanhai *et al.*, 2018) and deep-sea sediment (Bergmann *et al.*, 2017), as well as Arctic biota (Provencher *et al.*, 2010; Kühn *et al.*, 2018; Morgana *et al.*, 2018). A range of polymers have been identified in samples suggesting multiple sources, both local and long-range, with many particles being in the lower size ranges (<500 μm). These smaller particles are more readily available to marine biota (Lusher, 2015), although very few studies have been carried out on the effects of plastic ingestion on Arctic marine organisms. It is suggested that ingestion of smaller plastic particles, which can carry harmful toxins both on the surface and within the polymer matrix, can affect the mobility, reproduction and general functioning of the organisms that consume them (Kühn *et al.*, 2015). The ingestion of microplastics by blue mussels (*Mytilus edulis*) (Sundet *et al.*, 2016) and snow crabs (*Chionoecetes opilio*) (Sundet *et al.*, 2014) was investigated in Svalbard, with results showing that both species ingested microplastics, with 90% occurrence and an average of 9.5 items per individual for the mussels and 20% incidence for the snow crabs (PAME, 2019). Trevail *et al.* (2015) explain that plastic ingestion levels in the Arctic are generally minor compared to levels at lower latitudes. However, levels of ingestion are not as low as would be expected given the remoteness of some of the study regions. This raises questions about the sources, transport, and fate of plastic debris and microplastics in the Arctic as a whole.

Microplastics have been discovered in both surface and sub-surface marine waters, demonstrating that they are readily available for incorporation into sea ice (Lusher *et al.*, 2015), which has been identified as a sink for microplastic particles. Currently, there's a lack of knowledge concerning the processes and rates of microplastic incorporation within sea ice, and their possible effect on sea ice properties, such as influencing the albedo (proportion of incident sunlight reflected) of surface snow (Geilfus *et al.*, 2019). The Arctic is currently seeing record lows in sea ice extent, which could affect plastic particle dynamics and transport across the Arctic Ocean and see the release of large amounts of legacy microplastics that have previously been entrained in the ice (Obbard *et al.*, 2014). Arctic sea ice has been recognised as a repository for organic contaminants present in seawater but also deposited from the atmosphere (e.g., Pucko *et al.*, 2015; Garnett *et al.*, 2021), and microplastics have also been observed in sea ice and the ice-rafted snowpack. As sea ice forms, it will incorporate particles that are suspended in the water column, which will then be transported long distances as sea ice drifts. Contaminants will then be deposited when the sea ice melts, making them available to sympagic (ice-associated) and pelagic marine organisms (Obbard, 2018).

Research into plastic pollution in polar regions is lacking, especially regarding the fate of macroplastic litter and the formation and/or transport of microplastic particles. Studies have previously focussed on identifying the abundance, sources and accumulation patterns of microplastics, but the effects that plastic pollution are having on Arctic ecosystems is unknown (Horton & Barnes, 2020). Tirelli *et al.* (2020) explain that the first records of plastic pollution in the Arctic date back to the early 1960s, when the ingestion of plastic fragments was reported in Canadian seabirds (Threlfall, 1968; Provencher *et al.*, 2015), and more thoroughly in the 1970s, when the total amount of beached macrolitter items in the Bering Sea increased from over 2200 to more than 5300 in only 2 years — 1972–1974

(Merrell, 1980). Research on marine mammals and entanglement reached a peak during the 1980s and 1990s in Arctic waters of Alaska and adjacent areas (the Bering Sea and the Gulf of Alaska), but monitoring has since reduced, and in the rest of the Arctic, knowledge is fragmented and covers only some groups or species (PAME, 2019). The current knowledge on plastic pollution in the Arctic marine environment relates to areas where human activities are concentrated, such as the Barents Sea, Norwegian Sea, and Bering Sea, with little data available for the Central Arctic Ocean and the coastal areas around it in Siberia, Arctic Alaska, mainland Canada and the Canadian Arctic Archipelago (PAME, 2019). These regions are clearly more difficult to access, hence the lack of data, but if we are to get a homogenous understanding of marine plastic pollution in the Arctic, it is vital we include these areas. Against this backdrop of pollution pressures, the Arctic is warming on average more than twice the global mean (Jansen *et al.*, 2020). From 1971 to 2019, the annually averaged Arctic near-surface air temperature increased by 3.1°C, three times faster than the global average (AMAP, 2021), further demonstrating the importance of understanding the fate of plastic pollution in this rapidly changing environment. This rapid warming is leading to advanced thinning and melting of sea ice with implications for the global climate system. According to the US National Snow & Ice Data Center (NSIDC, 2022) and reported by AMAP, Arctic sea ice extent in September (the month with lowest sea ice cover) has declined by 43% between 1979 and 2019, and most areas are seeing a decline not only during the summer months but all year round. It is also reported that all Arctic regions are facing a net loss of land ice, with Greenland accounting for 51% of the total. These losses of land ice in the Arctic are a significant contributor to global sea-level rise (IPCC Special Report on the Ocean and Cryosphere in a Changing Climate). Arctic rivers are also experiencing shorter freezing periods, and ice thickness in the northernmost Arctic rivers is decreasing. The volume of water flowing through these rivers and into the Arctic Ocean has increased by 7.8% over the last five decades (AMAP, 2021) and has the potential to alter pollutant transport pathways to Arctic coastal seas and the wider Arctic Ocean. Increased anthropogenic pollution has the potential to exacerbate the negative effects of climate change on ecosystems (e.g., melting sea ice releasing large concentrations of microplastics back into the environment) and vice versa, for example, increased extreme weather as a result of climate change has the potential to worsen the spread of plastic pollution in the environment (Ford *et al.*, 2021).

2. Sources

Due to its low density, plastic debris can be transported long distances, meaning areas where the debris deposits are most likely not where the source of pollution is found. Sources of plastic pollution to the Arctic can be split into either terrestrial- or marine-based origins, for example, discarded fishing gear has been highlighted as one of the main contributors to floating plastic debris in the Arctic (PAME, 2019), which originates in the marine environment from both local and long-range sources. The remote nature of the Arctic would suggest that most plastic debris entering the region is from long-range sources, travelling via ocean currents, atmospheric pathways or riverine inputs.

2.1. *Marine inputs*

The main sectors of maritime activity in the Arctic include fishing, both commercial and recreational, aquaculture and shipping, including cruise tourism (PAME, 2019). Plastic debris associated with these activities such as nets, floats and ropes can enter the marine environment through accidental loss, poor waste management or illegal dumping (Sheavly and Register, 2007). These macroplastics create microplastics through fragmentation, by several processes including photolysis, thermal oxidation, hydrolysis and physical abrasion (Browne, 2015).

Bergmann *et al.* (2017) conducted a citizen science experiment which involved participants of a tourist cruise documenting and collecting plastic debris on Arctic beaches. Over the six beaches surveyed, a total of 991 kg of plastic debris was collected, with the highest quantities found on Reinstrandodden in Svalbard (524 gm^{-2}). The Svalbard beaches showed particularly high concentrations of fisheries-related plastic, with such items including ropes, buoys, floats, nets and smaller broken pieces of the aforementioned. These items accounted for 44–100% of the litter mass, with other plastic items such as packaging material, bottles, toys and other single-use plastics also present. Bergmann *et al.* (2017) explain that the dominance of fisheries-related litter on the beaches of Svalbard may be partly due to the unit used to express litter quantities (gm^{-2}), which is sensitive to the occurrence of single, particularly heavy, items such as fishing nets at Reinstrandodden, highlighting the importance of transparency when reporting methodologies. However, its strong dominance at all beaches examined clearly points to fisheries as a major source of litter pollution. Most fishing gears that end up on beaches and can be analysed are made of material that floats, meaning materials that do not float, such as set nets or crab pots, do not end up on beaches and remain in the sea (PAME, 2019). During the citizen science experiment, participants took images to demonstrate the effects of plastic litter on Arctic biota, which mostly portrayed wildlife entangled in fishing nets and ropes. Several Arctic species were affected, including Arctic whitlow-grass (*Draba bellii*), seabirds (specifically *Sterna paradisaea*), and terrestrial and marine mammals such as harbour seals (*Phoca vitulina*) and polar bears (*Ursus maritimus*).

2.2. *Terrestrial sources*

Plastic litter concentrations from local land sources should be relatively low in the Arctic due to its remote nature and relatively low population of 4 million, which are primarily concentrated in townships and capital cities, leaving the central Arctic area with a comparatively low population (Heleniak & Bogoyavlensky, 2015). However, increased discharge from the major Arctic rivers (North American and Eurasian rivers) that have very large catchments extending to sub-Arctic regions may be a significant and growing source of plastic debris to the coastal Arctic Seas. The remoteness of the central Arctic and the sparse coastal-based communities is unlikely to give rise to plastic waste from terrestrial sources on an Arctic-wide basis. Granberg *et al.* (2017) explain that the release of untreated sewage into coastal environments is one of the most significant unregulated local pollution sources in the Arctic, where sewage treatment is generally lacking, resulting in a continuous discharge of sewage and wastewater from households and commercial industries directly to coastal waters (PAME, 2019; Granberg *et al.*, 2017). Household waste is likely to contain microplastics, primarily fibres, originating from cosmetics and washing synthetic polymer clothing (Napper & Thompson, 2016). Kirkelund *et al.* (2017) found that in small Arctic communities, the traditional waste management solutions are uncontrolled waste dumps and simple incinerators with no or limited flue gas treatment. Magnusson *et al.* (2016) investigated wastewater outputs in various Nordic Sewage Treatment Plants (STPs) and found that those plants using only mechanical treatments such as those in Iceland had limited particle retention, and outputs of microplastics amounted to ~6,350,000 microplastics (of diameters

≥300 μm) per hour from Klettagarðar and ~2,230,000 microplastics per hour from Hafnarfjörður. An increase in tourism combined with rudimentary waste management will inevitably lead to higher levels of plastic pollution from local sources in the vicinity of Arctic communities and nearby coastal regions.

3. Transport

Contaminants such as macroplastic litter and microplastics can enter the Arctic via a combination of processes, such as surface oceanic currents, freshwater input from rivers and atmospheric processes, which all transport contaminants from long-range sources. Zarfl and Matthies (2010) considered the possibility that marine plastic particles could act as transport vectors for organic pollutants to the Arctic. The study used thermodynamically based model calculations to determine the relative importance of plastic debris as a source of persistent bioaccumulative toxins (PBTs) to the remote Arctic Ocean. Their models conclude that transport via atmospheric and ocean currents is orders of magnitude larger than via plastic particles, determining that the contribution of PBTs from plastic debris may be small compared with annual PBT flux from other global transport mechanisms.

3.1. *Ocean transport*

The central Arctic Ocean is surrounded by shallow coastal seas as well as passages and straits: the Barents Sea, Greenland Sea, Norwegian Sea and Fram Strait, located in the European region of the Arctic. The Labrador Sea, Davis Strait, Baffin Bay, Northern Passages, Hudson Strait, Hudson Bay and the Beaufort Sea are located around Canada. The Chukchi Sea and Bering Sea surround Alaska, and the East Siberian Sea, Laptev Sea and Kara Sea are located around northern Russia (Figure 1; PAME, 2013). Waters flowing into the Arctic basin include Atlantic waters entering via the Fram Strait as well as into the Barents Sea from the Norwegian Sea, and Pacific waters flowing from the Bering Sea via the Bering Strait and across the Chukchi Sea (Loeng *et al.*, 2005; Tirelli *et al.*, 2020).

Kanhai *et al.* (2018) report that the oceanic currents which supply the greatest water volumes to the Arctic Ocean, including the West Spitsbergen Current, a branch of the North Atlantic current which flows northwards

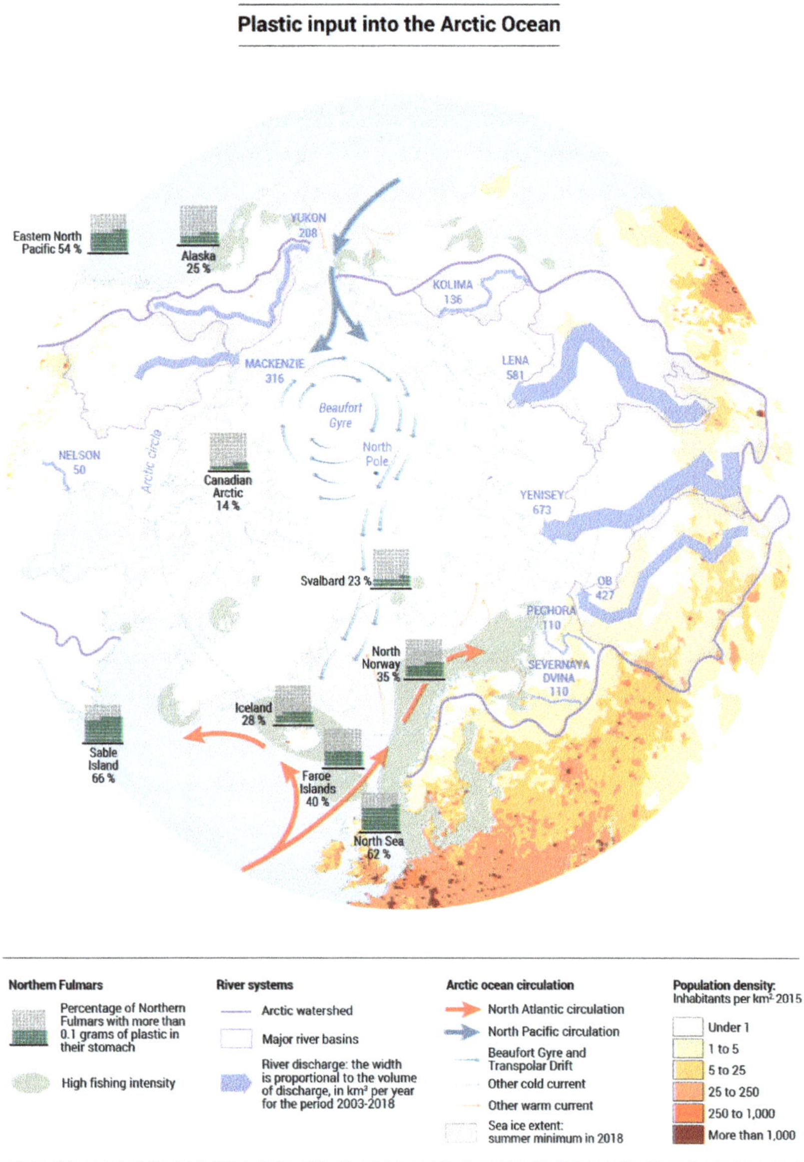

Figure 1. The Arctic region highlighting the major surface ocean currents, prevailing winds, sea ice drift patterns, terrestrial catchments and annual river basin water discharges to the Arctic coastal seas.

Source: GRID-ARENDAL (2019), https://www.grida.no/resources/13350.

through the eastern side of the Fram Strait, and the Bering Sea current bringing Pacific Ocean water into the Arctic and flows along the Siberian coastline, may also transport plastics, with the estimated flux ranging between 62,000 and 105,000 tonnes per year (Zarfl & Matthies, 2010). Onink *et al.* (2019) studied the role of various physical processes, such as wind-driven Ekman currents, geostrophic currents and wave-induced Stokes drift on the global surface distribution of floating microplastics. They predicted high concentrations in the Arctic seas compared to concentrations in the sub-tropics, suggesting an increase in transport northwards which they attribute to Stokes drift in particular. Because Stokes drift has not been used universally in studies investigating microplastic transport, the study suggests microplastic contamination in Arctic regions could be more severe than currently expected. Winds and currents play important roles in the advection of plastic particles both into and out of the Arctic (Cózar *et al.*, 2017), with macroplastic debris floating on the sea surface being affected by direct wind drag, which changes the drift dramatically compared to material that is fully submerged (Röhrs & Christensen, 2015). Obbard *et al.* (2014) suggest that a large amount of plastic pollution is entering the Arctic via the Pacific, where large concentrations of microplastics have been reported, and is subsequently being entrained within Arctic sea ice.

3.2. *River transport*

There have been very few studies thus far that have investigated Arctic rivers as a significant transport mechanism for plastic litter. Although there are numerous rivers that feed into Arctic coastal seas such as the Mackenzie and Yukon rivers in Canada and Northern America, and the Lena, Yenisey and Ob in Siberia (Holmes *et al.*, 2013), inputs from these rivers are unlikely to contribute large amounts of plastic pollution to the Arctic as they flow through sparsely populated areas (Obbard *et al.*, 2014). However, Yakushev *et al.* (2021) explain that the Arctic Ocean receives 11% of the global freshwater discharge, with the Ob (510 km^3/year), Yenisei (630 km^3/year) and Lena (530 km^3/year) rivers providing the most input, suggesting that even with low concentrations, these rivers could be a significant source of plastic debris. In a study on the effect of Atlantic waters and Siberian rivers on microplastic distribution in the Eurasian Arctic, Yakushev *et al.* (2021) found that smaller microplastics (with surface area <3 mm^2) dominated in the low-saline inner plumes adjacent to the river estuaries and deltas, whilst in the outer plume, microplastics had distinctly larger sizes (with surface area up to 10 mm^2). Small differences were found between the average abundance of microplastics in the inner and outer plumes (0.0045 particles m^{-3} and 0.0051 particles m^{-3}, respectively), however when looking at the mass concentration, a notable difference was detected (1.6 μg m^{-3} and 2.9 μg m^{-3}, respectively). The largest diversity in polymer types was found in the outer plume, consisting of five polymer types (13% fibres) compared to the inner plume which consisted of three polymer types (67% fibres). No microplastics were found outside the river plumes in the saline Polar surface water mass. Yakushev *et al.* (2021) explain that due to the absence of floating microplastics in this area, along with a higher abundance of sub-surface microplastic fragments, it is presumed that released microplastics do not remain at the sea surface but sink to the sub-surface layer after ice melting. This can be caused by physical and/or biological processes that modify floating properties of ice-trapped microplastics and explains the absence of sub-surface fragments of microplastics in river plumes. It is thought that floating microplastics in the plumes were discharged from the rivers to sea during an ice-free period and were not trapped by sea ice.

The study suggests that the lower abundance of microplastics reported in the results compared with other investigations is possibly due to a difference in sampling procedures and only considering particles >100 μm to be comparable to previous research. The study also found the FT-IR analysis decreased the final number of particles identified as plastics, which is something all investigations into plastic pollution must consider. Yakushev *et al.* (2021) go on to explain that the results found in this study are similar to levels reported for the Arctic Central Basin (ACB) in Kanhai *et al.* (2018), where similar sampling and identification methods were used, reinforcing the importance of consistent methodological approaches.

3.3. *Sea ice transport*

More recently, sea ice has been acknowledged as a significant sink for microplastic pollution, as well as a transport mechanism for moving plastic particles across the Arctic Ocean (Peeken *et al.*, 2018). Obbard *et al.* (2014) suggest that most of the plastic pollution entering the Arctic in surface currents is via the Pacific, where large concentrations of microplastics have been reported, and is subsequently being entrained to some extent within Arctic sea ice. Peeken *et al.* (2018) suggest that sea ice is one of the key compartments for transporting and redistributing microplastics in the Arctic, but this will be dependent on prevailing ice floe drift patterns. In general, pack ice passes through different regions of the Central Arctic Ocean but is eventually carried southward to the Fram Strait via the Transpolar Drift. In order to determine drift trajectories and source areas of sampled sea ice, Peeken *et al.* (2018) tracked sampled sea ice backwards in time using low-resolution ice drift via satellite observation, which showed that the sea ice originated from different source areas, namely, the Amerasian and Eurasian Basins. The estimated time between plastic contamination and ice arrival varies throughout the Arctic. For example, contamination from sources in the Laptev and Kara seas takes between 2–4 years to arrive in the Fram Strait, and up to 6–11 years from sources in the Amerasian Basin.

3.4. *Biotransport*

The role of biotransport poses two distinct issues: the first being marine biota acting as vectors for plastic and microplastic pollution, and the second being floating plastic debris acting as vectors for non-native species or pathogens that can enter the Arctic from other regions. For example, in Svalbard, non-native barnacles have been found on plastic debris (PAME, 2019). Macroplastic debris can act as "rafts" transporting toxins and invasive species, posing risks to ecosystems across the Arctic, from both regional (i.e., within Arctic) and long-range sources (Barnes, 2002). Provencher *et al.* (2010) studied ingested plastic in thick-billed murre (*Uria lomvia*), a migratory species to the Canadian Arctic, and reported the highest incidence of plastic ingestion to date for this species. Of the 186 thick-billed murres collected, 11% contained ingested plastics in their digestive tracts with a mean of 0.2 ± 0.8 microplastics found ingested per bird. The results of the study suggest that seasonal timing may determine how much plastic is ingested. For example, 13% of birds had ingested plastic early in the summer season when the birds had just arrived at the breeding colony. However, for birds sampled later in the season, after foraging in the remote area for several months, they had no detectable plastics in their digestive tracts. Thick-billed murres may be ingesting plastics in the waters of the Northwest Atlantic during the winter months, where plastic contamination is relatively higher. During seasonal migration, the birds will bring their burden of plastic contamination northwards to the Arctic. Presently, the quantity of plastic transported by biovectors such as migrating birds, fish and sea mammals into the Arctic is uncertain.

3.5. *Atmospheric transport*

Deposition with snowfall has been reported as a transport mechanism for smaller size ranges of microplastic particles. Bergmann *et al.* (2019) used FT-IR to quantify microplastic pollution in 21 snow samples taken from ice floes in the Fram Strait, and 20 contained microplastic particles. The concentration of microplastics in these samples ranged from 0 to 14.4×10^3 particles L^{-1} which is a substantial number but lower than concentrations reported in European snow where concentrations ranged from 0.19×10^3 to 154×10^3 particles L^{-1}. A variety of polymer types were observed in the snow samples, with acrylates/polyurethane-based varnish, rubber, polyethylene and polyamide particles making up most of the concentration. Eighty percent of all detected microplastics were ≤ 25 μm, and 98% of all particles were <100 μm. The detection limit in this study was 11 μm, indicating large numbers of smaller particles could be missed, however techniques to identify particles this small are lacking. The composition of the particles varied considerably, with 19 different polymer types found in total. The number of polymers per sample was significantly higher in European (mean, 8.63 ± 0.80) compared with Arctic (mean, 5.14 ± 0.79) samples, with acrylates/polyurethane-based varnish occurring most frequently (17 samples), followed by nitrile rubber, polyethylene and polyamide. There was a high contrast in the polymer composition from European and Arctic samples, with much higher abundances of polyamide, acrylates/polyurethane-based varnish, rubber type 3, nitrile rubber, ethylene-vinyl-acetate and PE in European samples. By contrast, polystyrene, polyvinyl chloride (PVC), polycarbonate, polylactic acid and polyimide occurred exclusively in Arctic snow.

4. Fate

Understanding contaminant fate within the Arctic Ocean is extremely difficult. This dynamic ecosystem is heavily influenced by the changing

Table 1. Studies on microplastic accumulation in Arctic and sub-Arctic snow and ice.

Reference	Location	Material	Sampling method	No. of samples	Size range (mm)	Lab methods	Microplastic concentrations			Reporting unit
							Min	Max	Mean	
Obbard *et al.* (2014)	Chuckchi Shelf	Ice	Sea ice core	1	0.22–5	Band saw, visual sorting and FTIR	n/a	n/a	~50	Particles L^{-1}
	Siberian Sea	Ice	Sea ice core	1	0.22–5	Band saw, visual sorting and FTIR	n/a	n/a	234	Particles L^{-1}
	Beaufort Gyre	Ice	Sea ice core	1	0.22–5	Band saw, visual sorting and FTIR	n/a	n/a	38	Particles L^{-1}
	Laptev Sea	Ice	Sea ice core	1	0.22–5	Band saw, visual sorting and FTIR	n/a	n/a	~125	Particles L^{-1}
Peeken *et al.* (2018)	Fram Strait, North of Svalbard, Nansen Basin	Sea ice	Ice coring	5	0.011 –0.15	Vacuum filtration and μFTIR (FPA detector)	$(1.1 \pm 0.8) \times 10_6$	$(1.2 \pm 1.4) \times 10_7$	n/a	Particles M^{-3}

(*Continued*)

Table 1. (*Continued*)

Reference	Location	Material	Sampling method	No. of samples	Size range (mm)	Lab methods	Microplastic concentrations			Reporting unit
							Min	Max	Mean	
Bergmann *et al.* (2019)	Svalbard and Fram Strait	Snow	Surface sample with a spoon	14	0.011–0.475	Vacuum filtration (1 mL aliquot) and μFTIR	ND	14,000	1151.7 ± 3816.4	Particles L^{-1}
Geilfus *et al.* (2019) field study	Baltic Sea	Sea ice	Sea ice core	9	0.063–1.4	Vacuum filtration and nile red	8	41	n/a	Particles L^{-1}
Kanhai *et al.* (2020)	Arctic Central Basin (ACB)	Sea ice	Sea ice core	25	0.1–4.99	Vacuum filtration, visual sorting, and FTIR	2	17	6.46 ± 3.41	Particles L^{-1}

Source: Modified from Tirelli *et al.* (2020).

climate, which inevitably affects the transport and fate of contaminants such as plastic. At present, there have been few studies on the accumulation of plastics in the Arctic, especially when regarding sea ice and snow, which are vital features of the Arctic environment. Understanding the sources, transport and fate of plastic pollution in the Arctic Ocean is critical to determining the potential threats posed by such contaminants to marine organisms that inhabit this ecosystem (Kanhai *et al.*, 2020).

4.1. *Sea ice*

The presence of plastic debris and microplastics in sea ice and the ice-rafted snowpack has been investigated in a few studies to date (Table 1). Bergmann *et al.* (2016) sampled ice from both land-fast and drifting ice floes, to differentiate between local entrainment sources and long-range transport. Mean concentrations of 2×10^6 particles m^{-3} in pack ice and 6×10^5 particles m^{-3} in land-locked ice were detected. Eleven different polymer types were identified, with polyethylene being the most abundant. Calculation of drift trajectories by backtracking of the ice floes sampled indicates multiple source areas, which explains the differences in the microplastic composition. Bergmann *et al.* (2016) explain that the results from this study exceed concentrations from the North Pacific by several orders of magnitude, which can be explained partly by the process of ice formation, during which (organic) particles tend to concentrate by 1–2 orders of magnitude compared with ambient seawater. However, the magnitude of the difference suggests that Arctic sea ice is a temporal sink for microplastics. Increasing quantities of small plastic litter items on the seafloor nearby corroborate the notion that melting sea ice releases entrained plastic particles and that sea ice acts as a vector of transport both horizontally and vertically. Kanhai *et al.* (2020) looked at microplastic concentrations in ice cores and seawater beneath ice floes and found that microplastic abundance in surface waters underlying ice floes (ND–18 particles m^{-3}) were orders of magnitude lower than microplastic concentrations in sea ice cores (0.002–0.017 particles m^{-3}). The majority of the microplastics that were found in the ice cores were fibres (79%) and there was no consistent pattern between the vertical distribution of microplastics within the various sea ice cores. The origins of the ice were estimated by using backward drift trajectories, which showed that the ice possibly originated from the Siberian shelves, western and central Arctic.

To investigate the distribution and impacts of microplastics in sea ice, Geilfus *et al.* (2019) conducted microcosm and field-based studies. Both approaches confirmed that sea ice does concentrate microplastics within its structure, with particles retained within the ice matrix. The microcosm study revealed that the presence of microplastics did not affect the growth and thickness of the sea ice, however, higher concentrations of microplastics were found where ice salinity was at its highest suggesting that brine mobility within the ice may affect the distribution of plastic particles within the ice column. The concentration of microplastics observed in the field study (8–41 particles L^{-1}), which took place in the Baltic Sea, was one order of magnitude less than those used in the microcosm study. Although the concentrations observed by Geilfus *et al.* (2019) were relatively low, there are concerns that higher concentrations of microplastics present in sea ice may alter surface sea ice properties such as the albedo. A reduction in ice/snow albedo due to the presence of particulate material like microplastics may promote surface melt.

Peeken *et al.* (2018) sampled ice from the Fram Strait and Central Arctic to quantify the concentration and composition of microplastics. Their study found extremely high microplastic concentrations, with the highest particle concentration ($1.2 \pm 1.4 \times 10^7$ particles m^{-3}) being observed in an ice core taken in the older pack ice of Fram Strait. These high concentrations are likely to reflect the age of the ice and the large geographic region through which the ice has drifted during its life cycle. The results from this study showed highly variable concentrations of microplastics in sea ice cores, with the second-highest microplastics concentration found in the Greenland land-fast ice of the Fram Strait (core A; $4.1 (\pm 2.0) \times 10^6$ particles m^{-3}). Concentrations in cores collected north of Svalbard and in the Nansen Basin, respectively, varied between $2.9 (\pm 2.4) \times 10^6$ particles m^{-3} and $2.4 (\pm 1.0) \times 10^6$ particles m^{-3}. The lowest concentration was found in a core taken from the north of Svalbard, $1.1 (\pm 0.8) \times 10^6$ particles m^{-3}. The values recorded in this study were 2–3 orders of magnitude higher than in a previous study from the Central Arctic, where microplastic concentrations in sea ice were 38–234 particles m^{-3} (Obbard *et al.*, 2014). The difference in results can largely be explained by the different methodologies used. In the study by Obbard *et al.* (2014), filters were first inspected using light microscopy, and suspected microplastic particles were then analysed individually using FT-IR microscopy. In the Peeken *et al.* (2018) study, Imaging FT-IR was used to scan entire areas of the filters, excluding the human bias which inevitably comes with visual selection of particles. Imaging FT-IR is able to detect very small particles (down to 11 μm) which are most likely missed during visual inspection and are subsequently unreported. Peeken *et al.* (2018) were able to show that the majority of microplastic particles identified in the sea ice cores were smaller than 50 μm, and on average, 67% of the particles were within the currently smallest-detectable size class of 11 μm, which would not have been considered in the Obbard *et al.* (2014) study.

It is important to note that when comparing microplastic abundances in various studies, the methods used for sampling, quantifying and identifying must be taken into consideration, as these can result in very different outcomes. Several studies have reported higher concentrations of microplastics in the smaller size ranges (e.g., Bergmann *et al.*, 2017, 2019; Peeken *et al.*, 2018), however certain sampling techniques may exclude these smaller size ranges, especially when looking at microplastics <100 μm, which may be leading to an underestimation of the abundance of microplastics in this region. Peeken *et al.* (2018) reported the highest concentrations thus far in Arctic sea ice, and the majority of microplastics identified using FTIR were <50 μm, however, fibres were excluded from the analysis. In comparison, Kanhai *et al.* (2020) report a pre-dominance of fibrous microplastics (79%) in sea ice from the Arctic Ocean but did not include particles <100 μm due to limitations from the chosen analytical techniques (visual identification followed by FT-IR spectroscopy of potential microplastics). It is, therefore, apparent that if either fibres or particles <100 μm are excluded from results, microplastic concentrations in sea ice will be underestimated (Kanhai *et al.*, 2020). As well as the methods used for identifying and quantifying, the units used to report plastic concentrations vary significantly, which adds to the difficulty of comparing results. As well as this, many studies do not report the procedures used to mitigate contamination in samples, through field sampling and lab processing, highlighting a pressing need for standardised reporting of quality control in plastic and microplastic samples (Hallanger & Gabrielsen, 2018). It is vital that we create a more standardised regime that is widely accessible and generate clear data that can be shared amongst researchers to ensure a global picture of plastic pollution in our oceans and beyond is presented.

4.2. *Surface and sub-surface waters*

Oceanographic models have been used to show that once in the ocean, plastic debris can accumulate in large gyres (Cole *et al.*, 2011). The fate of plastic particles there is not fully understood, but it is hypothesised that once in the gyres, plastic particles will be exposed to several processes such as biofouling, which may cause them to fragment until they are small enough to be ingested by marine biota. Cózar *et al.* (2017) point out that the sub-tropical ocean gyres are recognised as accumulation zones for plastic debris, however because of their remote nature, the polar regions have been overlooked. In the study, the Arctic Ocean was highlighted as a "dead end" for floating plastic debris. During a circumpolar expedition, the Arctic Ocean was extensively sampled for floating plastic debris, and although little debris was found in Arctic waters, high concentrations of 6.5 g km^2 were found in areas of Greenland and the Barents Sea. The study explains that this sector of the Arctic Ocean plays a key role in the global Thermohaline Circulation (THC) through the formation of deep water by cooling, and as the THC actively advects warm surface water from low to high latitudes across the North Atlantic Ocean to the Arctic, it could collect buoyant plastic from highly populated latitudes, leading to accumulation in the Greenland and Barents seas. Here, the landmasses and polar ice caps create a "dead end" for the surface transport of floating debris.

However, plastics have not only been found in surface seawater. Lusher *et al.* (2015) found microplastics in both the surface (top 16 cm) and sub-surface (6 m depth) samples taken from Arctic polar waters. Microplastics were found in more than 90% of samples, both in the surface waters using a manta net and at 6 m depth using the vessel's underway seawater pump. Surface sample concentrations ranged from ND to 1.31 particles m^{-3} and averaged 0.34 (±0.31 SD) particles m^{-3}. Microplastic abundance in sub-surface samples ranged between ND and 11.5 particles m^{-3} and averaged 2.68 (±2.95 SD) particles m^{-3}. Interestingly, the study found that although there were clear changes in polar fronts throughout the survey, microplastic abundances did not change with salinity, however there was a noticeable, although not statistically significant, difference with fewer particles in colder, less saline waters. Particle composition comprised of 95% fibres, which they suggest were coming from the breakdown of larger plastic items used for shipping activity or fishing equipment, recreation and offshore industries, and have likely arrived in the Arctic after long periods at sea, possibly transported over large distances. These microplastics may also be a result of the inputs from sewage and wastewater from more local sources, e.g., washing of textiles. There was a high proportion of fibrous microplastics (>90%) found in samples from Kanhai *et al.* (2018) study on the abundance, distribution and composition of microplastics in sub-surface waters of the ACB. Two independent sampling methods were used: underway sampling using the bow water system of ice-breaker Oden (single depth: 8.5 m), and conductivity, temperature, depth (CTD) profiler using a rosette sampler (multiple depths: 8–4369 m). Shallow, surface water samples contained microplastic accumulations ranging from non-detect to 7.5 particles m^{-3}, with CTD samples ranging from ND to 375 particles m^{-3}. The results show that microplastics are present throughout the water column in the ACB, and that in natural conditions microplastics are being vertically transported out of surface waters. The highest levels of microplastics were reported near the periphery of the ACB, in waters north of Svalbard. The study suggests this could be due to the lack of permanent sea ice cover in this region, allowing incoming Atlantic water to have a greater influence on near-surface waters.

4.3. *Deep-sea sediments*

Deep-sea sediments may be a permanent sink for microplastic particles that have deposited from surface waters following long-range transport via surface ocean currents or deposition from the atmosphere (Cózar *et al.*, 2017; Kanhai *et al.*, 2020). Bergmann *et al.* (2017) found high quantities of microplastics in Arctic deep-sea sediments taken from the Long-Term Ecological Research (LTER) HAUSGARTEN observatory in the Fram Strait at 79°N, west of Svalbard (Norway), which currently comprises 21 sampling stations along a latitudinal and a bathymetric gradient between 250 and 5,500 m water depth. Sediment samples were taken between 2340 and 5,570 m depth, where an overall mean number of 4,356 particles kg^{-1} sediment was found. The northernmost stations of this study, located within or close to the marginal ice zone, were found to have the highest microplastic abundances. A total of 18 different polymers were identified from all sediment samples, with almost 80% of particles being ≤20 μm diameter in size. The study found no significant correlation between depth and microplastic abundance, but a positive correlation was found between microplastic abundance and chlorophyll *a* content, suggesting vertical export via incorporation in sinking algal aggregates. The study suggests that the overlying sea ice affects the quantity and release of microplastics present in the water column and hence their deposition to deeper waters and presence in bottom sediments. The study also corroborates previously discussed conclusions that Arctic sea ice may act as a transport vehicle for microplastics, as after the ice breaks up in the Central

Arctic in spring, it is transported to the south as ice floes with the Transpolar Drift. On route, it encounters warmer Atlantic surface waters and continues to melt, possibly releasing entrained microplastics.

Kanhai *et al.* (2019) investigated deep-sea sediments of the ACB as a potential sink for microplastics. Gravity and piston corers were used to retrieve sediments from depths of between 855 and 4,353 m at 11 sites in the ACB. Surficial sediments from the various cores were taken and potential microplastics were isolated and analysed by Fourier Transform Infrared (FT-IR) spectroscopy. Of the surficial samples, 7 of the 11 samples contained synthetic polymers which included polyester, polystyrene, polyacrylonitrile, polypropylene, polyvinyl chloride and polyamide. The study suggests that the presence of low-density polymers such as polypropylene and polystyrene indicates that there are mechanisms operating within the ACB that are affecting the density of microplastics, driving the vertical transport of these particles through the water column. Once plastics reach deep-sea sediments, they are unlikely to become resuspended and redistributed, hence these areas being reported as significant sinks for microplastic pollution. There will likely be interactions between sinking particles and marine organisms inhabiting these areas, such as ingestion by benthic organisms. These organisms are highly sensitive to environmental change, relying on consistent and specific environmental conditions to survive, and so the addition of high concentrations of microplastics will inevitably have a detrimental effect on their health (Horton & Barnes, 2020).

5. Summary

It is clear that the Arctic can no longer be considered as a pristine environment. Plastic debris including microplastics are present in the remote marine environment of the Arctic, although detrimental effects on marine organisms and wildlife, particularly through exposure to microplastics, are not understood. The abundance of plastic pollution travelling to the Arctic from long-range sources is difficult to measure against more localised sources, but the connectivity of Arctic marine areas with the rest of the world's oceans undoubtedly provides input from a range of distant sources (PAME, 2019). Macroplastics and microplastics arising through long-range sources can be transported via pathways such as ocean and atmospheric currents and can be transported within the Arctic Circle through sea ice migration and biotransport. The fate of microplastics in the Arctic marine environment is still unclear, however their presence in seawater, sea ice, sediments and snow is unequivocal.

Unlike temperate regions, the cryosphere provides an additional environmental compartment that affects plastic distribution and fate in the Arctic. As highlighted in this chapter, sea ice has been recognised as a significant sink and potential transport vector for microplastic pollution, suggesting that high concentrations of microplastics are being transported across the Arctic and deposited in areas where we would not expect to find high abundances of plastic particles. Higher concentrations of microplastics in deep-sea sediments in regions with high ice cover (Bergmann *et al.*, 2017) support the view that ice is a significant source for microplastic pollution and may suggest these high concentrations in ice are exacerbating transfer to deep waters. However, at present, there is disparity in the number of plastic particles present in ice or seawater depending on the study in question and the analytical method involved. The reporting of ultra-small particles (e.g., $<10\ \mu$m in diameter) can result in much greater particle number concentrations compared to studies that do not, and hence an increase in the number of observational studies across the Arctic for different environmental media is called for.

Ice provides habitats for sympagic (ice-dwelling) organisms, and it is currently unclear whether these organisms are being exposed to high concentrations of microplastics present in the ice. If current trends in sea ice melt continue, then over the next decade, ~2.04 trillion m^3 of ice will melt (Comiso, 2012). Taking the lowest observed particle concentration from Obbard *et al.* (2014) (i.e., 38 particles per m^3) would result in the remobilisation of >1 trillion microplastic particles in Arctic Ocean surface waters. The entrapment and subsequent seasonal release of plastic debris and microplastic particles from sea ice have not been studied in any detail, and the impact of this "cycling" of plastic contamination on sympagic and pelagic organisms is unclear. One of the impacts from warming Arctic summers and the subsequent reduction in sea ice cover may be an increase in plastics debris in the Arctic, leading to an increase in plastic ingestion by marine biota (Mallory, 2008).

References

AMAP. (2019). *AMAP Climate Change Update 2019: An Update to Key Findings of Snow, Water, Ice and Permafrost in the Arctic (SWIPA) 2017*, 12. Oslo, Norway: Arctic Monitoring and Assessment Programme (AMAP).

AMAP. (2021). *Arctic Climate Change Update 2021: Key Trends and Impacts. Summary for Policy-Makers*, 16. Tromsø, Norway: Arctic Monitoring and Assessment Programme (AMAP).

Barnes, D. K. (2002). Invasions by marine life on plastic debris. *Nature, 416*(6883), 808–809.

Bergmann, M., Lutz, B., Tekman, M., & Gutow, L. (2017). Citizen scientists reveal: Marine litter pollutes Arctic beaches and affects wildlife. *Marine Pollution Bulletin, 125*(1–2), 535–540.

Bergmann, M., Mützel, S., Primpke, S., Tekman, M., Trachsel, J., & Gerdts, G. (2019). White and wonderful? Microplastics prevail in snow from the Alps to the Arctic. *Science Advances*, *5*(8), eaax1157.

Bergmann, M., Peeken, I., Beyer, B., Krumpen, T., Primpke, S., Tekman, M.B., Gerdts, G., 2017. Vast quantities of microplastics in arctic sea ice — A prime temporary sink for plastic litter and a medium of transport. In: Baztan, J., Jorgensen, B., Pahl, S., Thompson, R., Vanderlinden, J. (Eds.), *MICRO2016: Fate and Impact of Microplastics in Marine Ecosystems*. Elsevier, Amsterdam pp. 75–76.

Bergmann, M., Wirzberger, V., Krumpen, T., Lorenz, C., Primpke, S., Tekman, M., & Gerdts, G. (2017). High quantities of microplastic in Arctic deep-sea sediments from the HAUSGARTEN observatory. *Environmental Science & Technology, 51*(19), 11000–11010.

Browne, M.A., (2015). Sources and pathways of microplastics to habitats. In: Bergmann, L., Gutow, L., Klages, M. (Eds.), *Marine Anthropogenic Litter*. SpringerOpen, New York, pp. 229–244.

Cole, M., Lindeque, P., Halsband, C., & Galloway, T. S. (2011). Microplastics as contaminants in the marine environment: A review. *Marine Pollution Bulletin, 62*(12), 2588–2597.

Comiso, J. C. (2012). Large decadal decline of the Arctic multiyear ice cover. *Journal of Climate*, *25*(4), 1176–1193.

Cózar, A., Martí, E., Duarte, C., García-de-Lomas, J., van Sebille, E., Ballatore, T., Eguíluz, V., González-Gordillo, J., Pedrotti, M., Echevarría, F., Troublè, R., & Irigoien, X. (2017). The Arctic Ocean as a dead end for floating plastics in the North Atlantic branch of the thermohaline circulation. *Science Advances*, *3*(4), e1600582.

Ford, H. V., Jones, N. H., Davies, A. J., Godley, B. J., Jambeck, J. R., Napper, I. E., Suckling, C. C., Williams, G. J., Woodall, L., & Koldewey, H. J. (2021). The fundamental links between climate change and marine plastic pollution. *Science of the Total Environment*, *806*, 150392.

Garnett, J., Halsall, C., Thomas, M., Crabeck, O., France, J., Joerss, H., Ebinghaus, R., Kaiser, J., Leeson, A., & Wynn, P. M. (2021). Investigating the uptake and fate of poly-and perfluoroalkylated substances (PFAS) in sea ice using an experimental sea ice chamber. *Environmental Science & Technology*, *55*(14), 9601–9608.

Geilfus, N., Munson, K., Sousa, J., Germanov, Y., Bhugaloo, S., Babb, D., & Wang, F. (2019). Distribution and impacts of microplastic incorporation within sea ice. *Marine Pollution Bulletin, 145*, 463–473.

Granberg, M. E., Ask, A., & Gabrielsen, G. W. (2017). *Local Contamination in Svalbard. Overview and Suggestions for Remediation Actions*. Norsk Polarinstitutt. [S. l.].

Hallanger, I. G. & Gabrielsen, G. W. (2018). *Plastic in the European Arctic*. Tromsø: Norsk Polarinstitutt.

Heleniak, T. & Bogoyavlensky, D. (2015). Arctic populations and migration. In *Arctic Human Development Report: Regional Processes and Global Linkages*, Nordic Council of Ministers, Denmark, pp. 53–104.

Holmes, R.M., Coe, M.T., Fiske, G.J., Gurtovaya, T., McClelland, J.W., Shiklomanov, A.I., Spencer, R.G.M., Tank, S.E. and Zhulidov, A.V. (2012). Climate Change Impacts on the Hydrology and Biogeochemistry of Arctic Rivers. In Goldman, C.R. Kumagai, M., & Robarts, R. D. (Eds.), *Climatic Change and Global Warming of Inland Waters*, pp. 3–26. Chichester: Wiley-Blackwell. https://doi.org/10.1002/9781118470596.ch1.

Horton, A. A. & Barnes, D. K. (2020). Microplastic pollution in a rapidly changing world: Implications for remote and vulnerable marine ecosystems. *Science of the Total Environment, 738*, 140349.

Jansen, E., Christensen, J., Dokken, T., Nisancioglu, K., Vinther, B., Capron, E., Guo, C., Jensen, M., Langen, P., Pedersen, R., Yang, S., Bentsen, M., Kjær, H., Sadatzki, H., Sessford, E., & Stendel, M. (2020). Past perspectives on the present era of abrupt Arctic climate change. *Nature Climate Change, 10*(8), 714–721.

Kanhai, L., Gardfeldt, K., Krumpen, T., Thompson, R., & O'Connor, I. (2020). Microplastics in sea ice and seawater beneath ice floes from the Arctic Ocean. *Scientific Reports, 10*(1), 1–11.

Kanhai, L., Gårdfeldt, K., Lyashevska, O., Hassellöv, M., Thompson, R., & O'Connor, I. (2018). Microplastics in sub-surface waters of the Arctic Central Basin. *Marine Pollution Bulletin, 130*, 8–18.

Kanhai, L., Johansson, C., Frias, J., Gardfeldt, K., Thompson, R., & O'Connor, I. (2019). Deep sea sediments of the Arctic Central Basin: A potential sink for microplastics. *Deep Sea Research Part I: Oceanographic Research Papers, 145*, 137–142.

Kirkelund, G. M., Diez, L., Scheutz, C., & Eisted, R. (2017). Evaluating potentials for waste sorting in the Arctic: Waste separation studies from Greenland. *Waste Management & Research, 29*, 1064–1070.

Kühn, S., Bravo Rebolledo, E. L., & van Franeker, J. A. (2015). Deleterious effects of litter on marine life. In Bergmann, M., Gutow, L., & Klages, M. (Eds.), *Marine Anthropogenic Litter*. Cham: Springer. https://doi.org/10.1007/978-3-319-16510-3_4.

Kühn, S., Schaafsma, F., van Werven, B., Flores, H., Bergmann, M., Egelkraut-Holtus, M., Tekman, M., & van Franeker, J. (2018). Plastic ingestion by juvenile polar cod (*Boreogadus saida*) in the Arctic Ocean. *Polar Biology*, *41*(6), 1269–1278.

Loeng, H., Brander, K., Carmack, E., Denisenko, S., Drinkwater, K., Hansen, B. *et al.* (2005). Chapter 8, Marine systems. In *Arctic Climate Impact Assessment*, pp. 451–538. Cambridge: Cambridge University Press.

Lusher, A., Tirelli, V., O'Connor, I., & Officer, R. (2015). Microplastics in Arctic polar waters: The first reported values of particles in surface and sub-surface samples. *Scientific Reports*, *5*(1), 1–9.

Magnusson, K. Jörundsdóttir, H., Norén, F., Lloyd, H., Talvitie, J., Setälä, O. (2016). *Microlitter in Sewage Treatment Systems: A Nordic Perspective on Wastewater Treatment Plants as Pathways for Microscopic Anthropogenic Particles to Marine Systems*. Denmark: Nordic Council of Ministers, pp. 1–56.

Mallory, M. (2008). Marine plastic debris in northern fulmars from the Canadian high Arctic. *Marine Pollution Bulletin*, *56*(8), 1501–1504.

Meredith, M., Sommerkorn, M., Cassotta, S., Derksen, C., Ekaykin, A., Hollowed, A., Kofinas, G., Mackintosh, A., Melbourne-Thomas, J., Muelbert, M. M. C., Ottersen, G., Pritchard, H., & Schuur, E. A. G. (2019). Polar regions. In Pörtner, H.-O., Roberts, D. C., Masson-Delmotte, V., Zhai, P., Tignor, M., Poloczanska, E., Mintenbeck, K., Alegría, A., Nicolai, M., Okem, A., Petzold, J., Rama, B., & Weyer, N. M. (Eds.), *IPCC Special Report on the Ocean and Cryosphere in a Changing Climate*. pp. 203–320. Cambridge, New York: Cambridge University Press. DOI: 10.1017/9781009157964.005.

Merrell Jr, T. R. (1980). Accumulation of plastic litter on beaches of Amchitka Island, Alaska. *Marine Environmental Research*, *3*(3), 171–184.

Morgana, S., Ghigliotti, L., Estévez-Calvar, N., Stifanese, R., Wieckzorek, A., Doyle, T., Christiansen, J. S., Faimali, M., & Garaventa, F. (2018). Microplastics in the Arctic: A case study with sub-surface water and fish samples off Northeast Greenland. *Environmental Pollution*, *242*, 1078–1086.

Napper, I. E. & Thompson, R. C. (2016). Release of synthetic microplastic plastic fibres from domestic washing machines: Effects of fabric type and washing conditions. *Marine Pollution Bulletin*, *112*(1–2), 39–45.

National Snow and Ice Data Center (NSIDC). (2022). Retrieved from https://nsidc.org/. Accessed on 21/07/2022.

Obbard, R. W. (2018). Microplastics in polar regions: The role of long-range transport. *Current Opinion in Environmental Science & Health*, *1*, 24–29.

Obbard, R. W., Sadri, S., Wong, Y. Q., Khitun, A. A., Baker, I., & Thompson, R. C. (2014). Global warming releases microplastic legacy frozen in Arctic Sea ice. *Earth's Future*, *2*(6), 315–320.

Onink, V., Wichmann, D., Delandmeter, P., & Sebille, E. (2019). The role of Ekman currents, geostrophy, and Stokes drift in the accumulation of floating microplastic. *Journal of Geophysical Research: Oceans, 124*(3), 1474–1490.

PAME. (2019). *Desktop Study on Marine Litter including Microplastics in the Arctic.* Protection of the Arctic Marine Environment, 11th Arctic Council Ministerial Meeting, Rovaniemi, Finland (May 2019), pp. 1–114.

Peeken, I., Primpke, S., Beyer, B., Gütermann, J., Katlein, C., Krumpen, T., Bergmann, M., Hehemann, L., & Gerdts, G. (2018). Arctic sea ice is an important temporal sink and means of transport for microplastic. *Nature Communications*, *9*(1), 1–12.

Polasek, L., Bering, J., Kim, H., Neitlich, P., Pister, B., Terwilliger, M., Nicolato, K., Turner, C., & Jones, T. (2017). Marine debris in five national parks in Alaska. *Marine Pollution Bulletin, 117*(1–2), 371–379.

Provencher, J. F., Bond, A. L., & Mallory, M. L. (2015). Marine birds and plastic debris in Canada: A national synthesis and a way forward. *Environmental Reviews*, *23*(1), 1–13.

Provencher, J. F., Gaston, A., Mallory, M., O'hara, P., & Gilchrist, H. (2010). Ingested plastic in a diving seabird, the thick-billed murre (*Uria lomvia*), in the eastern Canadian Arctic. *Marine Pollution Bulletin, 60*(9), 1406–1411.

Pućko, M., Stern, G., Macdonald, R., Jantunen, L., Bidleman, T., Wong, F., Barber, D., & Rysgaard, S. (2015). The delivery of organic contaminants to the Arctic food web: Why sea ice matters. *Science of the Total Environment, 506–507*, 444–452.

Röhrs, J. & Christensen, K. H. (2015). Drift in the uppermost part of the ocean. *Geophysical Research Letters*, *42*(23), 10–349.

Sheavly, S. B. & Register, K. M. (2007). Marine debris & plastics: Environmental concerns, sources, impacts and solutions. *Journal of Polymers and the Environment*, *15*(4), 301–305.

Sundet, J. H. (2014). The snow crab (*Chionoecetes opilio*) in the Barents Sea (the lecture held at the workshop on red king- and snow crab, Tromsø, 2014). In *Workshop on King- and Snow Crabs in the Barents Sea, Tromsø.*

Sundet, J. H., Herzke, D., & Jenssen, M. (2016). Occurrence and sources of microplastics in sediment, and consequences for benthic fish and invertebrates on Svalbard. Sluttrapport No. Norway: 13.

Tekman, M. B., Krumpen, T., & Bergmann, M. (2017). Marine litter on deep Arctic seafloor continues to increase and spreads to the North at the HAUSGARTEN observatory. *Deep Sea Research Part I: Oceanographic Research Papers*, *120*, 88–99.

Threlfall, W. (1968). The food of three species of gulls in Newfoundland. *Canadian Field-Naturalist*, *82*, 176–180.

Tirelli, V., Suaria, G., & Lusher, A. L. (2022). Microplastics in polar samples. In *Handbook of Microplastics in the Environment*. pp. 281–322. Cham: Springer International Publishing.

Trevail, A. M., Kühn, S., & Gabrielsen, G. W. (2015). The state of marine microplastic pollution in the Arctic. *Norwegian Polar Institute, Tromso, 23*, 1–24.

United Nations Environment Programme (UNEP). (2021). Protection of Arctic Marine Environment. Retrieved from https://www.unep.org/explore-topics/oceans-seas/what-we-do/working-regional-seas/regional-seas-programmes/arctic-region. Accessed on 15/12/2021.

Yakushev, E., Gebruk, A., Osadchiev, A., Pakhomova, S., Lusher, A., Berezina, A., van Bavel, B., Vorozheikina, E., Chernykh, D., Kolbasova, G., Razgon, I., & Semiletov, I. (2021). Microplastics distribution in the Eurasian Arctic is affected by Atlantic waters and Siberian rivers. *Nature Communications Earth & Environment*, *2*(1), 1–10.

Zarfl, C. & Matthies, M. (2010). Are marine plastic particles transport vectors for organic pollutants to the Arctic? *Marine Pollution Bulletin*, *60*(10), 1810–1814.

https://doi.org/10.1142/9789811259111_0010

Chapter 9

Plastic Pollution in Antarctica and the Southern Ocean

Jack W. Buckingham[*,†,‖], Rebecca H. Peel[†,‡], Emily Rowlands[†,§] and Daniel R. Wilson[†,¶]

[*]*Energy and Environment Institute, University of Hull, Hull HU6 7RX, UK*

[†]*British Antarctic Survey, High Cross, Madingley Rd, Cambridge CB3 0ET, UK*

[‡]*Organic Geochemistry Unit, School of Chemistry, University of Bristol, Bristol BS8 1TS, UK*

[§]*College of Life and Environmental Science, University of Exeter, Streatham Campus, Exeter EX4 4PY, UK*

[¶]*Centre for Geography and Environmental Science, University of Exeter, Penryn Campus, Cornwall TR10 9FE, UK*

[‖]*j.buckingham-2018@hull.ac.uk*

1. Introduction

Antarctica and the Southern Ocean (ASO) together are a truly unique environment. Not just in terms of physical and geographical attributes, but also in governance, exploitation, and our historical and cultural relationships with it. The anthropogenic footprint on the Antarctic continent is unlike anywhere else on Earth. The nature of plastic pollution here is, therefore, worthy of independent consideration.

In this chapter, we provide background on the southern polar region, its geography, history and governance, and how they pertain to plastic pollution and waste regulation. Using a source to sink approach, we report what is currently known regarding the sources, transport vectors and sinks of plastic (macroplastic, microplastic and nanoplastic) in the region, and examine why plastic pollution has distinct implications for the marine system of ASO and its resident biota.

1.1. *Antarctica and the Southern Ocean region*

Many of the physical attributes of the Antarctic are well known: high latitudes, varied seasonal light levels, cold air and sea temperatures, and

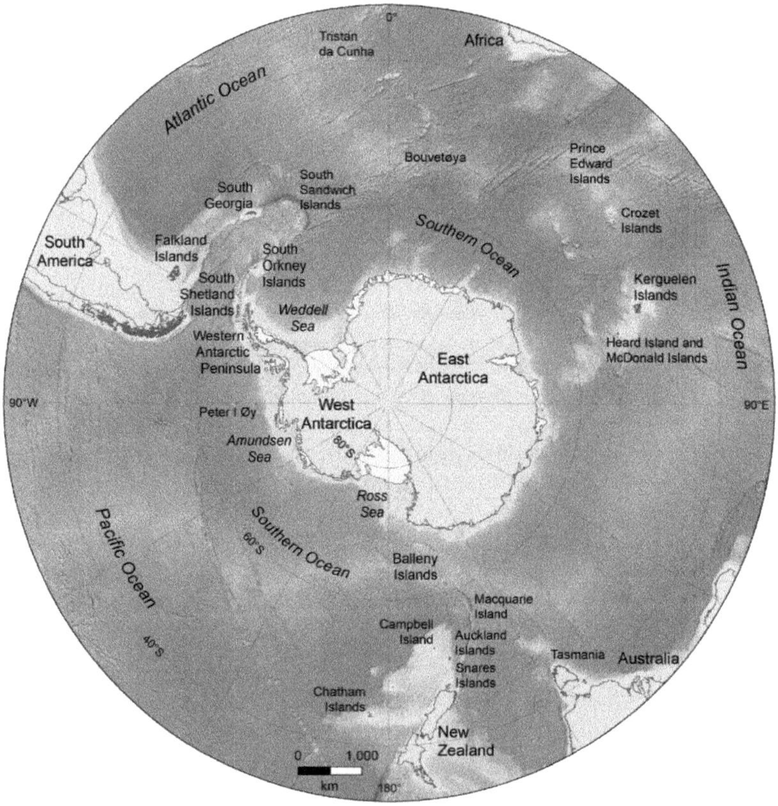

(a)

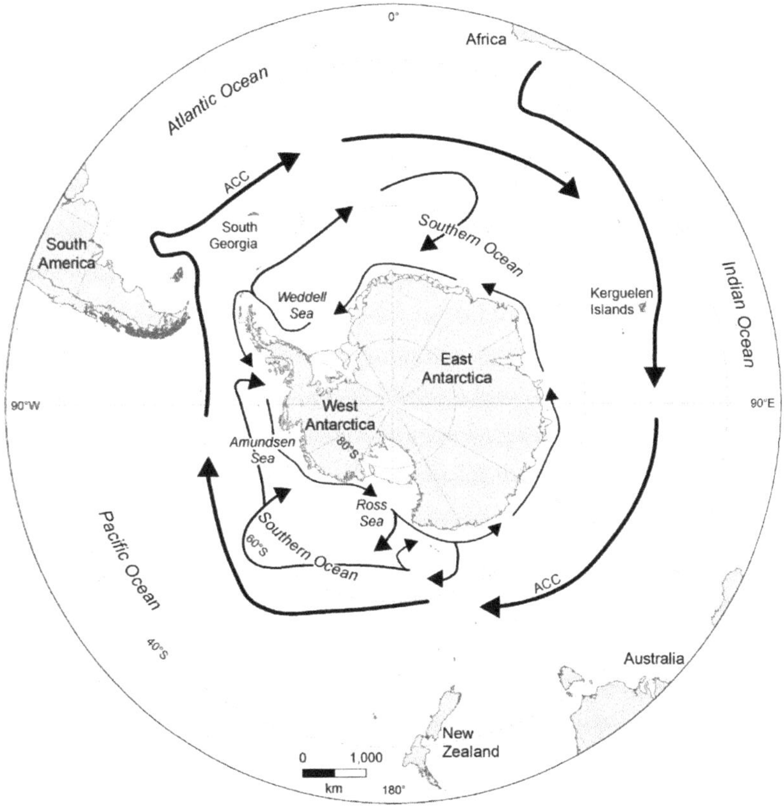

(b)

Figure 1. (a) The frozen continent of Antarctica surrounded by oceanic waters. (b) Major oceanic currents in ASO.

Source: Adapted from Armitage *et al.* (2018). Maps created with Esri ArcGIS version 10.6 courtesy of Laura Gerrish (British Antarctic Survey).

of course, ice. In these aspects, ASO bears similarities to the Arctic, yet the two poles are also immeasurably different. While the Arctic is an ocean surrounded by land, Antarctica is a frozen landmass surrounded by the Southern Ocean, which represents 5.4% of the world's oceans (Figure 1(a); Dayton *et al.*, 1994). Currents in the Southern Ocean are dominated by the Antarctic Circumpolar Current (ACC), an unbroken, 20,000 km long, eastwards-flowing mass of water (Figure 1(b)). It is the largest and fastest current on the planet, separating the Antarctic continent from any of its northern neighbours. A severe temperature discontinuity, the Polar Frontal Zone (PFZ), marks the boundary between the Southern Ocean and its adjacent oceans (Peck, 2005; Aronson *et al*., 2007). The temperature of the Southern Ocean is relatively stable, with total annual sea surface temperature variations of less than 1.5°C, much less than the variation experienced in the Arctic (NOAA, 2021), although sea ice in ASO varies extensively between seasons. Disconnected as it is from any other southern landmass, Antarctica's isolation has given rise to distinctive biota. Distance and oceanographic barriers inhibit migration as an evolutionary tactic (Peck *et al*., 2005) and, therefore, many Antarctic species belong to ancient genetic lineages (Dayton *et al*., 1994; Convey *et al*., 2009), making them susceptible to rapid contemporary change (Rowlands *et al*., 2021a). There are also high levels of endemism amongst microbial taxa (Vyverman *et al*., 2010), invertebrates (Pugh & Convey, 2008) and macrofauna (Eastman, 1997). This same isolation has prevented the settlement of an indigenous human population, unlike the Arctic which has been populated for thousands of years (McBride *et al*., 2014).

Figure 2. Demolition of the "plastic palace" at Signy Research Station, South Orkney Islands, Antarctica. Date: 2001/2002 summer season. Credit: John Shears (British Antarctic Survey). Image Ref: BAS 10006456.

Despite the remoteness of ASO, plastic has been present in the region for almost as long as it has been utilised by humans Figure 2. Marketed originally as a wonder-material, durable and lightweight, with unlimited applications, unsurprisingly it was used in early polar expeditions, for example, as parts of sledges during the Norwegian-British-Swedish expeditions of 1949 (Reece, 1953). However, it is difficult to discern from the scientific literature when plastic was first observed as a marine pollutant in the region.

1.2. *History and governance*

The Antarctic Treaty was signed in 1959, neither recognising nor annulling the previous, overlapping sovereignty claims made by any nation (Secretariat of the Antarctic Treaty (ATS), 1961), but instead setting the continent aside for peaceful, non-nuclear, scientific purposes only, with an eye to engendering international scientific cooperation (ATS, 1961). The only consideration of waste disposal at the time was the prohibition of nuclear waste disposal (ATS, 1961). In 1979, observations from the R.R.S Discovery cruises were published noting polyethylene (PE) and polypropylene (PP) pellets (3–5 mm, i.e., of a microplastic size by today's definition) in the South Atlantic, which is perhaps the first record of marine plastic pollution in ASO (Morris, 1980). Bonner and McCann (1982) also reported nylon collars on fur seals (*Arctocephalus gazella*), due to entanglement, at Bird Island, South Georgia, prior to 1982 (Figure 3). By this time, the unbridled optimism surrounding plastic use was becoming tarnished. The International Convention for the Prevention of Pollution from Ships (colloquially "MARPOL") brought in a ban on plastic disposed from shipping in 1988 (MARPOL, 1988) and the Commission for the Conservation of Antarctic Marine Living Resources (CCAMLR), set up in 1982 to preserve Antarctic marine life and environmental integrity, launched a Marine Debris Program to monitor marine litter levels in 1989 (CCAMLR, 2017a).

The issue of plastic pollution and waste disposal was addressed by the Antarctic Treaty with the enforcement of the Protocol on Environmental Protection (signed in 1991, enforced in 1998) Annex 3 of which concerns waste disposal and waste management (ATS, 1998). Polyvinyl chloride (PVC), polyurethane (PUR) foam, polystyrene (PS) foam and "all other plastic wastes, except low-density polyethylene containers (such as bags for storing waste)," were listed as waste

Figure 3. A mature male Antarctic fur seal (*Arctocephalus gazella*) collared by a plastic strap. The skin of the animal's neck was cut through for about three-quarters of its circumference. Image taken in 1982.

Source: Bonner and McCann (1982). UKRI British Antarctic Survey Copyright.

products which "shall be removed from the Antarctic Treaty area by the generator of such wastes" (ATS, 1998). Open burning of waste was phased out by the end of 1999, but sewage and domestic waste, large quantities of which are subject to maceration, could still be discharged directly into the sea if conditions allow for rapid dilution and dispersal. (ATS, 1998). At this time, microplastic had been observed in the marine environment elsewhere in the world but was not yet truly described or recognised as a pollutant with potentially adverse implications (Thompson *et al.*, 2004; Ryan *et al.*, 2015).

Today, there are 52 national parties to the Antarctic Treaty which must establish their own waste management plan and share them with other treaty parties (ATS, 1998). How plastic waste is dealt with, therefore, varies between stations and depends on a range of logistical factors, such as capacity for storage, waste disposal streams in home nations or gateway nations (i.e., Chile, Argentina, South Africa, UK Falkland Islands, Australia and New Zealand), the type of plastic material and its function in the field.

In 2019, the Scientific Committee on Antarctic Research (SCAR), an international organisation that coordinates scientific research in Antarctica and provides advice to the Antarctic Treaty Consultative Meetings, set up a new Action Group "Plastic in Polar Environments". The aims of this Action Group are to provide networking opportunities for plastic polar researchers, collate data, estimate current levels of plastic pollution in the region and its impacts, assess technology for plastic sampling and standardise procedures, and propose new measures to reduce or limit plastic pollution in polar environments (SCAR, 2019).

1.3. *The Antarctic Circumpolar Current*

Until recently, the ACC was thought to prevent the transfer of positively buoyant material, such as plastic pollution, to lower latitudes (Clarke *et al.*, 2005). This idea was reinforced by the geographic and genetic isolation of marine biota in the region (Clarke & Peck, 1991; Zane & Paternello, 2000) and the limited amount of plastic pollution that some studies, such as Waller *et al.* (2017), Kuklinski *et al.* (2019), Garcia-Garin *et al.* (2020) and Lacerda *et al.* (2019), report. However, this theory has now been called into question for numerous reasons:

- The discovery of non-native macrofauna, such as southern bull kelp *(Durvillaea antarctica*) endemic to Chile, on the Antarctic continent (Clarke *et al.*, 2005; Fraser *et al.*, 2018).
- Numerical modelling of the current that has revealed floating particles released north of the ACC can reach the inner Southern Ocean within three years (Fraser *et al.*, 2018).
- Estimations of microplastic output, based on an individual's microplastic production and the scale of the human footprint in ASO, proved five orders of magnitude lower than what is being observed in the field (Waller *et al.*, 2017).

The message is clear: ASO is neither completely isolated nor as pristine as once thought, posing significant questions regarding the sources and transport pathways of the plastic pollution observed there.

1.4. *Logistical and methodological challenges*

It is challenging and expensive to reach ASO, even before factoring in sampling in the harsh conditions. As such data on plastic pollution within the region are lacking compared to other oceans. Sampling is often carried out as part of multi-disciplinary research expeditions or opportunistically, limiting sample sizes and the geographical spread of observations. Research is also naturally clustered around regions, such as the Western Antarctic Peninsula (WAP), which have higher research ship traffic. Consequently, large areas of ASO remain unsampled for plastic pollution. This is in addition to the problems within the broader plastic pollution research field, such as a lack of standardisation when it comes to microplastic collection, extraction and identification methodologies.

Despite the difficulties of plastic research in ASO, there has been an upward swing in the past decade as plastic pollution weighs increasingly heavily on our collective psyche and the topic becomes more attractive for funding. However, due to the unique nature of policy interaction between and within signatory nations to the Antarctic Treaty, establishing best practice and coordinating any international or cross-disciplinary monitoring programme is complicated; and as a result, most action taken is isolated or disjointed.

2. Sources of Plastic to ASO

Primary plastic pollutants are those which enter the environment in their original form, be that a water bottle or a microbead from a household product. However, a large proportion of microplastic and nanoplastic originate from fragmentation in the environment. These secondary particles can form from physical abrasion, UV degradation or biological breakdown of larger plastic items. For example, Antarctic krill (*Euphausia superba*) have been observed to reduce microplastic to nanoplastic via digestive fragmentation within a laboratory environment, demonstrating a plausible source of nanoplastic in the water column of ASO (Dawson *et al.*, 2018), highlighting just one of many fragmentation routes.

The sources of plastic observed in the marine environment are notoriously difficult to determine. Whilst it is often not possible to work out what sort of item a fragment has come from, let alone where that parent item originated, transport models can be used to determine potential source regions for plastic debris. An item's function, the item's polymer type, concentration in the environment, and distribution on local and widespread scales can all be used to indicate the plastic debris' source.

This section examines the plastic footprint of ASO and examines the relative contributions of the three main industries operating in the region: fishing, tourism and scientific research.

2.1. *Fishing*

The Southern Ocean is home to only a handful of species which are commercially extracted, namely, krill, toothfish and icefish, but these are available only from the Southern Ocean making fishing there highly lucrative and desirable. Due to exploitation in the 1960s, fish stocks and populations of numerous species have declined (Hemmings, 2015), thus leading to the establishment of CCAMLR in 1982 (CCAMLR, 1979; Dodds, 2010). Today, CCAMLR regulates fisheries for Antarctic krill (*Euphausia superba*), mackerel icefish (*Champsocephalus gunnari*), Patagonian toothfish (*Dissostichus eleginoides*) and Antarctic toothfish (*Dissostichus mawsoni*), with limitations on what, where and how fishing can take place (CCAMLR, 2020). This includes catch limits and general protection zones, in which fishing is completely banned. Consequently, the number of vessels fishing around Antarctica is constrained, but annual numbers remain above 40 (CCAMLR, 2021a). Despite regulations, illegal, unreported and unregulated (IUU) fishing vessels are known to operate in the Southern Ocean, although steps are being taken to tackle this (CCAMLR, 2017b, 2021b).

Notwithstanding various waste management practices, some of the most notable macroplastic pollution in ASO originates from fishing. Since the late 19th/early 20th century, synthetic materials have been used for fishing. From the development of plastic lures (Gregg, 1877) to the first nylon fishing lines (Cuclo, 1964), plastic is now commonplace in the fishing industry. Items found in ASO include fishing nets, buoys, floats and rope, going as far back as 1989 (Convey *et al.*, 2002; Barnes *et al.*, 2010; Waluda *et al.*, 2020). Substantial contributions from fishing gear have also been observed in earlier (Eriksson *et al.*, 2013) and later (Suaria *et al.*, 2020a) studies, in other parts of Antarctica. Fishing gear is also thought to contribute to microplastic and nanoplastic pollution, through environmental degradation (Eriksson & Burton, 2003). This has been corroborated by several studies finding large contributions from polyethylene and polypropylene in the microplastic profiles observed — the two primary components in the manufacture of fishing gear (Cincinelli *et al.*, 2017; Jones-Williams *et al.*, 2020).

2.2. *Tourism*

Tourism is a growing industry in Antarctica, with more than 70,000 visitors in the 2019–2020 season (Dodds, 2010; IAATO, 2020a). This represents an increase of more than 100% since 2010, according to the International Association of Antarctica Tour Operators (IAATO, 2019a). The number of ships registered with IAATO has not increased as drastically, with around 50 vessels operating every year, conducting over 200 voyages annually. Not all tourist vessels are IAATO members, however, with an additional 28 vessels recorded in the 2017–2018 season (McCarthy *et al*., 2019). For those operators that are members of IAATO, various stipulations are in place to reduce their anthropogenic footprint (IAATO, 2020a). Only vessels carrying fewer than 500 passengers are allowed to dock in the Antarctic, with a maximum of 100 persons ashore at any one time. In addition, IAATO (and their Arctic partner) has developed a Plastic Reduction Programme, including guidelines for tourists about waste disposal and their consumption of single-use plastic (IAATO, 2019b, 2020b). For example, they are discouraged from using and packing disposable items, such as bottles and wrappers, as well as synthetic clothing and textiles, wherever possible.

2.3. *Scientific research*

There are currently 76 operational scientific research stations in Antarctica, belonging to 29 countries (COMNAP, 2020a). These vastly differ in size, accommodating anywhere between 6 and 1,200 people. While many were constructed around the time the Antarctic Treaty was signed (1959), a substantial number were also established in the late 1980s and early 1990s,with the latest being opened in 2014 (McCarthy *et al*., 2019). A limited number of these bases facilitate air travel, with the major route for access and resupply being via sea. Currently, there are 51 research vessels in service, which also aid in scientific research cruises around Antarctica and beyond (COMNAP, 2020b).

Research stations may contribute to plastic waste through scientific research, as well as day-to-day functioning (Urbina *et al*., 2015). Plastic use in science can be substantial, due to the need for specialist equipment, biosecurity and contamination control. In colder environments, extra synthetic clothing is also a necessity, which elevates the risk of contaminating the environment with microfibres, especially during laundering

(Tirelli *et al.*, 2020). Whilst the Protocol on Environmental Protection (Annex III) does legislate against discarding plastic pollution, research stations are not currently required to treat their wastewater before release. As of 2018, only 54% of research stations perform any kind of treatment to their wastewater, including that from laundry systems, and those with treatment methods in place are often not comprehensive (Reed *et al.*, 2018).

Investigating marine sediment samples around Rothera Research Station, Adelaide Island, Reed *et al.* (2018) discovered 31 microplastic particles from twenty 30 ml samples. The particles found, which were almost exclusively microfibres, were from samples close to the sewage outfall of the research station and contained the grey water from the washing machines. Rayon, a common component of synthetic clothing, was the most common constituent. Collectively, this led to the conclusion that laundering at the research base was the likely source of this microplastic contamination. Similar findings have been reported at other research bases (Cincinelli *et al.*, 2017; Munari *et al.*, 2017; Absher *et al.*, 2019). For example, Cincinelli *et al.* (2017) found higher concentrations of microplastic in the waters around the Mario Zucchelli Station (MZS) research base at Terra Nova Bay. Of fibres detected; the most common colours were those worn by the personnel on the base. Collectively, results indicated a point source of microfibres from the sewage outfall.

2.4. *Shipping*

In ASO, the aforementioned three activities are united by shipping. Ships are required by nearly everyone travelling to Antarctica, with an estimated 180 vessels and more than 500 voyages in ASO every year (McCarthy *et al.*, 2019). Aside from wastewater and ballast water discharge and plastic waste management stipulated by various international agreements (see History and Governance), a potential source of plastic from shipping is the vessels themselves. Amongst the microplastic particles discovered in the Southern Ocean, varnish and paint chips have also been found (Lacerda *et al.*, 2019; Jones-Williams *et al.*, 2020; Kelly *et al.*, 2020). These comprise polyurethane, polyacrylates and ethylene vinyl acetate, all of which are used industrially, particularly for the hulls of shipping vessels. These indicate ocean traffic as a source for this pollution, likely from the natural erosion of fishing, cruises and scientific research vessels in the area. Whilst many of these paint particles may not be considered plastic, the synthetic particles will have similar characteristics, behaviour and impacts (Lacerda *et al.*, 2019).

3. Transport

The ratio of plastic pollution that is being released within ASO compared to that which is transported across the ACC into the region from afar remains to be quantified. This segment of the chapter examines some of the known or hypothesised routes of plastic transportation into and around the region Figure 4.

3.1. *Transport to ASO*

We can attempt to better understand the marine transport of plastic pollution across the ACC using computer modelling, of which there are two main approaches: Lagrangian models that simulate individual particles and calculate the trajectory of each particle separately, and Eulerian models that simulate the concentration of particles and calculate the overall diffusion of particles (Saidi *et al.*, 2014).

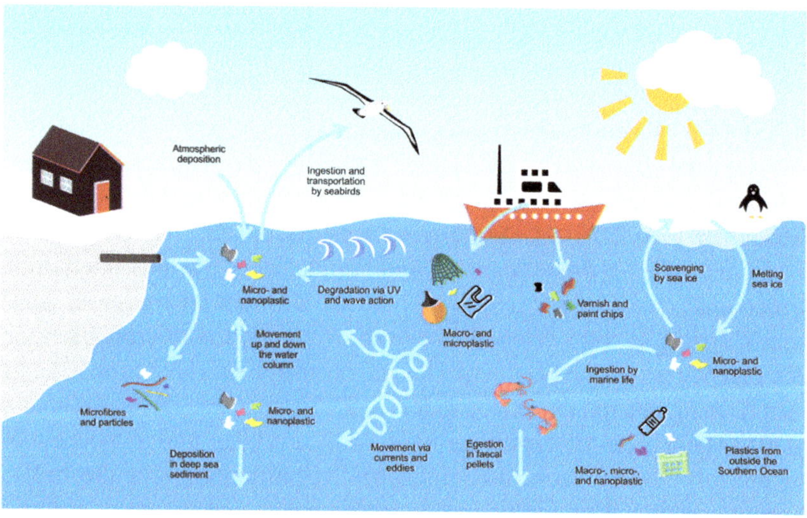

Figure 4. Potential sources, transport pathways and sinks of plastic pollution in Antarctica and the Southern Ocean.

High-resolution Lagrangian modelling has revealed that passively floating species can raft across the ACC when mesoscale eddies and Stokes drift are included in the model (Fraser *et al.*, 2018). Mesoscale eddies are swirling, energetic and time-limited circulations of water up to 100 km across that are often produced by the unstable meandering of intense currents such as the ACC (Rhines, 2001). Stokes drift is defined as the net drift that a floating particle experiences in the direction of wave propagation (Stokes, 1847), an effect that is observed in the top 1.8 m of the ocean (Clarke & van Gorder, 2018). Large, buoyant macroplastic is likely to be affected by similar transport processes to kelp and, therefore, drift to similar areas of the Southern Ocean. Overall, using the aforementioned model, between 0.0001% and 0.1915% of simulated kelp released outside the ACC crossed the current and reached the Antarctic coast within three years (Fraser *et al.*, 2018), which suggests a low proportion of the buoyant macroplastic present outside the ACC goes on to cross the ACC. Although low, in the context of the Southern Ocean which has extremely limited local sources of buoyant macroplastic pollution, it could represent a notable pathway for the transport of plastics to the inner Southern Ocean. The transport of buoyant macroplastics floating at the ocean surface is also affected by windage, which is the effect of surface winds blowing over the ocean and pushing against the area of plastic protruding from the surface of the ocean. This effect is highly dependent on the size, shape and buoyancy of each individual plastic item (van Sebille *et al.*, 2020).

Previous research has also shown that particularly strong mesoscale eddies can transport heat, carbon, oxygen and phosphates across the fronts which make up the ACC (Dufour *et al.*, 2015), suggesting smaller size fractions of plastic pollution such as microplastic and nanoplastic are also likely to be transported across the ACC, especially particles of plastic pollution which are positively buoyant, thus floating at the ocean surface where they are impacted by Stokes drift. Recent Eulerian modelling by Mountford and Morales Maqueda (2019) suggests that negatively buoyant microplastic particles begin to sink as soon as they enter the ocean, accumulating in the deepest areas of the global ocean and not making it to the Southern Ocean. Positively buoyant particles, that rise through the water column if submerged, were found to mostly accumulate in the five ocean gyres. However, their modelling of neutrally buoyant particles which have no rise velocity and, therefore, act as passive tracers suggests these neutrally buoyant particles could be potentially ubiquitous in the water column. Further research by Mountford and Morales Maqueda (2021) highlighted how neutrally buoyant microplastic particles are likely to be the dominant type of microplastic crossing the ACC into the Southern Ocean, travelling to the region through deep-water transport.

Modelling of the atmospheric transport of microplastics suggests there to be very little atmospheric microplastic transport to the Southern Ocean and negligible atmospheric microplastic transport to Antarctica (Evangeliou *et al.*, 2020). Recent observations of microplastic particles in the planetary boundary layer, however, favour a school of thought that microplastic and nanoplastic particles may be transported in the atmosphere for more than 1,000 kilometres and may be a notable source of the plastic observed in remote regions (Zhang *et al.*, 2019; González-Pleiter *et al.*, 2021). Further research is, therefore, necessary to assess the implications of this for ASO.

Another potential transport pathway for microplastic into ASO is through seabirds ingesting the particles when wintering at lower latitudes, which is then transferred to breeding areas within ASO through regurgitation or egestion (do Sul *et al.*, 2011). A study by van Franeker and Bell (1988), looking at the stomach contents of seabirds in the Antarctic continent, found 75% of Wilson's storm petrels (*Oceanires oceanicus*) and 33% of Cape petrels (*Daption capense*), which migrate out of ASO in winter, had microplastic in their stomach contents. Whereas for snow petrels (*Pagodroma nivea*) and Antarctic petrels (*Thalassoica antarctica*), which rarely migrate across the ACC, only 7% and 4% of these birds, respectively, had microplastic in their stomach contents. More recent research concurs that several seabird species, such as the grey-headed albatross (*Thalassarche chrysostoma*), could be transporting plastic from the more northerly areas of their distribution, that have higher plastic concentrations, into their breeding areas within ASO (Phillips & Waluda, 2020). The scale of this transport pathway remains unknown and further research is needed in this area.

3.2. *Transport within ASO*

Once inside the ACC, plastic pollution in the Southern Ocean is likely to be trapped for a significant period gradually fragmenting into smaller microplastic particles, and eventually to nanoplastics by physical stress from processes such as wave action and UV-initiated degradation (Jahnke *et al.*, 2017). For example, a backtracking dispersal model highlights how plastic sampled around the WAP in 2017 is unlikely to have originated north of

58°S in the preceding seven years (Lacerda *et al.*, 2019). Once within the Southern Ocean, there are several processes that can influence and/or determine the pathways that plastic pollution will take Figure 4.

First, the currents and fronts of the ACC transport plastic pollution in a clockwise direction around the Antarctic continent (Lacerda *et al.*, 2019). Plastic pollution that is present south of the ACC could also be transported towards the Antarctic coastline by the southward flowing limbs of Antarctic regional gyres such as those in the Weddell Sea and Ross Sea (Figure 1(b); Waller *et al.*, 2017). Eulerian modelling indicates there to be relatively high concentrations of neutrally buoyant microplastic particles in the Weddell Sea gyre, compared to surrounding areas of the Southern Ocean (Mountford & Morales Maqueda, 2021), suggesting buoyant microplastic may be accumulating in the Weddell Sea gyre.

A second potentially significant factor in the transport of plastic pollution in the Southern Ocean is sea ice. Microplastics have been reported in sea ice cores from both the Arctic (Peeken *et al.*, 2018) and the Antarctic (Kelly *et al.*, 2020), and once trapped, sea ice can transport microplastics laterally across the surface of the ocean (van Sebille *et al.*, 2020). Sea ice drift around Antarctica is largely determined by surface winds (Holland & Kwok, 2012) and often drifts in a different direction to underlying currents. Sea ice has been recorded off the coast of East Antarctica drifting at average speeds up to 19.8 km a day (Heil, 1999), highlighting how, with even a slight deviation from the underlying currents, plastic could be transported to areas they may not have reached if freely drifting in ocean surface waters.

Furthermore, whilst the long-range atmospheric transport of plastic particles to ASO is possible, but by current understanding likely limited (Evangeliou *et al.*, 2020), there is the potential for the transport of very small plastic particles within the region. This could include plastic particles being transported from terrestrial sources, such as scientific research stations, to the sea. Elsewhere, modelling of the atmospheric transport of microplastics within a remote area has suggested plastic fibres up to 750 μm long and fragments up to 300 μm in size could travel distances up to 95 km (Allen *et al.*, 2019).

The transport of plastic pollution within the Southern Ocean can also be influenced through interaction with marine life in a number of ways. First, plastic pollution in the ocean will undergo the process of biofouling, where organisms such as algae will build up on the surface of plastic over time, leading to an increase in the density of plastic pieces and subsequent decrease in their buoyancy (Chubarenko *et al.*, 2016). This then causes microplastic and smaller macroplastic items to sink, taking them below the depth at which transport is influenced by Stokes drift. Once particles have initially sunk, they are then likely to oscillate vertically in the water column as defouling and subsequent refouling take place as light availability, salinity, density and chlorophyll *a* concentration vary with depth (Kooi *et al.*, 2017).

Ingestion and egestion by organisms can also influence microplastic behaviour and fate Figure 4. Zooplankton present in the Southern Ocean, such as Antarctic krill (*Euphausia superba*), have been shown in laboratory studies to ingest microplastic (Dawson *et al.*, 2018). Given reported vertical migrations of up to 3,500 m (Clarke & Tyler, 2008), krill could act as a direct transport vector for microplastic by ingesting particles in surface waters and later egesting these at depths. Zooplankton faecal pellets are also a vector for microplastic and nanoplastic as they sink through the water column. When packed with positively buoyant plastics, faecal pellets can be less dense and, therefore, sink slower than pellets not containing plastic, making them more susceptible to being eaten by other organisms or to fragmentation of the pellet itself (Cole *et al.*, 2016). Alternatively, as suggested in the case of giant larvaceans (*Bathochordaeus stygius*), faecal pellets may sink all the way to the seafloor (Katija *et al.*, 2017), where the microplastic may subsequently be sequestered.

Concentrations of adhesive extracellular polymeric substances, which are emitted by unicellular and multicellular organisms including bacteria and phytoplankton, are particularly high in the Antarctic marine ecosystem (Meiners *et al.*, 2004) and may encourage aggregation of microplastic and nanoplastic particles, further altering their transport. Additionally, these abundant substances can impact the "ecocorona" of small plastic particulates, a coating adhered to the surface which comprises both organic material and contaminants (Galloway *et al.*, 2017). The aggregation of plastic particles and the composition of ecocoronas in ASO are largely unknown, though they are likely important parameters in determining the fate, transport and toxicity of microplastic and nanoplastic in the Antarctic marine ecosystem.

4. Sinks

In a scientific context, a sink is defined as "a vehicle for the removal of a substance from a system into a permanent or semi-permanent repository"

(European Environment Agency, 2021). Unquantified plastic sinks are a key explanation for the observed discrepancy between known plastic inputs into natural systems and observed levels (van Sebille *et al.*, 2015; Erni-Cassola *et al.*, 2019).

ASO may be considered a plastic sink, in that plastic is geographically removed from other systems. But, as discussed previously, the region is not as isolated as it is generally perceived. Plastic debris has been identified in many marine Antarctic and SO environments, some potentially constituting notable, and surprising, sinks Figure 4.

4.1. *Plastic in seawater*

The only long-term, rigorous study of macrodebris (including plastic) in ASO has been conducted over 30 years (1989–2019) at Bird Island, South Georgia, where 90 kg of plastic has been recovered from a single 291 m-long stretch of beach (Waluda *et al.*, 2020). These levels are comparable with other islands in the South Atlantic (Ryan & Watkins, 1988) but might not represent ASO as a whole. South Georgia sits in the path of the AAC which, as discussed, is not an impenetrable barrier to positively buoyant items (Barber *et al.*, 1959; Coombs & Landis, 1966; Fraser *et al.*, 2018) and appears to be a buffer zone where these items have some residence time. This is evidenced by the higher counts of plastic on western shorelines of South Georgia, exposed to the eastwards flowing AAC (Torres & Jorquera, 1996; Waluda *et al.*, 2020). A study of macrodebris on another sub-Antarctic island, Signy Island, conducted less frequently but over a similar period (1991–2019) found notably fewer debris items, however the ratios of plastic compared to other forms of debris were similar (Waluda *et al.*, 2020).

Generally, macroplastics in the ASO region are low in quantity and certainly lower than other oceanic basins. Hypotheses for this are twofold. First, we don't yet have a holistic picture of macroplastic levels due to the nature of existing reports which are often singular (Gregory *et al.*, 1984; Gregory, 1987; Slip & Burton, 1990; Convey *et al.*, 2002; Eriksson *et al.*, 2013), incidental (Barnes & Fraser, 2003) or opportunistic (Barnes *et al.*, 2010), or at least, not from rigorous, systematic or targeted surveys. Second, macroplastic levels are generally low due to sea ice which may shield continental Antarctica from macroplastic deposits (Barnes & Fraser, 2003; Waluda *et al.*, 2020). Additionally, in line with other regions of the world, macroplastic observations in ASO are generally limited to the sea surface and what is washed ashore, which we know to be a fraction of what sinks through the water column or is redistributed into smaller fragments (Thompson *et al.*, 2004; Barnes *et al.*, 2009; Ryan, 2020).

There are marginally more records of microplastics than macroplastics in ASO (Waller *et al.*, 2017; Tirelli *et al.*, 2020) and definitively more than nanoplastics (Rowlands *et al.*, 2021a). Using reliable methods during the same year, Cincinelli *et al.* (2017) and Isobe *et al.* (2017) report concentrations of 0.003–1.18 particles m^{-3} (excluding fibres) and 0.003–0.099 particles m^{-3}, respectively. These low figures contradict earlier findings which equate concentrations in the Southern Ocean to those in other ocean basins, notably the South Atlantic (Eriksen *et al.*, 2014; Cózar *et al.*, 2014). Across the region, microplastics are not ubiquitous; several studies have found negligible (Ryan, 1990; Suaria *et al.*, 2020a) or no (Kuklinski *et al.*, 2019) microplastic fragments in surface water samples. Waller *et al.* (2017) suggest that anthropogenic sources only influence microplastic concentrations on a local scale, which may partially explain this uneven distribution. It is worth noting that all aforementioned studies only cover surface samples (down to ~5 m), and the mesh size of nets deployed in surface trawls tend to differ. Cózar *et al.* (2014) used a net with a mesh size of 200 μm, whilst Isobe *et al.* (2017) used nets with a mesh size of 350 μm, and Cincinelli *et al.* (2017) used a saltwater intake pump and passed the water under pressure through a glass fibre filter with a pore size of 1 μm. However, we know that microplastic is present in high quantities at mid-ocean depths (Pabortsava & Lampitt, 2020) and in deep-water currents (Kane *et al.*, 2020) in other regions of the world but as yet have no data from ASO. Nanoplastics have yet to be quantified in the SO, as with other regions globally, due to analytical limitations.

4.2. *Plastic in marine sediments*

The depth of the Antarctic continental shelf is, on average, 500 m, notably deeper than the global average (~100 m), and in many regions, is quite narrow (in width from the coast). There have been few examinations of sub-marine sediments for plastic in ASO and even fewer beyond the continental shelf. A single microplastic fragment (with a maximal Feret diameter of 118 μm) was retrieved from a single sample at a depth of 2,749 m in the Weddell Sea, though at a location north of the Polar Frontal Zone (Van Cauwenberghe *et al.*, 2013). The situation is similar for other records from the region, namely, most studies are largely limited to a single geographic area or depth range.

Munari *et al.* (2017), Waller *et al.* (2017) and Reed *et al.* (2018) all report microplastics in sub-tidal sediment, although only from samples in localised areas and less than 120 m deep. The only Antarctic or SO-focused study of sub-marine sediments across multiple areas, with sampling from across different depth zones (136–3633 m), found concentrations of microplastic up to 9.52 particles g^{-1}. These results are comparable with recent results from the Arctic, Scapa Flow and Canadian riverine sediments, all of which have a higher degree of anthropogenic activity nearby than Antarctic and the SO (Cunningham *et al.*, 2020). Across surveys, there was only a single recorded instance of non-microplastic size plastic debris (a 22 mm fragment) in SO sub-marine sediment. Yet, in the same study, macrodebris accounted for only 1.7% of plastic retrieved (Munari *et al.*, 2017).

4.3. *Plastic in sea ice*

Antarctic sea ice extends to approximately 7 million square miles in winter and shrinks to a minimum of 1 million square miles in summer (NOAA, 2021 — data from 2020). Just 15% of Antarctic sea ice is considered multiyear ice and is likely to be thicker than one metre (NOAA, 2021). There has been little work on sea ice and plastic interactions in ASO, but records and studies from the Arctic indicate the potential significance of sea ice as a transport vector and temporary sink for plastic <5 mm (Obbard *et al.*, 2014; Peeken *et al.*, 2018).

The relationship between sea ice formation and microplastic is complex and likely varies between locations, season, speed of the freeze and abundance of associated biota. Microplastics have been observed in sea ice at both poles correlating with levels of chlorophyll *a*, suggesting living biota could assist microplastic incorporation, particularly via the production of adhesive extracellular polymeric substances (Hoffmann *et al.*, 2020; Kelly *et al.*, 2020). Arctic examples indicate that sea ice is extremely variable as a sink, with concentrations of 33–75,143 particles L^{-1} (melted volume) reported across the region (Peeken *et al.*, 2018). Although upper concentrations detected are higher in some places than the surface waters of the Pacific gyre "garbage patch", they only represent localised findings (Obbard *et al.*, 2014; Peeken *et al.*, 2018; van Sebille *et al.*, 2020).

The only record of microplastic in sea ice in Antarctica comes from East Antarctic, coastal, land-fast ice, with 11.71 particles L^{-1} (Kelly *et al.*, 2020). This is relatively low compared to records from Arctic sea ice, yet higher than any record from Antarctic surface seawater. The passive uptake of floating particles during freezing, a process known as "scavenging", is hypothesised to be the method of microplastic incorporation (Hoffmann *et al.*, 2020; Kelly *et al.*, 2020). Sea ice, therefore, constitutes a temporary sink for microplastic with particles being taken up in the austral autumn and winter when sea ice grows and released again in the austral summer when sea ice cover reduces. The Southern Ocean favours frazil ice formation (60–80% of total ice formation), a fast method of ice growth more adept at scavenging particulates than the slow congelation ice crystal growth more prevalent in the Arctic (Thoman & Dieckmann, 2008; Obbard *et al.*, 2014). To our knowledge, there are no physical records of macroplastic or nanoplastic accumulation or transport via sea ice in ASO.

5. ASO-Specific Considerations

The distinctiveness of the physical environment and biota in the ASO region renders its relationship with plastic pollution unique.

5.1. *The physical environment*

The ASO physical environment is noted for the following: (1) extremely low temperatures (falling to an average of −2°C in winter), (2) the presence of sea ice and terrestrial ice, (3) high levels of UV radiation and (4) the ACC barrier.

While sea ice is a known sink for plastics and vector for plastic transport, terrestrial ice also increases plastic transport when it melts. Almost four gigatonnes of Antarctic terrestrial ice have melted into the Southern Ocean since the 1990s (Adusumilli *et al.*, 2020). This influx of fresh water reduces local salinity and alters the relative buoyancy of plastic, which, in turn, affects particle transport and aggregation (Koelmans *et al.*, 2015; van den Berk *et al.*, 2019). The low temperatures of the Southern Ocean can also impact the oxidative degradation of plastic polymers, lessening the chemical reactions which lead to fragmentation (Gewert *et al.*, 2015; Min *et al.*, 2020). Conversely, UV radiation potentially enhances plastic degradation (Gewert *et al.*, 2015). This may be amplified in the region due to intense irradiation in longer hours of daylight during the austral summer coupled with the seasonal thinning of ozone.

When considering the physical environment, the remoteness of ASO and the physical oceanographic buffer of the AAC means that it takes longer for plastic items to penetrate the region; hence a larger proportion of plastic will be weathered to a greater degree (Obbard, 2018). It is, therefore, hypothesised that the region contains a larger proportion of plastics in the smaller size classifications than other regions.

5.2. *The ASO biosphere*

Now consider the sensitive biota of the region's ecosystem, renowned for the following: (1) slow growth and metabolic rates, (2) high levels of endemism and (3) stenothermal traits (i.e., only able to tolerate a small range of temperature). The Antarctic and SO ecosystems are unique, comprising short food chains where a large proportion of resident fish, squid, seals, whales, penguins, albatrosses and petrels predominantly rely on krill. Antarctic krill is the keystone species of the Antarctic marine ecosystem, and the largest biomass on earth. Nonetheless, a range of key zooplankton including salps (barrel-shaped gelatinous tunicates), amphipods and copepods are also critical to the ecosystem. As a planktonic primary producer, diatoms are also a key taxon on which the region hinges. Collectively, the Antarctic marine ecosystem is home to the largest concentration of endemic species globally and may be considered more vulnerable than others to anthropogenic pollutants, including plastic pollution (Horton & Barnes, 2020). Slow growth rates, typical of Antarctic marine organisms, are a particularly important consideration for pollutants given that early life stages are often the most vulnerable to anthropogenic stressors (Clarke & Peck, 1991; Peck, 2002; Kurihara, 2008). The aforementioned high levels of endemism, across various taxa, result in weakened genetic differentiation (Zane & Patarnello, 2000), while the stenothermal nature of many Antarctic organisms makes them vulnerable to predicted temperature increases from ocean warming (Peck, 2005). Combined, these factors make the biosphere sensitive to environmental change by reducing overall ecosystem resilience and its capacity for recovery following change or disturbance (Sommer, 1995; Dial & Roughgarden, 1998).

6. Impact

As the largest form of plastic debris observable *in situ*, macroplastic has the most visible consequences. Entanglement of megafauna in discarded or lost fishing gear has been recorded amongst fur seals in South Georgia (Bonner & McCann, 1987; Garcia-Garin *et al.*, 2020). Incidental ingestion of macroplastic by seabirds is another frequent occurrence globally, and marked increases of ingestion in seabirds from South Georgia have been observed since the 1990s (Phillips & Waluda, 2020). Ingested macroplastic can result in partial gut obstruction, blockage of enzyme secretion, appetite suppression and weight reduction in seabirds or gut perforation (in the case of ingested sharp objects), although the likelihood of population-level impacts is low (Phillips & Waluda, 2020).

For microplastics, while there is a paucity of data relating to their effects on polar species, we can make assumptions from laboratory studies of non-polar species. Data collected thus far suggest that the effects of microplastic are rarely lethal, though sub-lethal effects impact survival, feeding, oxidative status and contaminant uptake (Cole *et al.*, 2013, 2019; Wright *et al.*, 2013; Coppock *et al.*, 2019). In the only study to date on species of the Antarctic and SO ecosystem, Dawson *et al.* (2018) exposed Antarctic krill to polyethylene spheres sized 27–32 μm in a laboratory environment and suggest that acute toxicity of ingestion is limited, however sub-lethal chronic end points are likely and require quantification. Many microplastic particles fall within the optimum feeding range for numerous Antarctic zooplankton species. Jones-Williams *et al.* (2020) provide the only insight into microplastic ingestion in ASO zooplankton (pelagic amphipods), observing actual encounter rates (0.8%) to be higher than encounter rates predicted if random chance were the only factor. This highlights that, even at low concentrations, microplastic is preferentially consumed by Antarctic zooplankton. There are records of microplastic internalised and excreted by penguin species (Bessa *et al.*, 2019; Le Guen *et al.*, 2020), seabirds (Suaria *et al.*, 2020b), benthic invertebrates (Sfriso *et al.*, 2020) and demersal fish, including the economically important Antarctic toothfish (*Dissostichus mawsoni*) which inhabits depths of 2,200 m (Cannon *et al.*, 2016). These examples highlight the ubiquity of microplastic, how Antarctic organisms interact with plastics and the future work required to determine the impact of microplastic in the region.

Finally, only a handful of studies focus on the impact of nanoplastic in ASO and the toxicological impacts of plastic debris of this size range on Antarctic species. Due to their physical characteristics (small size, high surface curvature and large surface area to volume ratio), nanoplastics are hypothesised to be potentially more hazardous than microplastics (Koelmans *et al.*, 2015). Polystyrene spheres (40–50 nm) have been shown to impair the immune response of Antarctic sea urchins (*Sterechinus neumayeri*) (Bergami *et al.*, 2019), an important species of the Antarctic benthic community and a model organism in the field of toxicology. With the same polymer (50–60 nm) sub-lethal effects such as reduced swimming behaviour and increased exuviae production were observed in juvenile Antarctic krill (Bergami *et al.*, 2020). Whilst the behaviour of nanoplastic has been observed in ambient Antarctic seawater in a laboratory environment (Bergami *et al.*, 2019, 2020), the behaviour of small plastic particles and their toxicity to biota in the Southern Ocean itself are yet to be determined. Early indications based on observations from other regions in the world suggest adverse implications are possible.

Although we discuss plastic as a single stressor, like all marine ecosystems, ASO are subjected to multiple stressors concomitantly and indeed plastics are not one type of material. Multistress scenarios are of particular relevance in the region since ocean warming and acidification are exacerbated, and the region is considered a dead end for a range of chemical pollutants. In a recent review, Rowlands *et al.* (2021a) highlighted potential mixture effects of microplastic and nanoplastic coupled with chemical and climate related stressors in the polar regions and reiterate the need to consider multistress scenarios in future plastic ecotoxicity research on Antarctic species (Horton & Barnes, 2020). Only one study to date explores the impact of plastic pollution from a multistress experiment on Antarctic and SO biota. Rowlands *et al.* (2021b) demonstrated that nanoplastic in conjunction with a future ocean acidification scenario for the SO hindered development of Antarctic krill embryos. They suggest this to be a result of ocean acidification altering the chemical equilibrium of plastic particles and consequently their ability to leach chemicals as seen in other studies (Piccardo *et al.*, 2020), as well as due to increased energy needs to maintain homeostasis in a multistress environment.

Despite the field of plastic toxicity on species of the Antarctic marine ecosystem being relatively young, the region's organisms are as susceptible to the adverse impacts of marine plastic pollution as anywhere else. Whether Antarctic species may be more vulnerable to plastic pollution due to their unique ecosystem is yet to be quantified and potential multistressor implications are largely unknown.

7. Summary, Conclusions and Future Directions

Antarctica and the Southern Ocean benefit from the lowest human presence on Earth, making potential point source emissions of plastic pollution within the region comparatively easy to pinpoint. Possibilities are limited to fishing, tourism, scientific research and the transport vehicles associated with each of them. Yet, the issue is not as clear-cut as once believed. Despite long-standing hypotheses to the contrary, floating items (including biota and plastic) likely cross the AAC, and given the minimal human footprint, long-range sources are likely to be key contributors to plastic pollution within the Southern Ocean. Establishing the sources of plastic found in ASO is as difficult as it is in less-remote regions.

Whilst plastic is now considered ubiquitous in the world's oceans, the Southern Ocean generally has lower levels of surface plastic pollution compared to global trends. However, owing to the harsh polar conditions, logistical constraints and expense of conducting research, a paucity of data are apparent when compared to other regions.

The need to maintain environmentally responsible exploitation in ASO is like no other region, as the endemic biota are arguably the most sensitive on the planet to changes brought about by anthropogenic activities. The polar ecosystem is particularly vulnerable in its reliance on keystone species, such as Antarctic krill (*Euphasia superba*) which are known to ingest microplastic with sub-lethal effects. Plastic pollution is a complex, multifaceted threat, but its effects cannot be considered in isolation. The Antarctic ecosystem is subject to a range of stressors — climate change, ocean acidification, chemical pollution — and it is essential to bear this in mind when analysing and addressing anthropogenic impacts.

Producing high-quality, comparable data may be even more essential in ASO due to the difficulties in sampling, aligned with a desire to minimise activities in the region, though this could also be said for many other regions. It is partly this reason which prompted the formation of the SCAR "Plastics in the Polar Environments" Action Group with the aim to enhance cross-disciplinary and international collaboration in plastic research (SCAR, 2019). Many questions remain for science to answer:

- How much microplastic is present in various layers of the Southern Ocean?
- How does sea ice contribute to plastic transport in the region?

- How will the ecotoxicological implications of microplastics and nanoplastics impact Antarctic biota?
- How much plastic actually crosses the AAC?

In ASO, there is definite potential for these answers to come about collaboratively. The desire to reduce our plastic footprint is present in the tourism industry, e.g., IAATO and fishing industry regulator, e.g., CCAMLR. A partnership which expands beyond scientific communities to all industries with unique access to the region can be hugely beneficial, enabling tourist and fishing vessels to assist with monitoring campaigns, minimising the number of vessels in the Southern Ocean and aiding the collection of much needed long-term datasets; though care must be taken to avoid convenience and accessibility taking precedence overtargeting areas of scientific interest.

Since ASO benefit from the unique governance of the Antarctic Treaty, including 29 consultative countries and 25 non-consultative parties (ATS, 1961), it presents a rare opportunity for international cooperative research and action (though the Antarctic Treaty only covers latitudes higher than 60° South and lower latitudes must rely on other international regulating bodies). As research in the region increases, findings can be incorporated into new policies to aid further protection. For example, targeting waste treatment of grey water from bases and vessels and moving towards non-synthetic clothing. However, in some respects, many challenges are still to come. Impacts will likely worsen before improving as plastic production increases and steadily finds its way from distant locations into ASO, and as the human footprint in the region increases. Nonetheless, plastic pollution research is still relatively young and, in ASO, still in its infancy. Perhaps with the many advantages the region has in terms of interdisciplinary collaborative opportunities, it can become a barometer for plastic monitoring and control in other regions, with standardised, long-term monitoring, effective citizen science through tourism, and opportunities to incorporate findings into policy.

Acknowledgements

The authors would like to acknowledge the assistance of a number of people at the British Antarctic Survey, High Cross, Madingley Rd, Cambridge, CB3 0ET, UK: Laura Gerrish, Pete Bucktrout, Kevin Roberts,

Clara Manno and Stephen Roberts. Authors RP, ER and DW are funded by NERC GW4+ DTP scholarships. Author JB is funded by the Energy and Environment Institute of the University of Hull.

References

Absher, T. M., Ferreira, S. L., Kern, Y., Ferreira, A. L., Christo, S. W., & Ando, R. A. (2019). Incidence and identification of microfibers in ocean waters in Admiralty Bay, Antarctica. *Environmental Science and Pollution Research, 26*(1), 292–298.

Adusumilli, S., Fricker, H. A., Medley, B., Padman, L., & Siegfried, M. R. (2020). Interannual variations in meltwater input to the Southern Ocean from Antarctic ice shelves. *Nature Geoscience, 13*(9), 616–620.

Allen, S., Allen, D., Moss, K., Le Roux, G., Phoenix, V. R., & Sonke, J. E. (2020) Examination of the ocean as a source for atmospheric microplastics. (Mukherjee, A. (Ed.)). *PLoS One, 15*(5), e0232746.

Aronson, R. B., Thatje, S., Clarke, A., Peck, L. S., Blake, D. B., Wilga, C. D., & Seibel, B. A. (2007). Climate change and invasibility of the Antarctic benthos. *Annual Review of Ecology, Evolutionary Systems, 38*, 129–154.

Barber, H. N., Dadswell, H. E., & Ingle, H. D. (1959). Transport of driftwood from south America to Tasmania and Macquarie Island. *Nature, 184*(4681), 203–204.

Barnes, D. K. & Fraser, K. P. (2003). Rafting by five phyla on man-made flotsam in the Southern Ocean. *Marine Ecology Progress Series, 262*, 289–291.

Barnes, D. K. A., Galgani, F., Thompson, R. C., & Barlaz, M. (2009). Accumulation and fragmentation of plastic debris in global environments. *Philosophical Transactions of the Royal Society B: Biological Sciences, 364*(1526), 1985–1998.

Barnes, D. K. A., Walters, A., & Goncalves, L. (2010). Macroplastics at sea around Antarctica. *Marine Environmental Research, 70*(2), 250–252.

Bergami, E., Emerenciano, E., Krupinski, A., González-Aravena, M., Cárdenas, C. A., Hernández, P., Silva, J. R. M. C., & Corsi, I. (2019). Polystyrene nanoparticles affect the innate immune system of the Antarctic sea urchin Sterechinus neumayeri. *Polar Biology, 42*, 743–757.

Bergami, E., Manno, C., Cappello, S., Vannuccini, M. L., & Corsi, I. (2020). Nanoplastics affect moulting and faecal pellet sinking in Antarctic krill (*Euphausia superba*) juveniles. *Environmental International, 143*, 105999.

Bernhard, G. & Stierle, S. (2020). Trends of UV radiation in Antarctica. *Atmosphere, 11*(8), 795.

Bessa, F., Ratcliffe, N., Otero, V., Sobral, P., Marques, J. C., Waluda, C. M., Trathan, P. N., & Xavier, J. C. (2019). Microplastics in gentoo penguins from the Antarctic region. *Scientific Reports, 9*(1), 1–7.

Bonner, W. N. & McCann, T. S. (1982). Neck collars on fur seals, Arctocephalus gazelle, at South Georgia. *British Antarctic Survey Bulletin, 57*, 73–77.

Cannon, S. M., Lavers, J. L., & Figueiredo, B. (2016). Plastic ingestion by fish in the Southern Hemisphere: A baseline study and review of methods. *Marine Pollution Bulletin, 107*(1), 286–291.

CCAMLR. (1979). Text of the Convention for the Conservation of Antarctic Marine Living Resources. [online] Retrieved from https://www.ccamlr.org/en/document/publications/convention-conservation-antarctic-marine-living-resources. Accessed on 07/05/2021.

CCAMLR. (2017a). *Marine Debris* [online]. Retrieved from https://www.ccamlr.org/en/science/marine-debris#:~:text=The%20CCAMLR%20Marine%20Debris%20program,regard%20to%20fishing%20debris%20items.&text=Protection%20of%20the%20Antarctic%20marine,species%20covered%20by%20the%20Convention. Accessed on 16/04/2021.

CCAMLR. (2017b). Conservation Measure 10-08. [online] Retrieved from https://www.ccamlr.org/en/measure-10-08-2017. Accessed on 07/05/2021.

CCAMLR. (2020). Statistical Bulletin, Vol. 32. [online] Retrieved from https://www.ccamlr.org/en/publications/statistical-bulletin. Accessed on 07/05/2021.

CCAMLR. (2021a). List of Authorised Vessels. [online] Retrieved from https://www.ccamlr.org/en/compliance/list-vessel-authorisations. Accessed on 07/05/2021.

CCAMLR. (2021b). Non-Contracting Party IUU Vessel List. [online] Retrieved from https://www.ccamlr.org/en/compliance/iuu-vessel-lists. Accessed on 07/05/2021.

Chubarenko, I., Bagaev, A., Zobkov, M., & Esiukova, E. (2016). On some physical and dynamical properties of microplastic particles in marine environment. *Marine Pollution Bulletin, 108*(1–2), 105–112.

Cincinelli, A., Scopetani, C., Chelazzi, D., Lombardini, E., Martellini, T., Katsoyiannis, A., Fossi, M.C., & Corsolini, S. (2017). Microplastic in the surface waters of the Ross Sea (Antarctica): Occurrence, distribution and characterization by FTIR. *Chemosphere, 175*, 391–400.

Clarke, A. & Peck, L. S. (1991). The physiology of polar marine zooplankton. *Polar Research, 10*(2), 355–370.

Clarke, A. J., Barnes, D. K. A., & Hodgson, D. A. (2005). How isolated is Antarctica? *Trends in Ecology & Evolution, 20*(1), 1–3.

Clarke, A. J. & Tyler, P. A. (2008). Adult Antarctic krill feeding at abyssal depths. *Current Biology, 18*(4), 282–285.

Clarke, A. J. & van Gorder, S. (2018). The relationship of near-surface flow, Stokes drift and the wind stress. *Journal of Geophysical Research: Oceans, 123*(7), 4680–4692.

Cole, M., Coppock, R., Lindeque, P. K., Altin, D., Reed, S., Pond, D. W., Sørensen, L., Galloway, T. S., & Booth, A. M. (2019). Effects of nylon microplastic on feeding, lipid accumulation, and moulting in a coldwater copepod. *Environmental Science & Technology, 53*(12), 7075–7082.

Cole, M., Lindeque, P., Fileman, E., Halsband, C., Goodhead, R., Moger, J., & Galloway, T. S. (2013). Microplastic ingestion by zooplankton. *Environmental Science & Technology*, *47*(12), 6646–6655.

Cole, M., Lindeque, P. K., Fileman, E., Clark, J., Lewis, C., Halsband, C., & Galloway, T. S. (2016) Microplastics alter the properties and sinking rates of zooplankton faecal pellets. *Environmental Science and Technology*, *50*(6), 3239–3246.

COMNAP. (2020a). COMNAP Antarctic Facilities. [online] Retrieved from https://github.com/PolarGeospatialCenter/comnap-antarctic-facilities (Accessed: 07/05/2021).

COMNAP. (2020b). COMNAP Antarctic Vessels. [online] Retrieved from https://github.com/PolarGeospatialCenter/comnap-antarctic-vessels/. Accessed on 07/05/2021.

Convey, P., Barnes, D., & Morton, A. (2002). Debris accumulation on oceanic island shores of the Scotia Arc, Antarctica. *Polar Biology*, *25*(8), 612–617.

Convey, P., Stevens, M. I., Hodgson, D. A., Smellie, J. L., Hillenbrand, C. D., Barnes, D. K., Clarke, A., Pugh, P. J., Linse, K., & Cary, S. C. (2009). Exploring biological constraints on the glacial history of Antarctica. *Quaternary Science Reviews*, *28*(27–28), 3035–3048.

Coombs, D. S. & Landis, C. A. (1966). Pumice from the south sandwich eruption of March 1962 reaches New Zealand. *Nature*, *209*(5020), 289–290.

Coppock, R. L., Galloway, T. S., Cole, M., Fileman, E. S., Queirós, A. M., & Lindeque, P. K. (2019). Microplastics alter feeding selectivity and faecal density in the copepod, Calanus helgolandicus. *Science of the Total Environment*, *687*, 780–789.

Cózar, A., Echevarría, F., González-Gordillo, J. I., Irigoien, X., Úbeda, B., Hernández-León, S., Palma, Á. T., Navarro, S., García-de-Lomas, J., Ruiz, A., & Fernández-de-Puelles, M. L. (2014). Plastic debris in the open ocean. *Proceedings of the National Academy of Sciences*, *111*(28), 10239–10244.

Cuclo, J. A. (1964). United States of America Patent No. 3,156,750. U. S. P. Office.

Cunningham, E. M., Ehlers, S. M., Dick, J. T., Sigwart, J. D., Linse, K., Dick, J. J., & Kiriakoulakis, K. (2020). High abundances of microplastic pollution in deep-sea sediments: Evidence from Antarctica and the Southern Ocean. *Environmental Science & Technology*, *54*(21), 13661–13671.

Dawson, A. L., Kawaguchi, S., King, C. K., Townsend, K. A., King, R., Huston, W. M., & Bengtson Nash, S. M. (2018). Turning microplastics into nanoplastics through digestive fragmentation by Antarctic krill. *Nature Communications*, *9*(1), 1–8.

Dayton, P. K., Mordida, B. J., & Bacon, F. (1994). Polar marine communities. *American Zoologist*, *34*(1), 90–99.

Dial, R. & Roughgarden, J. (1998). Theory of marine communities: The intermediate disturbance hypothesis. *Ecology*, *79*(4), 1412–1424.

do Sul, J. A. I., Barnes, D. K. A., Costa, M. F., Convey, P., Costa, E. S., & Campos, L. (2011). Plastics in the Antarctic environment: Are we looking only at the tip of the iceberg? *Oecologia Australis*, *15*(1), 150–170.

Dodds, K. (2010). Governing Antarctica: Contemporary challenges and the enduring legacy of the 1959 Antarctic Treaty. *Global Policy, 1*(1), 108–115.

Dufour, C. O., Griffies, S. M., de Souza, G. F., Frenger, I., Morrison, A. K., Palter, J. B., Sarmiento, J. L., Galbraith, E. D., Dunne, J. P., Anderson, W. G., & Slater, R. D. (2015). Role of mesoscale eddies in cross-frontal transport of heat and biogeochemical tracers in the Southern Ocean. *Journal of Physical Oceanography*, *45*(12), 3057–3081.

Eastman, J. T. (1997). Comparison of the Antarctic and Arctic fish faunas. *Cybium (Paris)*, *21*(4), 335–352.

Eriksen, M., Lebreton, L. C., Carson, H. S., Thiel, M., Moore, C. J., Borerro, J. C., Galgani, F., Ryan, P. G., & Reisser, J. (2014). Plastic pollution in the world's oceans: More than 5 trillion plastic pieces weighing over 250,000 tons afloat at sea. *PloS One*, *9*(12), e111913.

Eriksson, C. & Burton, H. (2003). Origins and biological accumulation of small plastic particles in fur seals from Macquarie Island. *Ambio, 32*(6), 380–384.

Eriksson, C., Burton, H., Fitch, S., Schulz, M., & van den Hoff, J. (2013). Daily accumulation rates of marine debris on sub-Antarctic island beaches. *Marine pollution bulletin*, *66*(1–2), 199–208.

Erni-Cassola, G., Zadjelovic, V., Gibson, M. I., & Christie-Oleza, J. A. (2019). Distribution of plastic polymer types in the marine environment; A meta-analysis. *Journal of Hazardous Materials*, *369*, 691–698.

European Environment Agency (2021). TERM: Pollution Sink. Retrieved from https://www.eea.europa.eu/help/glossary/gemet-environmental-thesaurus/pollution-sink. Accessed on 08/02/2021.

Evangeliou, N., Grythe, H., Klimont, Z., Heyes, C., Eckhardt, S., Lopez-Aparicio, S., & Stohl, A. (2020). Atmospheric transport is a major pathway of microplastics to remote regions. *Nature Communications*, *11*(1), 1–11.

Fang, X., Park, S., Saito, T., Tunnicliffe, R., Ganesan, A. L., Rigby, M., Li, S., Yokouchi, Y., Fraser, P. J., Harth, C. M., & Krummel, P. B. (2019). Rapid increase in ozone-depleting chloroform emissions from China. *Nature Geoscience*, *12*(2), 89–93.

Fraser, C. I., Morrison, A. K., Hogg, A. M., Macaya, E. C., van Sebille, E., Ryan, P. G., Padovan, A., Jack, C., Valdivia, N., & Waters, J. M. (2018). Antarctica's ecological isolation will be broken by storm-driven dispersal and warming. *Nature Climate Change, 8*(8), 704–708.

Galloway, T. S., Cole, M., & Lewis, C. (2017). Interactions of microplastic debris throughout the marine ecosystem. *Nature, Ecology & Evolution*, *1*, 0116.

Garcia-Garin, O., García-Cuevas, I., Drago, M., Rita, D., Parga, M., Gazo, M., & Cardona, L. (2020). No evidence of microplastics in Antarctic fur seal scats from a hotspot of human activity in Western Antarctica. *Science of the Total Environment, 737*, 140210.

Gewert, B., Plassmann, M. M., & MacLeod, M. (2015). Pathways for degradation of plastic polymers floating in the marine environment. *Environmental Science: Processes & Impacts, 17*(9), 1513–1521.

Gregg, W. H. (1877). United States of America Patent No. 185,914. U. S. P. Office.

Gregory, M. R., Kirk, R. M., & Mabin, M. C. G. (1984). Pelagic tar, oil, plastics and other litter in surface waters of the New Zealand sector of the Southern Ocean, and on Ross Dependency shores. *Antarctic Records. N. Z., 6*(1), 12–28.

Gregory M. R. (1987). Plastics and other seaborne litter on the shores of New Zealand's Subantarctic islands. *NZ Antarctic Records, 7*, 32–47.

González-Pleiter, M., Edo, C., Aguilera, Á., Viúdez-Moreiras, D., Pulido-Reyes, G., González-Toril, E., Osuna, S., de Diego-Castilla, G., Leganés, F., Fernández-Piñas, F., & Rosal, R. (2021). Occurrence and transport of microplastics sampled within and above the planetary boundary layer. *Science of the Total Environment, 761*, 143213.

Heil, P. & Allison, I. (1999). The pattern and variability of Antarctic sea-ice drift in the Indian Ocean and western Pacific sectors. *Journal of Geophysical Research: Oceans, 104*(C7), 15789–15802.

Hemmings, A. D. (2015). Commercial harvest in Antarctica. In Liggett, D., Storey, B., Cook, Y., & Meduna, V. (Eds.), *Exploring the Last Continent: An Introduction to Antarctica*, pp. 413–428. Cham: Springer International Publishing.

Hoffmann, L., Eggers, S. L., Allhusen, E., Katlein, C., & Peeken, I. (2020). Interactions between the ice algae Fragillariopsis cylindrus and microplastics in sea ice. *Environment International, 139*, 105697.

Holland, P. R. & Kwok, R. (2012). Wind-driven trends in Antarctic sea-ice drift. *Nature Geoscience, 5*(12), 872–875.

Horton, A. A. & Barnes, D. K. (2020). Microplastic pollution in a rapidly changing world: Implications for remote and vulnerable marine ecosystems. *Science of the Total Environment, 738*, 140349.

IAATO. (2019a). Overview of Antarctic Tourism: 2018–19 Season and Preliminary Estimates for 2019–20 Season. [online] Retrieved from https://iaato.org/wp-content/uploads/2020/03/IP140-IAATO-Overview-of-Antarctic-Tourism-2018-19-Season-and-Preliminary-Estimates-for-2019-20-Season.pdf. Accessed on 07/05/2021.

IAATO. (2019b). Reducing Single-Use Plastic and Waste Generated by Polar Tourism. [online] Retrieved from https://iaato.org/wp-content/uploads/2020/03/IP099-Reducing-Single-Use-Plastic-and-Waste-Generated-by-Polar-Tourism.pdf. Accessed on 07/05/2021.

IAATO. (2020a). Antarctic Visitor Figures 2019-2020. [online] Retrieved from https://iaato.org/wp-content/uploads/2020/07/IAATO-on-Antarctic-visitor-figures-2019-20-FINAL.pdf. Accessed on 07/05/2021.

IAATO. (2020b). Reducing Waste — Guidelines for Visitors to Antarctica. [online] Retrieved from https://iaato.org/wp-content/uploads/2020/04/Reducing-waste-visitor-guidelines.pdf. Accessed on 07/05/2021.

Isobe, A., Uchiyama-Matsumoto, K., Uchida, K., & Tokai, T. (2017). Microplastics in the Southern Ocean. *Marine Pollution Bulletin, 114*(1), 623–626.

Jahnke, A., Arp, H. P. H., Escher, B. I., Gewert, B., Gorokhova, E., Kühnel, D., Ogonowski, M., Potthoff, A., Rummel, C., Schmitt-Jansen, M., Toorman, E., & MacLeod, M. (2017). Reducing uncertainty and confronting ignorance about the possible impacts of weathering plastic in the marine environment. *Environmental Science and Technology Letters*. American Chemical Society.

Jones-Williams, K., Galloway, T., Cole, M., Stowasser, G., Waluda, C., & Manno, C. (2020). Close encounters-microplastic availability to pelagic amphipods in sub-Antarctic and Antarctic surface waters. *Environment International, 140*, 105792.

Kane, I. A., Clare, M. A., Miramontes, E., Wogelius, R., Rothwell, J. J., Garreau, P., & Pohl, F. (2020). Seafloor microplastic hotspots controlled by deep-sea circulation. *Science, 368*(6495), 1140–1145.

Katija, K., Choy, C. A., Sherlock, R. E., Sherman, A. D., & Robison, B. H. (2017). From the surface to the seafloor: How giant larvaceans transport microplastics into the deep sea. *Science Advances, 3*(8), e1700715.

Kelly, A., Lannuzel, D., Rodemann, T., Meiners, K. M., & Auman, H. J. (2020). Microplastic contamination in east Antarctic sea ice. *Marine Pollution Bulletin, 154*, 111130.

Koelmans, A. A., Besseling, E., & Shim, W. J. (2015). Nanoplastics in the aquatic environment. Critical review. In: Bergmann, L., Gutow, L., Klages, M. (Eds.), Marine Anthropogenic Litter. SpringerOpen, New York, pp. 229–244

Kooi, M., van Nes, E. H., Scheffer, M., & Koelmans, A. A. (2017). Ups and downs in the ocean: Effects of biofouling on vertical transport of microplastics. *Environmental Science and Technology, 51*(14), 7963–7971.

Kuklinski, P., Wicikowski, L., Koper, M., Grala, T., Leniec-Koper, H., Barasiński, M., Talar, M., Kamiński, I., Kibart, R., & Małecki, W. (2019). Offshore surface waters of Antarctica are free of microplastics, as revealed by a circum-Antarctic study. *Marine Pollution Bulletin, 149*, 110573.

Lacerda, A. L. d. F., Rodrigues, L. dos S., van Sebille, E., Rodrigues, F. L., Ribeiro, L., Secchi, E. R., Kessler, F., & Proietti, M. C. (2019). Plastics in sea surface waters around the Antarctic Peninsula. *Scientific Reports, 9*(1), 3977.

Le Guen, C., Suaria, G., Sherley, R. B., Ryan, P. G., Aliani, S., Boehme, L., & Brierley, A. S. (2020). Microplastic study reveals the presence of natural and synthetic fibres in the diet of King Penguins (Aptenodytes patagonicus) foraging from South Georgia. *Environment International*, *134*, 105303.

MARPOL (Adoption: 1973 (Convention), 1978 (1978 Protocol), 1997 (Protocol-Annex VI); Entry into force: October 2, 1983 (Annexes I and II). Accessed on 16/04/2021. Retrieved from https://www.imo.org/en/About/Conventions/Pages/International-Convention-for-the-Prevention-of-Pollution-from-Ships-(MARPOL).aspx#:~:text=The%20International%20Convention%20for%20the,from%20operational%20or%20accidental%20causes.

McBride, M. M., Dalpadado, P., Drinkwater, K. F., Godø, O. R., Hobday, A. J., Hollowed, A. B., Kristiansen, T., Murphy, E. J., Ressler, P. H., Subbey, S., & Hofmann, E. E. (2014). Krill, climate, and contrasting future scenarios for Arctic and Antarctic fisheries. *ICES Journal of Marine Science*, *71*(7), 1934–1955.

McCarthy, A. H., Peck, L. S., Hughes, K. A., & Aldridge, D. C. (2019). Antarctica: The final frontier for marine biological invasions. *Global Change Biology*, *25*(7), 2221–2241.

Meiners, K., Brinkmeyer, R., Granskog, M., & Lindfors, A. (2004). Abundance, size distribution and bacterial colonization of exopolymer particles in Antarctic sea ice (Bellingshausen Sea). *Aquatic Microbial Ecology*, 35, 283–296.

Min, K., Cuiffi, J. D., & Mathers, R. T. (2020). Ranking environmental degradation trends of plastic marine debris based on physical properties and molecular structure. *Nature Communications*, *11*(1), 1–11.

Morris, R. J. (1980). Plastic debris in the surface waters of the South Atlantic. *Marine Pollution Bulletin*, *11*(6), 164–166.

Mountford, A. S. & Morales Maqueda, M. A. (2019). Eulerian modeling of the three-dimensional distribution of seven popular microplastic types in the global ocean. *Journal of Geophysical Research: Oceans*, *124*(12), 8558–8573.

Mountford, A. S. & Morales Maqueda, M. A. (2021). Modeling the accumulation and transport of microplastics by sea ice. *Journal of Geophysical Research: Oceans*, *126*(2), e2020JC016826.

Munari, C., Infantini, V., Scoponi, M., Rastelli, E., Corinaldesi, C., & Mistri, M. (2017). Microplastics in the sediments of Terra Nova Bay (Ross Sea, Antarctica). *Marine Pollution Bulletin*, *122*(1–2), 161–165.

NOAA. (2021). Understanding climate: Antarctic sea ice extent. [online] Retrieved from https://www.climate.gov/news-features/understanding-climate/understanding-climate-antarctic-sea-ice-extent. Accessed on 18/03/2021.

Obbard, R. W., Sadri, S., Wong, Y. Q., Khitun, A. A., Baker, I., Richard, C., & Thompson, R. C. (2014). Global warming releases microplastic legacy frozen in Arctic sea ice. *Earth's Future*, *2*(6), 315–320.

Obbard, R. W. (2018). Microplastics in polar regions: The role of long range transport. *Current Opinion in Environmental Science & Health*, *1*, 24–29.

Pabortsava, K. & Lampitt, R. S. (2020). High concentrations of plastic hidden beneath the surface of the Atlantic Ocean. *Nature Communications*, *11*(1), 1–11.

Peck, L. S. (2002). Ecophysiology of Antarctic marine ectotherms: Limits to life. In *Ecological Studies in the Antarctic Sea Ice Zone*, pp. 221–230. Berlin, Heidelberg: Springer.

Peck, L. S. (2005). Prospects for survival in the Southern Ocean: Vulnerability of benthic species to temperature change. *Antarctic Science*, *17*(4), 497.

Peeken, I., Primpke, S., Beyer, B., Gütermann, J., Katlein, C., Krumpen, T., Bergmann, M., Hehemann, L., & Gerdts, G. (2018). Arctic sea ice is an important temporal sink and means of transport for microplastic. *Nature Communications*, *9*(1), 1–12.

Phillips, R. A. & Waluda, C. M. (2020). Albatrosses and petrels at South Georgia as sentinels of marine debris input from vessels in the southwest Atlantic Ocean. *Environment International*, *136*, 105443.

Pugh, P. J. & Convey, P. (2008). Surviving out in the cold: Antarctic endemic invertebrates and their refugia. *Journal of Biogeography*, *35*(12), 2176–2186.

Reece, A. (1953). Sledges of the Norwegian-British-Swedish Antarctic Expedition, 1949–1952. *Polar Record*, *6*(46), 775–787.

Reed, S., Clark, M., Thompson, R., & Hughes, K. A. (2018). Microplastics in marine sediments near Rothera Research Station, Antarctica. *Marine Pollution Bulletin*, *133*, 460–463.

Rhines, P. B. (2001). Mesoscale eddies. In *Encyclopedia of Ocean Sciences*. Amsterdam: Elsevier. DOI: 10.1006/rwos.2001.0143.

Rowlands, E., Galloway, T., & Manno, C. (2021a). A Polar outlook: Potential interactions of micro-and nano-plastic with other anthropogenic stressors. *Science of the Total Environment*, *754*, 142379.

Rowlands, E., Galloway, T., Cole, M., Lewis, C., Peck, V., Thorpe, S., & Manno, M. (2021b). The effects of combined ocean acidification and nanoplastic exposures on the embryonic development of Antarctic krill. *Frontiers in Marine Science (Marine Pollution)*, *8*, 709763.

Ryan, P. G. & Watkins, B. P. (1988). Accumulation of stranded plastic objects and other artefacts at Inaccessible Island, Central South Atlantic Ocean. *South African Journal of Antarctic Research*, *18*(1), 11–13.

Ryan, P. G. (1990). The marine plastic debris problem off Southern Africa: Types of debris, their environmental effects, and control measures. In *Proceedings of the Second International Conference on Marine Debris* (Washington, DC: NOAA), pp. 85–102.

Ryan, P. G. (2015). A brief history of marine litter research. In *Marine Anthropogenic Litter*, pp. 1–25. Cham: Springer.

Ryan, P. G. (2020). The transport and fate of marine plastics in South Africa and adjacent oceans. *South African Journal of Science*, *116*(5–6), 1–9.

Saidi, M. S., Rismanian, M., Monjezi, M., Zendehbad, M., & Fatehiboroujeni, S. (2014). Comparison between Lagrangian and Eulerian approaches in predicting motion of micron-sized particles in laminar flows. *Atmospheric Environment*, 89, 199–206.

SCAR. (2019). Plastic in Polar Environments (PLASTIC-AG). [online] Retrieved from https://www.scar.org/science/plastic/about/. Accessed on 14/05/2021.

Secretariat of the Antarctic Treaty (ATS). (1961). The Antarctic Treaty (Signed: December 1, 1959; Entry into force: June 23, 1961). Accessed on 16/04/2021. Retrieved from https://www.ats.aq/e/antarctictreaty.html.

Secretariat of the Antarctic Treaty (ATS). (1998). Protocol on Environmental Protection to the Antarctic Treaty (Signed: October 4, 1991; Entry into force: January 14, 1998). Accessed on 16/04/2021. Retrieved from https://www.ats.aq/e/protocol.html.

Sfriso, A. A., Tomio, Y., Rosso, B., Gambaro, A., Sfriso, A., Corami, F., Rastelli, E., Corinaldesi, C., Mistri, M., & Munari, C. (2020). Microplastic accumulation in benthic invertebrates in Terra Nova Bay (Ross Sea, Antarctica). *Environment International*, *137*, 105587.

Slip, D. J. & Burton, H. R. (1990). The composition and origin of marine debris stranded on the shores of subantarctic Macquarie Island. In *Proceedings of the Second International Conference on Marine Debris*, pp. 403–415. US Dept. of Commerce.

Sommer, U. (1995). An experimental test of the intermediate disturbance hypothesis using cultures of marine phytoplankton. *Limnology and Oceanography*, *40*(7), 1271–1277.

Stokes, G. G. (1847). On the theory of oscillatory waves. *Transactions of the Cambridge Philosophical Society*, *8*(4), 441–455.

Suaria, G., Achtypi, A., Perold, V., Lee, J. R., Pierucci, A., Bornman, T. G., Aliani, S., & Ryan, P. G. (2020b). Microfibres in oceanic surface waters: A global characterization. *Science Advances*, *6*(23), 8493.

Suaria, G., Perold, V., Lee, J. R., Lebouard, F., Aliani, S., & Ryan, P. G. (2020a). Floating macro- and microplastics around the Southern Ocean: Results from the Antarctic Circumnavigation Expedition. *Environment International, 136*, 105494.

Thomas, D. N. & Dieckmann, G. S. (Eds.). (2008). Sea ice: An introduction to its physics, chemistry, Wiley-Blackwell, Oxford.

Thompson, R. C., Olsen, Y., Mitchell, R. P., Davis, A., Rowland, S. J., John, A. W., McGonigle, D., & Russell, A. E. (2004). Lost at sea: Where is all the plastic? *Science (Washington)*, *304*(5672), 838.

Tirelli, V., Suaria, G., & Lusher, A. L. (2020). Microplastics in polar samples. In Rocha-Santos, T., Costa, M., & Mouneyrac, C. (Eds.), *Handbook of*

Microplastics in the Environment, pp. 1–42. Cham: Springer International Publishing.

Torres D, Jorquera D (1996) Monitoring results of marine debris at Cape Shirreff, Livingston Island, South Shetland Islands, during the Antarctic season 1995/96. Serie Cientifica INACH, *46*,121–132.

Urbina, M. A., Watts, A. J. R., & Reardon, E. E. (2015). Labs should cut plastic waste too. *Nature, 528*(7583), 479–479.

Van Cauwenberghe, L., Vanreusel, A., Mees, J., & Janssen, C. R. (2013). Microplastic pollution in deep-sea sediments. *Environmental Pollution, 182*, 495–499.

van den Berk, J., Drijfhout, S. S., & Hazeleger, W. (2019). Atlantic salinity budget in response to Northern and Southern Hemisphere ice sheet discharge. *Climate Dynamics*, *52*(9), 5249–5267.

van Franeker, J. A., & Bell, P. J. (1988). Plastic ingestion by petrels breeding in Antarctica. *Marine Pollution Bulletin*, *19*(12), 672–674.

van Sebille, E., Wilcox, C., Lebreton, L., Maximenko, N., Hardesty, B. D., van Franeker, J. A., Eriksen, M., Siegel, D., Galgani, F., & Law, K. L. (2015). A global inventory of small floating plastic debris. *Environmental Research Letters*, *10*(12), 124006.

van Sebille, E., Aliani, S., Law, K. L., Maximenko, N., Alsina, J. M., Bagaev, A. *et al*. (2020). The physical oceanography of the transport of floating marine debris. *Environmental Research Letters*, *15*(2), 023003.

Vyverman, W., Verleyen, E., Wilmotte, A., Hodgson, D.A., Willems, A., Peeters, K., van de Vijver, B., De Wever, A., Leliaert, F., & Sabbe, K. (2010). Evidence for widespread endemism among Antarctic micro-organisms. *Polar Science*, *4*(2), 103113.

Waller, C. L., Griffiths, H. J., Waluda, C. M., Thorpe, S. E., Loaiza, I., Moreno, B., Pacherres, C. O., & Hughes, K. A. (2017). Microplastics in the Antarctic marine system: An emerging area of research. *Science of The Total Environment*, *598*, 220–227.

Waluda, C. M., Staniland, I. J., Dunn, M. J., Thorpe, S. E., Grilly, E., Whitelaw, M., & Hughes, K. A. (2020). Thirty years of marine debris in the Southern Ocean: Annual surveys of two island shores in the Scotia Sea. *Environment International, 136*, 105460.

Wright, S. L., Thompson, R. C., & Galloway, T. S. (2013). The physical impacts of microplastics on marine organisms: A review. *Environmental Pollution*, *178*, 483–492.

Zane, L. & Patarnello, T. (2000). Krill: A possible model for investigating the effects of ocean currents on the genetic structure of a pelagic invertebrate. *Canadian Journal of Fisheries and Aquatic Sciences*, *57*(S3), 16–23.

Zhang, Y., Gao, T., Kang, S., & Sillanpää, M. (2019). Importance of atmospheric transport for microplastics deposited in remote areas. *Environmental Pollution*, *254*, 112953.

https://doi.org/10.1142/9789811259111_0011

Chapter 10

Current Research on Microbe — Plastic Interactions in the Marine Environment

Mira Latva*, Vinko Zadjelovic* and Robyn J. Wright†,‡

**School of Life Sciences, University of Warwick, Coventry, UK*

†*Department of Pharmacology, Faculty of Medicine, Dalhousie University, Halifax, Nova Scotia, Canada*

‡*robyn.wright@dal.ca*

1. Introduction

The quantity of plastic pollution entering the oceans annually is increasing year-on-year (Borrelle *et al.*, 2020), but the ultimate fate and longevity of plastic in the oceans are unknown (Ward & Reddy, 2020). With some studies suggesting a persistence of hundreds of years (Avio *et al.*, 2016) or fragmentation rates of only 1–5% per year (Gerritse *et al.*, 2020), this has led us to look towards microbes for a solution to this problem. When plastics enter the environment, they are rapidly covered by organic matter, known as the ecocorona (Galloway *et al.*, 2017), and are colonised by microbes within minutes (Harrison *et al.*, 2014). These plastic-colonising microbes — bacteria, fungi and single-celled eukaryotes as well as macro and other organisms — are collectively termed the "plastisphere" (Zettler *et al.*, 2013). While marine plastics have long been observed to have a colonising biofilm (e.g., Carpenter and Smith, 1972), the first specific characterisation of the plastisphere was published by Zettler *et al.*

(2013), following a call for research into microbial communities on plastics by Harrison *et al.* (2011). Some earlier studies investigating biofilm formation in the marine environment did also include plastic materials (e.g., Dang *et al.*, 2008) or investigated the biomass colonising plastics in relation to plastic degradation (e.g., Artham *et al.*, 2009), but the specific focus on the taxonomic or functional characterisation of plastisphere communities has only been within the last approximately eight years.

A literature search carried out on 4th January 2021 for the search terms "plastics plastisphere", "plastics microbial community" and "plastics microbial degradation" yielded a total of 1069 unique results. This was supplemented with our own literature collections and filtered manually to include only studies and reviews that fit the following criteria: (i) were available online by the beginning of 2021; (ii) examined at least one aspect of the colonisation or degradation of surfaces in the marine environment; and (iii) the surfaces used included at least one recalcitrant, petrochemical plastic. This resulted in 140 primary research articles (Figures 1 and 2 and Table S1) and 65 review articles (Figure S1 and Table S2). Remarkably, 34% and 45% of the research articles and reviews, respectively, published to date were made available online in 2020.

In the following sections, we summarise current knowledge on the taxonomic and functional capacity of the plastisphere as well as the future directions suggested by the plastisphere reviews included here.

2. The Plastisphere as a Microbial Biofilm on Plastics

The microbial communities on plastics are typically dominated by *Proteobacteria* (~70% relative abundance), *Bacteroidetes* (~10%) and Cyanobacteria (~20%). There are usually not large differences in alpha diversity indices between plastics, control marine biofilms and planktonic samples, although some control marine biofilms have been found to have larger proportions of *Bacteroidetes* (~20%) and *Planctomycetes* (~5%), while planktonic samples have a larger proportion of *Verrucomicrobia* (~5%) (Wright *et al.*, 2021b). The composition of macromolecules likely drives colonisation in the first few minutes of incubation (Siboni *et al.*, 2007; Dang and Lovell, 2016), but much of the variation in plastisphere composition is driven by stochasticity (Wright *et al.*, 2019) as well as environmental variables (Amaral-Zettler *et al.*, 2020; Oberbeckmann and

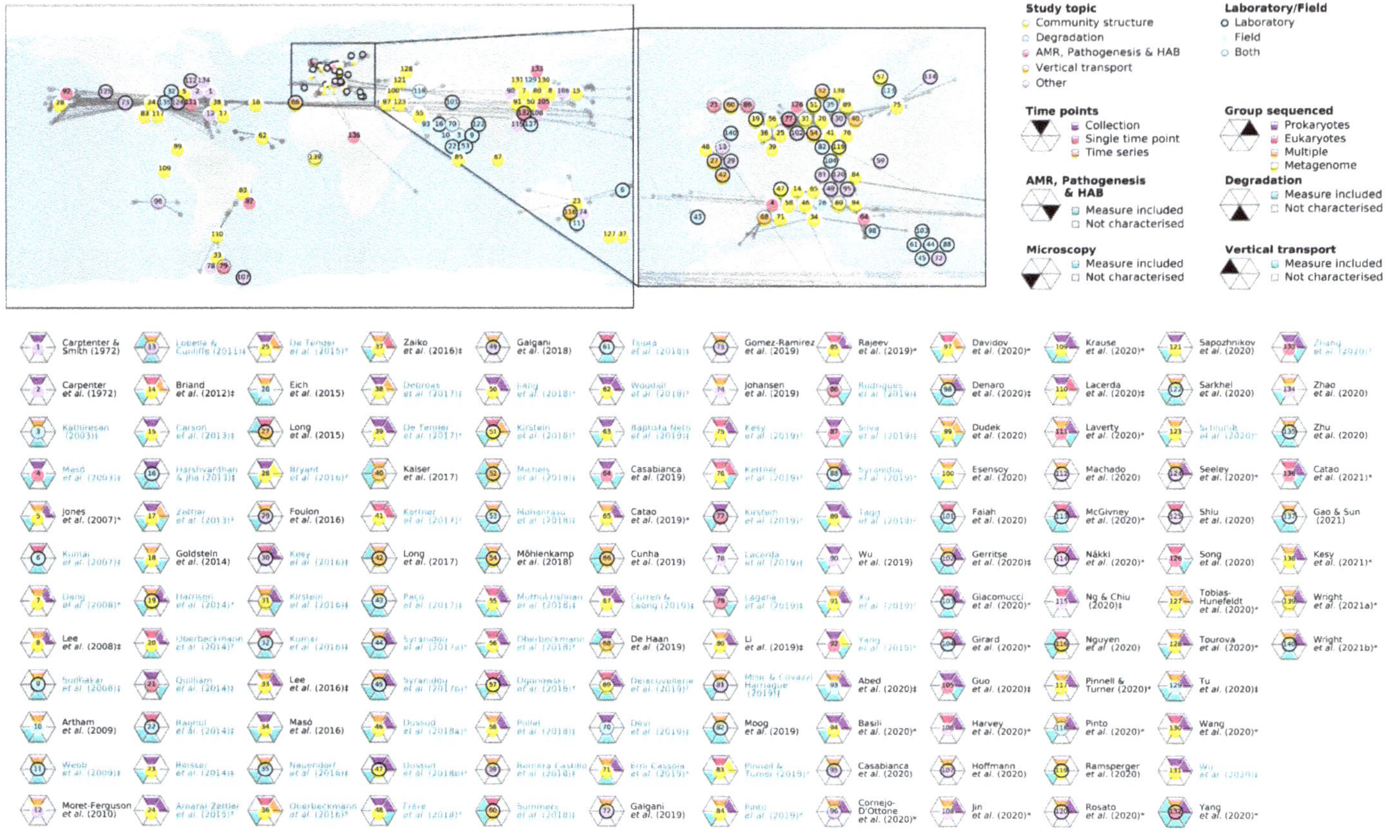

Study topic
Community structure
Degradation
AMR, Pathogenesis & HAB
Vertical transport
Other
Laboratory/Field
Laboratory
Field
Both
Time points
Collection
Single time point
Time series
Group sequenced
Prokaryotes
Eukaryotes
Multiple
Metagenome
AMR, Pathogenesis & HAB
Measure included
Not characterised
Degradation
Measure included
Not characterised
Microscopy
Measure included
Not characterised
Vertical transport
Measure included
Not characterised
Carpenter & Smith (1972)
Carpenter et al. (1972)
Jones et al. (2007)*
Lee et al. (2008)‡
Artham et al. (2009)
Moret-Ferguson et al. (2010)
Briand et al. (2012)‡
Goldstein et al. (2014)
Eich et al. (2015)
Long et al. (2015)
Foulon et al. (2016)
Lee et al. (2016)‡
Masó et al. (2016)
Zaiko et al. (2016)‡
Kaiser et al. (2017)
Long et al. (2017)
Galgani et al. (2018)
Möhlenkamp et al. (2018)
Casabianca et al. (2019)
Catao et al. (2019)*
Cunha et al. (2019)
De Haan et al. (2019)
Galgani et al. (2019)
Gomez-Ramirez et al. (2019)
Johansen et al. (2019)
Li et al. (2019)‡
Moog et al. (2019)
Rajeev et al. (2019)*
Wu et al. (2019)
Abed et al. (2020)‡
Basili et al. (2020)*
Casabianca et al. (2020)
Cornejo-D'Ottone et al. (2020)*
Davidov et al. (2020)*
Denaro et al. (2020)‡
Dudek et al. (2020)
Esensoy et al. (2020)
Faiah et al. (2020)
Gerritse et al. (2020)‡
Giacomucci et al. (2020)*
Girard et al. (2020)*
Guo et al. (2020)‡
Harvey et al. (2020)*
Hoffmann et al. (2020)
Jin et al. (2020)*
Krause et al. (2020)*
Lacerda et al. (2020)‡
Laverty et al. (2020)*
Machado et al. (2020)
McGivney et al. (2020)*
Näkki et al. (2020)*
Ng & Chiu (2020)‡
Nguyen et al. (2020)
Pinnell & Turner (2020)*
Pinto et al. (2020)*
Ramsperger et al. (2020)
Rosato et al. (2020)*
Sapozhnikov et al. (2020)
Sarkhel et al. (2020)
Seeley et al. (2020)*
Shiu et al. (2020)
Song et al. (2020)
Tobias-Hunefeldt et al. (2020)*
Tourova et al. (2020)*
Tu et al. (2020)‡
Wang et al. (2020)*
Yang et al. (2020)*
Zhao et al. (2020)
Zhu et al. (2020)
Catao et al. (2021)*
Gao & Sun (2021)
Kesy et al. (2021)*
Wright et al. (2021a)*
Wright et al. (2021b)*

Figure 1. (*see figure on the previous page*) An overview of the sampling locations and methodologies used by all relevant primary research articles returned by our literature search. All sampling locations are marked by grey points with lines linking them to the numbered study that they were from (top; coloured by the topic that the study is focussed on, with black, white and grey edge colours being for studies carried out in the laboratory, field or both, respectively) with each study being represented by a hexagon (bottom) with six triangles summarising the experimental methodology used (legend top right). White points within the "Group sequenced" triangle indicate that this sequencing was not high throughput, e.g., by Denaturing Gradient Gel Electrophoresis (DGGE). Study names shown in green were included in our previous review. * after the study name denote that sequencing data are accessible, while ‡ denotes that data are not accessible. A table with full summaries of the methodology and key findings for each study is in Table S1.

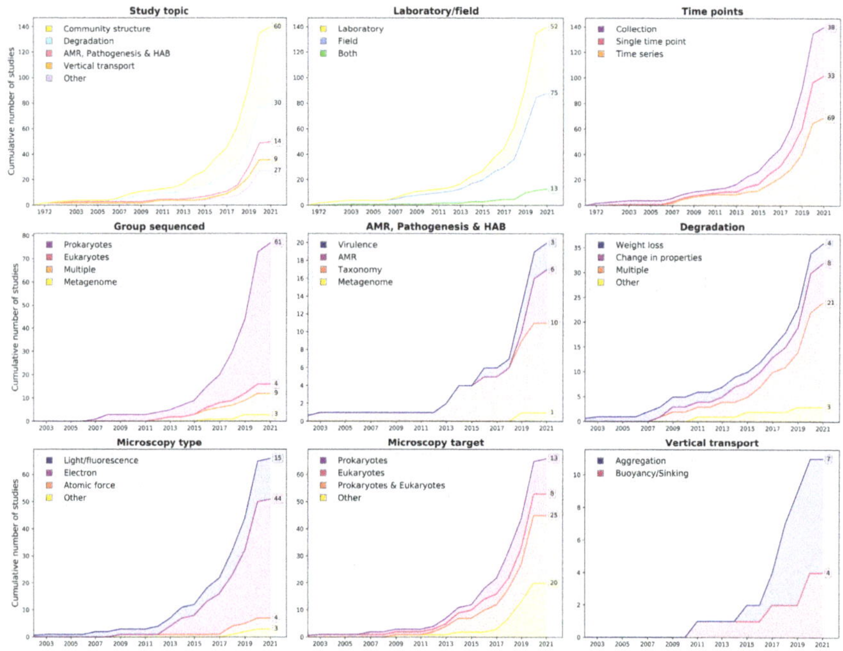

Figure 2. Cumulative number of plastisphere primary research articles published per year relating to different study topics (as in Figure 1).

Labrenz, 2020; Wright *et al.*, 2021b). The microbial communities colonising plastics likely also go through different successional stages — as with any biofilm (Datta *et al.*, 2016; see Wright, *et al.*, 2020a) — although our current understanding of these successional states on plastics is complicated by the long timescales favoured by many plastic studies (in our recent meta-analysis, the mean incubation time for all samples was 100 days; Wright *et al.*, 2021b). While results fluctuate depending on the individual study, when looking at coarse timescales, current research suggests that the *Alphaproteobacteria* are generally more abundant at early (below one week) than late (above one week) time points, while the opposite is true for the *Gammaproteobacteria, Bacteroidetes* and *Actinobacteria* (Wright *et al.*, 2021b). However, of the 69 studies that include a time series (Figures 1 and 2), only 15 include these early time points, with only two studies including sequencing of the microbial community before one day of incubation (Lee *et al.*, 2008; Kesy *et al.*, 2021). Interestingly, one of these actually suggests that at the earliest time points (3 or 9 h incubation), the *Gammaproteobacteria* dominate, while their results from 24 or 36 h of incubation were consistent with other studies where the *Alphaproteobacteria* dominate (Lee *et al.*, 2008), although further studies are needed to confirm patterns of succession, particularly at early time points.

Our knowledge of the functional potential as well as the functional capacity of the plastisphere community is currently very limited, although the abilities of plastisphere isolates have been more comprehensively tested. There are just three metagenomic datasets available (Bryant *et al.*, 2016; Pinnell and Turner, 2019; Bhagwat *et al.*, 2021) as well as several studies that measure one or several aspects of plastisphere function, e.g., alkane degradation, antimicrobial resistance or pathogenesis (Table S1). The metagenomic datasets, collected by Bryant *et al.* (2016), Pinnell and Turner (2019) and Bhagwat *et al.* (2021), were from plastics collected from the Northern Pacific Gyre, incubated in coastal Gulf of Mexico sediments or incubated in an Australian estuary, respectively. While Pinnell and Turner and Bhagwat *et al.* included both control biofilm samples (ceramic or wood, respectively) and a biodegradable polyester (polyhydroxyalkanoate [PHA] or polycaprolactone [PCL], respectively), Bryant *et al.* compared the plastisphere only with planktonic samples. It is likely for this reason that while Bryant *et al.* found that functions including nitrogen fixation, chemotaxis, type IV and VI secretion systems and xenobiotics degradation were significantly enriched on plastics (polyethylene [PE] and polypropylene [PP]), Pinnell and Turner found that recalcitrant plastic biofilms (polyethylene terephthalate [PET]) were not distinguishable from control biofilms (ceramic; although biodegradable plastic biofilms were) and Bhagwat *et al.* did not find significant functional differences between any material types [PCL, PP, polystyrene (PS), polyvinyl chloride (PVC) and wood]. It is, therefore, important that we continue to design experiments that test whether the functional capacity of plastic biofilms differs from that of other biofilms rather than planktonic samples.

3. Importance of Eukaryotes in Shaping the Plastisphere

Despite almost 10 years of Plastisphere research, the eukaryotic community fraction has consistently received far less scientific attention compared with prokaryotes (Figures 1 and 2). Among the 140 studies in our literature review, 77 papers characterised plastisphere communities using PCR-based methods, and only 16 of these included a eukaryotic fraction. Furthermore, while there were 29 studies that included the prokaryotes published during 2020 (up from 14 published in 2019), there were only four that included the eukaryotic fraction (up from three published in 2019, two of which were metagenomes that did not comment on the eukaryotes). Out of the 65 publications included in our list of plastisphere reviews (Table S2), only one was focussed on eukaryotes, reviewing the potential of fungi to degrade petrochemical plastics (Sánchez, 2020). The low inclusion rate of eukaryotes in plastisphere research, however, does not correlate with their prevalence or ecological importance in these biofilms: metagenomic investigations in the North Pacific Gyre revealed that 40–99% of reads from each of the 12 samples analysed were identified as eukaryotic sequences (Bryant *et al.*, 2016). Even though, due to the highly variable rRNA gene copy number in eukaryotes, the number of reads usually does not directly correspond to the number of actual individuals, these findings still highlight the significant contribution of eukaryotes to the overall community composition of the plastisphere (Amaral-Zettler *et al.*, 2020).

Plastisphere metabarcoding studies that target the eukaryotic fraction have typically used the 18S rRNA gene as a marker, but some have focussed on the fungal plastisphere fraction, in some cases employing fungal-specific ITS or ITS2 marker genes (Figure 2 and Table S1). Additional markers for marine microalgae (*tufA*) and metazoans (COI) have also been included on one occasion (Davidov *et al.*, 2020). Taxa including members of the diatoms, dinoflagellates, ciliates, fungi, brown, green and red algae, as well as a range of metazoan taxa are often reported as the dominant eukaryotes in the plastisphere (e.g., Zettler *et al.*, 2013;

Kettner *et al.*, 2019; Davidov *et al.*, 2020; reviewed by Rogers *et al.*, 2020). Indeed, diatoms are consistently described as an abundant and diverse group of colonisers on plastics, particularly during the early stages of biofilm formation (Carpenter and Smith, 1972; Carson *et al.*, 2013; Esensoy *et al.*, 2020), while macroscopic and microscopic forms of metazoan taxa (e.g., bryozoans, crustaceans, molluscs, anthozoans, sponges, nematodes and annelids) often also make up a major proportion of the eukaryotic communities likely, at least in part, because of their multicellular nature (Zaiko *et al.*, 2016; Kirstein *et al.*, 2018; Kettner *et al.*, 2019; Amaral-Zettler *et al.*, 2020). Apart from PCR-based methods, eukaryotic plastisphere members have been described in several publications using microscopy techniques (Figure 2). These techniques, however, are not high throughput and do not usually provide sufficient coverage or taxonomic resolution for estimation of the whole microbial community (De Tender *et al.*, 2017a). They can, however, be useful for revealing structural and spatial details unattainable via sequencing approaches, particularly when used in combination with phylogenetic labelling with fluorescent probes (De Tender *et al.*, 2017a; Davidov *et al.*, 2020; Schlundt *et al.*, 2020).

Eukaryotic microorganisms play several important roles in community regulation, dynamics and functioning in marine biofilms, ranging from photosynthesis and primary production (e.g., diatoms; Zettler *et al.*, 2013) to predation (e.g., predatory ciliates (Kettner *et al.*, 2019) and amoebae (Dudek *et al.*, 2020), parasitism (e.g., dinoflagellates and fungi; Kettner *et al.*, 2019) and other symbiotic interactions (e.g., *ectosymbiotic* bacteria on ciliates (Zettler *et al.*, 2013; Kettner *et al.*, 2019) or coral-symbiont dinoflagellates (Bryant *et al.*, 2016); reviewed by Amaral-Zettler *et al.*, 2020). Several studies have also reported that certain fungal taxa are able to degrade plastics, however only three of these strains are marine (*Zalerion maritimum* (Paço *et al.*, 2017), *Aspergillus tubingensis* and *A. flavus* (Devi *et al.*, 2015); reviewed by Rogers *et al.*, 2020 and Sánchez, 2020). Nevertheless, the search for marine plastic degraders has prompted a special interest in fungi, and, in recent years, the fungal fraction of the plastisphere — which has been found to comprise approximately 2.8–3% of eukaryotic sequences in field and lab incubations in Baltic (along an estuarine continuum) and North Sea waters, respectively (Kettner *et al.*, 2017; Kirstein *et al.*, 2018) — has been specifically targeted by at least four studies (De Tender *et al.*, 2017b; Kettner *et al.*, 2017; Davidov *et al.*, 2020; Lacerda *et al.*, 2020).

Perhaps the most interesting aspect about eukaryotes in the plastisphere, however, is their potential to shape the composition of the prokaryotic community inhabiting the biofilm through eukaryote–prokaryote interactions and close associations, some of which can be species-specific (Mühlenbruch *et al.*, 2018; Kettner *et al.*, 2019; Schlundt *et al.*, 2020). Bacteria can benefit from phytoplankton by gaining access to their cell exudates that provide a source of carbon, energy and nutrients (Amin *et al.*, 2012; Gutierrez, 2018), as well as oxygen produced *via* photosynthesis (reviewed by Stocker and Seymour, 2012). Bacteria are often observed either colonising (Amin *et al.*, 2012) or living in close proximity (Schlundt *et al.*, 2020) to phytoplankton cells, where they may end up as a result of random encounters or active chemotaxis towards the variety of chemical cues originating from the microenvironment surrounding individual phytoplankton cells — the "phycosphere" (reviewed by Stocker and Seymour, 2012 and Seymour *et al.*, 2017). Similar associations, particularly between diatoms and bacteria, have been observed in biofilms on plastics incubated in coastal Atlantic Ocean and Wadden Sea waters *via* microscopy (Schlundt *et al.*, 2020), but also by inference from sequencing data from North Sea samples (Oberbeckmann *et al.*, 2016). A co-occurrence network analysis conducted on samples incubated along a river–Baltic Sea continuum further suggested that prokaryotic community dynamics on plastic substrates (PE and PS) were largely driven by interactions with the eukaryotic community fraction. Wood substrates, on the other hand, were mainly characterised by connections between bacteria (Kettner *et al.*, 2019), probably because bacteria were able to use the wood as a carbon source while inert plastics often support carbon-producing phototrophic communities (Bryant *et al.*, 2016; Tobias-Hünefeldt *et al.*, 2020). In addition to plastic surfaces and diatoms, prokaryotes can also attach to and interact with a variety of other eukaryotic organisms, such as encrusting bryozoans or ciliates (Zettler *et al.*, 2013; Bryant *et al.*, 2016; Schlundt *et al.*, 2020). Diatom-associated bacterial groups have often been found dominating bacterial communities on marine biofilms, including those on plastics (Oberbeckmann *et al.*, 2016; Schlundt *et al.*, 2020; Wright *et al.*, 2021b).

Dudek *et al.* (2020) reported that in their Caribbean Sea time series investigations, eukaryotes — diatoms in particular — were observed to exhibit substrate specificity between the different plastic types incubated, in contrast to prokaryotes. This observation was supported by earlier findings where diatom communities differed between PE and a starch-based biodegradable plastic (Eich *et al.*, 2015). Phytoplankton are a known source of hydrocarbons, both aliphatic and aromatic, and may also adsorb them from the surrounding seawater, accumulating these compounds in the phycosphere (Gutierrez, 2018). The potential ramifications of these hydrocarbons for degradation assessments are discussed further in the following, but further investigations are needed that experimentally address the dynamics of eukaryote–prokaryote interactions in the plastisphere and differentiate them from microbe–plastic interactions (see Bryant *et al.*, 2016). Schlundt *et al.* (2020) recently conducted the first study that incorporates the analysis of the spatial structure and eukaryote–prokaryote interactions in the plastisphere by combining phylogenetic labelling [Combinatorial Labelling And Spectral Imaging–Fluorescence In Situ Hybridization (CLASI-FISH)] with confocal microscopy, as recommended earlier (Oberbeckmann *et al.*, 2016). We hope that more studies like this will follow in the future, ideally in combination with high-throughput sequencing of all community fractions, elucidating the most relevant drivers behind plastisphere community composition.

4. Plastic Degradation

Biodegradation refers to the breakdown of organic chemicals by the action of living organisms (i.e., biotic degradation), usually carried out through enzymatic catalysis (Crawford, 2011). However, this process in nature is never isolated from the action of abiotic factors. Hence, biodegradation can be understood as a chemical breakdown of organic molecules by biotic and abiotic factors (Lucas *et al.*, 2008). There are several key stages in the biodegradation of polymeric hydrocarbons, such as plastics: (i) biodeterioration — a decline in physicochemical properties mediated by microbial activity on the surface of the polymer; (ii) biofragmentation — a lytic process reducing polymer molecular weight; (iii) assimilation — integration of atoms resulting from the fragmentation process into biomass, in this case carbon uptake; and (iv) mineralisation — the process of carbon uptake and complete transformation into biomass, CO_2/H_2O in oxic environments and $CO_2/CH_4/H_2O$ in anoxic environments (Lucas *et al.*, 2008; Shah *et al.*, 2008; Wright *et al.*, 2020a). While assimilation and mineralisation are the most significant biological factors, abiotic factors also play a role in biodegradation. The abiotic degradation of polymers produces weathered materials by the action of either temperature, light, gases, mechanical forces, water or a combination of these factors (Crawford and Quinn, 2017). Depending on the type of weathering, polymers' modifications can take place through different routes. For instance, sunlight can influence the weathering of polymers by the action of ultraviolet (UV; UV-A and UV-B) radiation (i.e., photooxidation) and by heating (i.e., thermooxidation) which is mediated by the visible section of sunlight and infrared radiation (Shah *et al.*, 2008). The oxidation of polymers can be observed by the formation of different chemical groups, such as ketones, carboxylic acids and esters (Gardette *et al.*, 2013). The biological degradation of chemicals is an intricate process in which the abiotic contribution is essential as the initial step of the degradation of many recalcitrant polymers (Hakkarainen and Albertsson, 2004), which we focus on here.

Physicochemical properties inherent to the polymers impact the (bio) degradability of a specific material (Min *et al.*, 2020). Among these properties, molecular weight, glass transition temperature, hydrophilic/hydrophobic behaviour and crystallinity are relevant features to look at when determining biodegradability (Tokiwa *et al.*, 2009; Min *et al.*, 2020). For instance, highly crystalline polymers have shown to be resistant to biodegradation compared with polymers with more amorphous regions — irregular polymer structures that are more susceptible to microbial/enzymatic attack (Restrepo-Flórez *et al.*, 2014; Balani *et al.*, 2015). Furthermore, the chemical structure of the polymer itself can indicate how prone to degradation it may be; polyesters such as PET and polycaprolactone (PCL) contain functional ester groups that serve as susceptible points for the catalytic action of enzymes with esterase activity (Zheng *et al.*, 2005). On the other hand, polymers with no evident functional groups are less susceptible to enzymatic attack (i.e., more recalcitrant polymers); some examples are typical olefins, such as PE and PP. Expectedly, for the potential degradation of polyolefins, it seems to be essential to have abiotic pre-oxidation, introducing weak points for enzymatic attack: functional groups. This oxidation will generate degradation products of lower molecular weight that can be taken up by cells and metabolised (i.e., mineralised), completing the degradation process (Hakkarainen and Albertsson, 2004; Gewert *et al.*, 2018). Overall, biodegradation requires: (i) the presence of microbes able to access, uptake and mineralise the substrate, (ii) appropriate environmental conditions (e.g., temperature, salinity and pH) to initiate microbial development and (iii) a polymer susceptible to enzymatic degradation.

Microbes' potential ability to degrade plastics seems to be well distributed across a wide range of taxa. Furthermore, the majority of putative degraders reported in degradation studies were found in soil (27.8%), followed by plastic waste dumping sites (9.6%) and finally compost (5.3%) (Gambarini *et al.*, 2021). The isolation or characterisation of aquatic plastic degraders is less frequent (Zrimec *et al.*, no date), which is probably related to the extra challenge that aquatic environments, such as the oceans, represent (e.g., low nutrient concentration and temperature) when compared with soil and compost. In soil and compost, high nutrient availability, humidity and temperature are often conditions in which microorganisms thrive, positively impacting any potential biodegradation processes (Haider *et al.*, 2019). As previously explained, functional groups within a polymer (e.g., ester carbonyls and ketones) will serve as a target for the enzymatic breakdown of the polymer. In this regard, lipases and esterases have been connected to the degradation of polyesters, such as PET, initially in the terrestrial environment (Yoshida *et al.*, 2016; Danso *et al.*, 2019) but now also in the marine environment (Table S3; Gao and Sun, 2021; Wright *et al.*, 2021a). Conversely to polyester degradation, the identification of enzymes related to the degradation of polyolefins has been more challenging. So far, we do not think there is sufficient evidence to determine whether this is an active process (i.e., polymer break down mediated by enzymes) or only a passive uptake of the released carbon from the process of weathering. However, the degradation of plastics has been repeatedly linked to hydrocarbon degradation, particularly alkane degradation, due to their chemical semblance to polyolefins — relatively inert carbon-based backbone chains (Jeon and Kim, 2015). Thus, several marine studies have targeted alkane degradation genes (Table S3) (Syranidou *et al.*, 2017a, 2017b; Tsiota *et al.*, 2018).

To date, there is not a unified fashion to determine plastic degradation in the literature (Zumstein *et al.*, 2019). On the contrary, there are several methods to determine degradation, with the most commonly applied being mass loss, spectroscopic profiles (addition and removal of functional groups) and scanning electron microscopy to visualise changes in surface structure (SEM; Figure 2 and Table S3; see also Chamas *et al.*, 2020). Though these techniques on their own are not enough to properly screen for biodegradation, a combination of these can provide a better idea of the potential of putative plastic degraders. Some other more informative but less frequently applied techniques involve monitoring the incorporation of isotopically labelled carbon, e.g., ^{13}C-labelled carbon, into biological structures and monitoring CO_2 evolution (Rogers *et al.*, 2020). Using a combination of methods to evaluate plastic degradation is especially relevant when the degradation rate is minimal or when plastics contain a high concentration of plasticisers (e.g., PVC contains around 30–80% plasticisers by weight; Rahman and Brazel, 2004).

Even though plasticisers are chemically complex structures, they are degradable molecules with several described degradation pathways (Wright *et al.*, 2020b). Furthermore, hydrocarbons produced by phytoplankton may provide sustenance for hydrocarbon-degrading bacteria when other sources of hydrocarbons are scarce (see Gutierrez, 2018), and it has been suggested that diatoms could attract and concentrate hydrocarbon degraders in the plastisphere (Dudek *et al.*, 2020). These additional carbon sources are possibly more readily available for microbes to degrade than the plastics themselves. It is, therefore, necessary to rule out any preliminary "positive" results of biodegradation studies that may be attributed to the potential underlying degradation of plasticisers or phytoplankton-produced hydrocarbons. Performing comprehensive studies that monitor the integration of labelled carbon into biomass and characterise the spatial structure of the plastisphere will allow us to go beyond just noting the presence of taxa of interest and will hopefully shed light on potentially conflicting speculations that higher abundances of hydrocarbonoclasts could be: (i) degrading the plastics (Delacuvellerie *et al.*, 2019; Devi *et al.*, 2019); (ii) due to UV-induced leaching of degradation products (Pinto *et al.*, 2019); or (iii) in response to phototrophs (Dudek *et al.*, 2020).

5. Potential Pathogens Associated with Plastics

After the description of the plastisphere by Zettler *et al.* (2013), increasing attention has been directed toward the identification and understanding of the potential role of the microbial community drifting along with plastic material in several environments (e.g., oceans, rivers and wastewater). Apart from describing the microbial community on plastic marine debris, Zettler, Mincer and Amaral-Zettler also noted the presence of a *Vibrio* Operational Taxonomic Unit (OTU). Although the pathogenicity of this OTU could not be determined, this has sparked multiple enquiries surrounding the concept of "harm" posed by pathogens on plastics. In this sense, the occurrence of potentially pathogenic taxa on plastics has been extensively reviewed, and several investigations have reported examples of these microorganisms (Keswani *et al.*, 2016; Bowley *et al.*, 2020; Zhang *et al.*, 2020). These are usually referred to as "potential" pathogens (Table S4), with findings being dominated by microorganisms related to *Escherichia coli*, *Vibrio* spp. and occasionally dinoflagellates (e.g., *Alexandrium* spp.). However, it is essential to highlight that the actual pathogenicity of these microorganisms is not usually assessed, and their classification as pathogens is frequently attributed indirectly by mere taxonomic association (Figure 2). For instance, *Vibrio* strains are frequently found on marine plastic debris and are automatically associated with the pathogenic members of the genus (Baker-Austin *et al.*, 2018). Here, we give an overview of the genetic material associated with the pathogenic process and antimicrobial resistance in prokaryotes as well as the potential mode of pathogenicity, or toxicity, for microeukaryotes.

To gain insight into the harm posed by potentially pathogenic prokaryotic taxa colonising plastic debris to marine organisms, it is useful to look at the primary examples of potential pathogens, such as *E. coli* and *Vibrio* spp., both found on these materials (Muniz Silva *et al.*, 2019). Even though there are many examples of pathogenic *E. coli* and *Vibrio* spp. associated with human disease (Kaper *et al.*, 2004; Baker-Austin *et al.*, 2018), not all strains are necessarily pathogenic (Julie *et al.*, 2010; Lorenz *et al.*, 2020). Thus, it is necessary to have substantial evidence to catalogue a particular strain as a pathogen. In this context, several pathogenic or virulence factors are usually used as markers for pathogenicity. For instance, the encoding of adherence factors, toxins, type III and IV secretion systems, among others, are typically targeted as genetic features encoded by pathogenicity islands (Schmidt and Hensel, 2004). Of equal importance, pathogenic microorganisms are defined by the presence of a susceptible host — an organism vulnerable to the infection process. The existence of a susceptible host will enable a pathogen to successfully display different infectious traits, determined by genes associated with pathogenicity (i.e., pathogenic factors), which establish the type and infectious impact on the host — human or not (Casadevall and Pirofski, 2018). However, despite the relevance of identifying pathogenicity islands and the set of pathogenic factors encoded by these distinct genomic regions, such characterisations of microbial isolates retrieved from plastic debris or metagenomic analysis targeting these genetic elements are not frequently performed. One example is the investigation carried out by Muniz Silva *et al.* (2019), where it was possible to detect several virulence genes from different *E. coli* and *Vibrio* spp. isolates obtained from plastic debris. Similarly, Kirstein *et al.* (2016) also tried to identify virulence-related genes in *Vibrio* spp. isolates retrieved from marine plastics, but none of their isolates encoded any of the targeted virulence genes (Tables S4 and S5).

As noted earlier, microorganisms' capacity to produce infectious disease relies on the expression of virulence factors. When the natural immune response of the host is insufficient, the application of antibiotics may diminish and terminate the infectious process of virulent microbes. Unfortunately, microbes have also evolved several strategies to elude antibiotics' effect, a capacity provided by various antibiotic resistance genes (ARGs). The spread of antibiotic-resistant bacteria is considered a matter of global concern (IACG, 2019). ARGs in virulent microbes can complicate infections in humans potentially leading to fatal results; it is estimated that annually in the U.S.A., >35,000 people die as a consequence of >2.8 million antibiotic resistant infections (CDC, 2019). Furthermore, it is expected that antimicrobial resistant diseases could push 24 million people into extreme poverty by 2030 and reach a staggering 10 million deaths per year by 2050 (IACG, 2019). Therefore, the elucidation of such resistance genes and the physio-pathological activity of microbes carrying ARGs have been extensively studied (Peterson, 1996; Foster, 2017). This includes investigations to determine the potential for plastics to harbour microbes containing such genetic elements. Some prevalent ARGs on plastics are tetracycline (*tet*) and sulfonamide (*sul*), macrolide-lincosamide-streptogramin (MLS) and multidrug resistance genes; these last two examples have been detected by metagenomic analyses (Table S4). Interestingly, extended-spectrum beta-lactamases (ESBLs) are not frequently identified on plastics despite their widespread global distribution (Amos *et al.*, 2014; Hawkey, 2015; Song *et al.*, 2020).

Alongside potentially pathogenic bacteria, plastics have also been reported to host potentially pathogenic or Harmful Algal Bloom (HAB) forming microeukaryotes (see Audrézet *et al.*, 2021); these taxa include members of toxin-producing dinoflagellates (e.g., the genera *Alexandrium, Amphidinium, Coolia, Heterocapsa, Ostreopsis, Pfiesteria* and *Prorocentrum*) and diatoms (*Pseudo-nitzschia*) as well as other protists that comprise potential seagrass (*Labyrinthulaceae* sp(p).) and coral (*Halofolliculina* sp.) pathogens (Table S4). As with the potential pathogenicity of plastic-associated bacteria, the hypothetical threats from these microeukaryotic rafters are mostly just noted but are rarely tested in any experimental manner, and the harmfulness of microeukaryotes may not always be easily determined by amplicon sequencing or taxonomic approaches alone. For example, the toxin production (or lack thereof) of dinoflagellates from the genus *Alexandrium*, responsible for causing paralytic shellfish poisoning in humans, varies considerably in both quantity as well as quality between — and even within — species and can also be affected by a number of factors related to growth cycle phase and environmental conditions (reviewed by Murray *et al.*, 2015). Likewise, several ecological and environmental factors are likely to regulate their bloom-forming potential in marine waters (Murray *et al.*, 2015). To our knowledge, the only marine plastic related study that took this a step further and quantitatively characterised the toxin-production profiles of harmful dinoflagellates (*Alexandrium pacificum*) isolated from marine plastic debris was that of Casabianca *et al.* (2019), who found that all of the 10 dinoflagellate strains produced 4 or more of the 12 paralytic shellfish toxins tested. They further hypothesised that marine plastic debris could accumulate these toxins and aid their transfer through the food web (Casabianca *et al.*, 2019).

Overall, the analysis of plastic samples collected from different environments reveals that potentially virulent — or harmful — microbes and ARGs are present within the plastisphere (Tables S4 and S5). Caution is suggested when analysing and interpreting results because of both the uncertainty in determining whether plastisphere microbes are pathogenic and the issues related to transport and plastic residence times, discussed in the following. What is clear, however, is the necessity to have more robust evidence when suggesting the presence of pathogens on plastic debris.

6. Effects of Microbe–Plastic Interactions on Plastic Buoyancy and Transport

Even though in many cases naturally occurrent substrates (e.g., wood, leaves and rocks) are also able to concentrate high abundances and diversity of microorganisms, including potential pathogens (Muniz Silva *et al.*, 2019; Wu *et al.*, 2019; Song *et al.*, 2020), plastics are lightweight and highly resistant to both biotic and abiotic degradation. This means that — whereas naturally occurring materials such as wood, leaves or other biogenic substrates will be eventually degraded (Kooi *et al.*,

2018) — plastics can be easily transported in the environment, from one ecosystem to another. This can occur via rivers, currents and winds (Hale *et al.*, 2020) as well as rain (Brahney *et al.*, 2020), snowfall (Bergmann *et al.*, 2019) and melting sea ice (Peeken *et al.*, 2018) and is often referred to as a dynamic process: stranded plastic particles — potentially colonised by resident microbiota — can be subsequently washed back into the water flow and transported downstream or back out to sea. This extended life afloat enables plastic debris to be transported across long distances and to reach remote locations that may host vulnerable ecosystems (e.g., Thiel *et al.*, 2005; Gregory, 2009; Barnes *et al.*, 2018).

Several non-indigenous taxa have been documented in association with marine plastic debris (see Audrézet *et al.*, 2021), prompting concerns regarding the ecological impacts caused by the spread of potentially invasive species to new areas as a result of rafting on plastic waste (Gregory, 2009; Barnes *et al.*, 2018). Caution regarding these potentially harmful rafters seems justified, especially as global warming may further support the spread of non-indigenous or harmful species to new areas, particularly in polar regions (Barnes, 2002; Harvey *et al.*, 2020). Moreover, recent research indicates that ecologically and economically important seagrass meadows (Jones *et al.*, 2020; de Smit *et al.*, 2021) and other habitat-forming taxa, such as hard corals (de Smit *et al.*, 2021), can trap and accumulate microplastics, while potential seagrass (Dudek *et al.*, 2020) and coral (Goldstein *et al.*, 2014) pathogens have been found on plastics and plastic debris has been associated with an increased likelihood of disease on coral reefs (Lamb *et al.*, 2018). Other taxa that are not generally regarded as invasive, toxic or pathogenic as such, but that can cause nuisance via mucilage events, such as the diatom species *Ceratoneis Closterium* (Masó *et al.*, 2016), have also been reported as plastisphere members. Alongside the potential toxicity of the plastic polymer itself, or plastic additives and sorbed contaminants (see Pham *et al.*, 2021), plastics' potential to be considered harmful clearly also depends on the interaction with pathogens and non-native taxa present in the environment (see above), especially in polluted environments, such as wastewater effluents and water bodies in the proximity of densely populated urbanised areas (Bastaraud *et al.*, 2020).

The role of biofouling has long been acknowledged as a factor affecting the buoyancy and potential sedimentation of marine plastics, i.e., vertical transport. Observations of low-density plastics resting on the seafloor or buried in sediments have been reported since the mid-1970's

(Holmström, 1975; Schwarz *et al.*, 2019). Despite that most plastics produced as well as most plastic debris found in the marine environment are less dense than seawater, recently these should-be-buoyant polymers have also been collected from the deeper layers of the water column, demonstrating the "fallout" of plastics from surface waters (Geyer *et al.*, 2017; Schwarz *et al.*, 2019; Egger *et al.*, 2020). Due to the rapid colonisation of marine plastics — including by calcareous animals and/or algae that can cover the plastics with thick, filamentous growth — the overall density of biofouled particles may exceed that of seawater, causing them to lose buoyancy and sink (Ye and Andrady, 1991; Barnes, 2002; Fazey and Ryan, 2016; Kaiser *et al.*, 2017). Both experimental field studies and mathematical sinking models have found that biofouled plastics that sink to deeper layers of the water column can also undergo rapid de-fouling due to a number of factors (Ye and Andrady, 1991; Wright *et al.*, 2020a). This could potentially cause them to continue oscillating up and down in the water column as a result of recurring fouling and de-fouling events, thereby retaining the highest microplastic concentrations at mid-depths of the water body (Kooi *et al.*, 2017; see Wright *et al.*, 2020a for further discussion).

Among the relatively few field-based incubation studies and other observations on the topic, buoyancy-loss and biofouling-induced sinking of plastics ($\geq$1 mm) in the marine environment have been mainly associated with colonisation by macroorganisms and/or calcareous animals (Holmström, 1975; Ye and Andrady, 1991; Fazey and Ryan, 2016; Kaiser *et al.*, 2017). It is, therefore, currently unclear whether natural microbial biofouling alone can cause individual, low-density microplastics to sink in seawater (Fazey and Ryan, 2016; Kaiser *et al.*, 2017; Wright *et al.*, 2020a). While some studies have concluded that macroorganisms are likely necessary for the transport of buoyant plastics through the water column, others have noted that some macroorganisms (e.g., barnacles) are unlikely to colonise sub-millimetre particles (Fazey and Ryan, 2016; Kaiser *et al.*, 2017); Assuming this statement can be extended to most macrocolonisers, the potential for vertical transport of individual sub-millimetre microplastic particles should depend mainly on microbial colonisation, although aggregation with inorganic matter and detritus is likely to play a role as well Kaiser *et al.*, 2017; Möhlenkamp *et al.*, 2018; Wright *et al.*, 2020a; Wu *et al.*, 2020a). A recent modelling study concluded that small (radius range 10 μm–1 mm) spherical microplastics did not sink within 90 days in oligotrophic ocean waters, where algal growth is scarce

(Lobelle *et al.*, 2021). To our knowledge, however, no studies exist that experimentally investigated the effect of microbial biofouling on the buoyancy of individual, sub-millimetre, low-density microplastics in seawater in either lab or field settings, likely due to the difficulties in the handling, recovery and analysis of small particles while retaining natural exposure conditions, as noted by several studies (Kaiser *et al.*, 2017; Arias-Andres *et al.*, 2018; Tu *et al.*, 2020).

The effects of interactions between biogenic material (such as marine snow and marine microalgae) and very small, sub-millimetre microplastics (down to 2 μm) have in fact been investigated in a number of laboratory studies (Long *et al.*, 2015, 2017; Möhlenkamp *et al.*, 2018; Porter *et al.*, 2018; Cunha *et al.*, 2019). These studies, however, have either focussed on heteroaggregation, rather than the colonisation of individual microplastic particles, or polymers that are denser than seawater (that would readily sink without biofouling) (Nguyen *et al.*, 2020). These studies have found that heteroaggregation can either increase or decrease the settling or rising velocity of microplastics or even change the direction of their movement altogether (i.e., making buoyant polymers sink and, according to a brackish water model, non-buoyant polymers rise/float), depending on the polymer type and the composition of the heteroaggregates (Long *et al.*, 2015; Möhlenkamp *et al.*, 2018; Nguyen *et al.*, 2020). Biofilm coating has further been shown to promote aggregation between microplastics and natural particles (Michels *et al.*, 2018), while field-collected marine aggregates have also been found to contain microplastics (Zhao *et al.*, 2018; de Haan *et al.*, 2019). Despite a number of studies having been published on the sinking and rising behaviour of microplastics incorporated into heteroaggregates, the experimental data needed to enable comparisons between empirical data and mathematical model predictions are lacking, specifically: (i) on the effects of microbial biofouling on the buoyancy of individual sub-millimetre microplastics; and (ii) on natural marine biofilm parameters, namely, biofilm thickness and density (Wright *et al.*, 2020a). Further impacts of microbial colonisation on the vertical transport of plastics are related to the increased ingestion and subsequent egestion of plastics by macroorganisms, although microeukaryotes, including marine dinoflagellates and ciliates, have also been observed to ingest small microplastics, possibly making microplastics more bioavailable for marine organisms at higher trophic levels (Galloway *et al.*, 2017; Long *et al.*, 2017; Zhang *et al.*, 2021).

7. Effects of Plastic Pollution on Biogeochemical Cycling in the Oceans

Despite being included in the call for research by Harrison *et al.* (2011), the potential impacts of plastic pollution and microbe–plastic interactions on ecologically critical biogeochemical processes have remained under-discussed and largely understudied (Figure S1). This topic is beginning to gain more attention, with a number of studies published so far in 2021 (e.g., Birarda *et al.*, 2021; Boldrini *et al.*, 2021) as well as reviews and other works that either focused solely on biogeochemical effects of marine microbe–plastic interactions or included it as one of several topics (e.g., Galgani and Loiselle, 2021; Wang *et al.*, 2021b).

Although often referred to as "inert" materials, plastics can release many chemical compounds (Tetu *et al.*, 2019; Galgani and Loiselle, 2021). This happens when they are immersed in seawater — particularly when exposed to solar irradiation — and the compounds range from additives, metals, greenhouse gases (e.g., methane and ethylene) and other degradation and weathering products to pollutants adsorbed from the surrounding environment (Gewert *et al.*, 2018; León *et al.*, 2018; Royer *et al.*, 2018; Tetu *et al.*, 2019; Capolupo *et al.*, 2020). Depending on the integrity and chemical composition of the plastic, these leachates can either stimulate or inhibit microbial activity (Romera-Castillo *et al.*, 2018; Sarker *et al.*, 2020; Zhu *et al.*, 2020). It has been estimated that dissolved organic carbon (DOC) leaching from marine plastics (i.e., plastic leachates) can amount up to 23,600 metric tonnes per year and, in areas with high levels of plastic pollution, could comprise as much as 10% of all DOC present in the marine surface microlayer (Romera-Castillo *et al.*, 2018). Note that the term DOC is used by Romera-Castillo *et al.* (2018) to include all potential leachates from plastics and has previously been defined as the organic carbon present in water after filtering out particles of approximately >0.45 μm in size (Zsolnay, 2003). This definition is problematic given that nanoplastics (1–1,000 nm) could, therefore, be included within this and we, therefore, suggest that future research makes a distinction between the "dissolved" organic carbon that is particulate and that which is in suspension. Nevertheless, this organic carbon is also available for microbial utilisation, enhancing the growth of heterotrophic marine microbes (Romera-Castillo *et al.*, 2018; Zhu *et al.*, 2020). Since marine microbes play a crucial role in global biogeochemical processes, including carbon and nutrient cycling as well as oxygen production and climate regulation, while also forming the basis of marine food webs via primary production (Cavicchioli *et al.*, 2019), the potential threat of plastic pollution disturbing these processes is not trivial.

The presence of microplastics has been shown to alter the composition of the overall pool of marine dissolved organic matter (DOM) by accumulating light-absorbing and more recalcitrant forms of DOM (i.e., chromophoric DOM [CDOM]) as a result of increased microbial processes (Galgani *et al.*, 2018; Galgani and Loiselle, 2021). This accumulation of CDOM in surface waters, along with the potential shading effect due to the accumulation of buoyant plastics (Arias-Andres *et al.*, 2018), could hamper light availability in deeper layers of the water column and have implications for phototrophic primary production as well as overall marine carbon dynamics (Galgani *et al.*, 2018). Microplastics can also stimulate microbial production of organic carbon and marine gel particles which, alongside the demonstrated capacity of microplastics to agglomerate with and alter the sinking rates of natural marine aggregates, may potentially affect the functioning of the marine biological pump (Long *et al.*, 2015; Möhlenkamp *et al.*, 2018; Cunha *et al.*, 2019; Galgani *et al.*, 2019; Galgani and Loiselle, 2021). Several studies have previously suggested that similar effects would be caused by any inert or inorganic particles, and this hypothesis was recently confirmed (Galgani *et al.*, 2018, 2019; Boldrini *et al.*, 2021). Boldrini *et al.* compared the effects of PS microplastics to those of silica gel particles and found that the silica gel particles actually increased the production and transformation of CDOM by microbes even more than microplastics (Boldrini *et al.*, 2021). Indeed, it has long been known that solid particles immersed in seawater adsorb and concentrate nutrients as well as organic matter on their surfaces, transforming them into "hotspots" of microbial activity (Zobell, 1943; Dang and Lovell, 2016; Mincer *et al.*, 2016). This phenomenon has been shown to apply to plastics as well, demonstrated, for example, by higher bacterial activity on different polymer substrates compared with seawater (Dussud *et al.*, 2018a) and by the ability of bacterial communities to remain viable over several months in cultures amended with PET particles, while similar cultures without PET soon died out (Webb *et al.*, 2009; Mincer *et al.*, 2016). Furthermore, microplastic abundance in the environment is only predicted to increase, thereby having the potential to considerably and irreversibly increase the number of particles affecting microbial activities in the (otherwise largely oligotrophic) oceans

(Bryant *et al.*, 2016; Galgani *et al.*, 2018; Oberbeckmann *et al.*, 2018; Borrelle *et al.*, 2020).

In addition to carbon cycling, some studies have demonstrated that microplastics can interfere with nitrogen cycling in marine sediments (Cluzard *et al.*, 2015; Seeley *et al.*, 2020). PVC microplastics were found to inhibit nitrification and denitrification in coastal salt marsh sediments in a microcosm study, while microparticles of polyurethane foam (PUF), polylactic acid (PLA) and PE enhanced these processes (Seeley *et al.*, 2020). Contrastingly, PE microbeads extracted from a cosmetic product were found to impair microbial ammonium removal via denitrification in intertidal sediment microcosms, leading to the accumulation of ammonium in the overlying water, potentially leading to eutrophication and HAB incidents in the natural environment (Cluzard *et al.*, 2015). Residues of antimicrobial compounds in the cosmetic PE microbeads may have adversely affected the microbial communities responsible for ammonium removal, whereas the disruptions to nitrogen cycling by PVC were possibly caused by sulfide producing bacteria, whose significant enrichment was potentially linked to an unidentified additive, highlighting the importance of the chemical additive composition of plastics (Cluzard *et al.*, 2015; Seeley *et al.*, 2020).

Marine plastics harbour microbial communities that are taxonomically and functionally (e.g., enrichment of genes coding for nitrogen fixation) distinct from planktonic communities, and conservative estimates reveal that marine plastic biofilms comprise approximately 1% of the total biomass in the surface microlayer of the oceans (Bryant *et al.*, 2016; Zhao *et al.*, 2020; Wright *et al.*, 2021b). As discussed earlier, it is unclear whether marine microbes biodegrade and utilise the recalcitrant polymer backbones of all plastics for growth, although evidence for microbial mineralisation and assimilation of some plastics exists (Gao and Sun, 2021; Wright *et al.*, 2021a). Concerns have, however, been raised regarding the ecological and biogeochemical impacts of alterations to the overall marine microbiome as plastic surfaces provide new niches for and enrich taxa that would otherwise be scarce in abundance in the planktonic community fraction (Bryant *et al.*, 2016; Oberbeckmann *et al.*, 2018; Oberbeckmann and Labrenz, 2020; Zhao *et al.*, 2020). Furthermore, high concentrations of plastic debris are accumulating in oligotrophic open ocean gyres, with unknown consequences to the functioning of these ecosystems (Cózar *et al.*, 2014; Mincer *et al.*, 2016; Oberbeckmann *et al.*, 2018). Increased (heterotrophic) microbial activity is often coupled with increased oxygen consumption, which could potentially lead to oxygen depleted conditions in areas with high levels of plastic pollution, including the seafloor (e.g., Arias-Andres *et al.*, 2018; Krause *et al.*, 2020). Moreover, the impacts of plastic pollution on primary production by microbial photoautotrophs in the marine environment need further clarifying as floating plastic debris has been described as hotspots for net autotrophic production, while adverse effects on these organisms have also been reported (Bryant *et al.*, 2016; Yokota *et al.*, 2017; Tetu *et al.*, 2019; Sarker *et al.*, 2020). The plastisphere has also been found to contribute to greenhouse gas (CO_2 and N_2O) fluxes in the Eastern South Pacific Ocean, albeit the contribution was rather small (less than 1% of the overall ocean contribution) (Cornejo-D'Ottone *et al.*, 2020). Experimental evidence also indicates that exposure to plastics and related leachates can negatively affect biocalcification by marine forams, resulting in possible impacts on oceanic carbon storage and consequently climate regulation (Birarda *et al.*, 2021).

8. Conclusions and Future Directions

We now have a wealth of information on the taxonomic composition of the prokaryotic (particularly bacterial) component of the plastisphere and, while we are not advocating for studies to stop characterising the prokaryotes, there are several other areas that need further investigation. We have, therefore, brought together some of the key themes that came up in the conclusions and future directions suggested by many of the reviews returned by our literature search (Figure S1 and Table S2), namely, that future studies should:

(i) use multi-OMIC and interdisciplinary approaches to understand the role of the plastisphere in plastic biodegradation as well as in determining the mechanisms and pathways used for biodegradation;
(ii) examine the impacts of plastic waste in concert with other ecological threats, such as climate change and antibiotic resistance;
(iii) assess the consequences of the microbial communities on plastics on plastic toxicity and interactions with other organisms and their environments;
(iv) empirically investigate the ability of the potentially harmful microbes (including bacteria and microeukaryotes) already identified on plastics to induce infection or cause harm;

(v) use isotopically labelled plastics or other methods to confirm plastic degradation more comprehensively alongside assessments of plastic degradation in the environment.

There are also suggestions by several studies to standardise methodologies in order to increase the ability to draw fair comparisons between different locations and environmental conditions (e.g., De Tender *et al.*, 2017a; Rogers *et al.*, 2020; Table S2). While we agree that standard methodologies or test substrates could be useful tools in order to compare the plastisphere on a global scale and could be incorporated with data collected for the Earth Microbiome Project (e.g., Roager and Sonnenschein, 2019), this must be balanced against the rapid development of new and better methodologies, alongside the fact that there are few universally agreed upon best practice methods for microbiome or plastics research (De Tender *et al.*, 2017a; Knight *et al.*, 2018). Furthermore, a standardised methodology is only useful if the authors of studies guarantee that their data are made available upon publication of their study. Even in 2020, 24% of published plastisphere studies that used next-generation sequencing did not make their data accessible (Figure 1). We, therefore, urge authors and reviewers to confirm that these data are made accessible at the time of publication, ensuring that these data may be used for other purposes, such as in other meta-analyses (Wright *et al.*, 2021b), the construction of databases containing known genes for plastic degradation (Gan and Zhang, 2019; Gambarini *et al.*, 2021), or for the mining of metagenomes for plastic degradation genes (Danso *et al.*, 2018).

As discussed earlier, we are also currently lacking information on the non-bacterial plastisphere constituents; to our knowledge, there are currently no studies reporting on or investigating viruses in the plastisphere, and we have relatively little information on the eukaryotic plastisphere fraction (Figure 2). We hope that with the decreasing cost and, therefore, increasing availability of metagenomic sequencing, this approach will be used more extensively, and future studies will, therefore, include information on all taxonomic groups. This will hopefully help to elucidate the role of eukaryotes in shaping the plastisphere. It is currently not clear whether it is "cost efficient" for hydrocarbonoclastic taxa to degrade plastic polymers (if theoretically capable) when more accessible forms of hydrocarbons may be available from other biofilm members (e.g., diatoms). Therefore, future research should aim to determine whether prokaryotic community composition differs if eukaryotes are excluded from the colonisation process (Bryant *et al.*, 2016; Gutierrez, 2018; Kettner *et al.*, 2019). There are further key research questions around biogeochemical cycling and the impacts plastics may have on this with increasing levels of plastic litter and under changing environmental conditions.

One suggested solution to the problems caused by the environmental accumulation of traditional (i.e., petrochemical) recalcitrant plastics is the use of biodegradable plastic types (Wang *et al.*, 2021a). However, the risks and benefits of such replacement must be carefully weighed in light of microbial processes and biogeochemical cycling (Pinnell and Turner, 2019; Wang *et al.*, 2021a), particularly as high concentrations of bioplastics also have the potential to affect marine carbon, nitrogen and sulfur cycles (Pinnell and Turner, 2019; Seeley *et al.*, 2020). Together with further implications to marine methanogenesis and cycling of natural PHAs (Pinnell and Turner, 2019), it seems that replacing traditional plastics with biodegradable ones could lead to an equally bad — or even worse — situation. Therefore, the greatest ways of mitigating the effects of plastic pollution are to radically decrease the production and use of single-use plastics — degradable or not — and to improve waste management practices globally.

Supplemental Materials

All supplementary figures, tables and code used are on Figshare: doi.org/10.6084/m9.figshare.14096912.

Figure S1: Summary of all plastisphere review articles reviewed.
Table S1: Summary of all primary research articles reviewed.
Table S2: Summary of all plastisphere review articles reviewed.
Table S3: The methods used and degradation found by studies including a measurement of biodegradation.
Table S4: Potential pathogens (prokaryotic and eukaryotic) found on plastics by the studies reviewed here.
Table S5: Antibiotic resistance genes found on plastics in aquatic environments.

References

Abed, R. M. M. *et al.* (2020). Degradability and biofouling of oxo-biodegradable polyethylene in the planktonic and benthic zones of the Arabian Gulf.

Marine Pollution Bulletin, *150* (September 2019), 110639. https://doi.org/10.1016/j.marpolbul.2019.110639.

Amaral-Zettler, L. A., Zettler, E. R., & Mincer, T. J. (2020). Ecology of the plastisphere. *Nature Reviews Microbiology*, *18*, 139–151. Springer US. DOI: 10.1038/s41579-019-0308-0.

Amaral-Zettler, L. A. *et al.* (2015). The biogeography of the plastisphere: Implications for policy. *Frontiers in Ecology and the Environment*, *13*(10), 541–546.

Amin, S. A., Parker, M. S., & Armbrust, E. V. (2012). Interactions between diatoms and bacteria. *Microbiology and Molecular Biology Reviews*, *76*(3), 667–684. DOI: 10.1128/MMBR.00007-12.

Amos, G. C. A. *et al.* (2014). Waste water effluent contributes to the dissemination of CTX-M-15 in the natural environment. *Journal of Antimicrobial Chemotherapy*, *69* (May), 1785–1791. DOI: 10.1093/jac/dku079.

Arias-Andres, M. *et al.* (2018). Microplastic pollution increases gene exchange in aquatic ecosystems. *Environmental Pollution*, *237*, 253–261. Elsevier Ltd. DOI: 10.1016/j.envpol.2018.02.058.

Artham, T. *et al.* (2009). Biofouling and stability of synthetic polymers in sea water. *International Biodeterioration & Biodegradation*, *63*(7), 884–890. Elsevier Ltd. DOI: 10.1016/j.ibiod.2009.03.003.

Audrézet, F., Zaiko, A., Lear, G., Wood, S. A., Tremblay, L. A., & Pochon, X. (2021). Biosecurity implications of drifting marine plastic debris: Current knowledge and future research. *Marine Pollution Bulletin*, *162*, 111835.

Avio, C. G., Gorbi, S., & Regoli, F. (2016). Plastics and microplastics in the oceans: From emerging pollutants to emerged threat. *Marine Environmental Research*, *128*, 2–11.

Baker-Austin, C. *et al.* (2018). Vibrio spp. infections. *Nature Reviews Disease Primers*, *4*(8), 1–19. Springer US. DOI: 10.1038/s41572-018-0005-8.

Balani, K. *et al.* (2015). Physical, thermal, and mechanical properties of polymers. In *Biosurfaces: A Materials Science and Engineering Perspective*, pp. 329–344. Hoboken: John Wiley & Sons.

Baptista Neto, J. A. *et al.* (2019). Microplastics and attached microorganisms in sediments of the Vitória bay estuarine system in SE Brazil. *Ocean and Coastal Management*, *169* (September 2018), 247–253. Elsevier. DOI: 10.1016/j.ocecoaman.2018.12.030.

Barnes, D. K. A. (2002). Invasions by marine life on plastic debris. *Nature*, *416*(4), 808–809. DOI: 10.1080/00785236.1995.10422039.

Barnes, D. K. A. *et al.* (2018). Marine plastics threaten giant Atlantic marine protected areas. *Current Biology*, *28*(19), R1137–R1138. DOI: 10.1016/j.cub.2018.08.064.

Basili, M. *et al.* (2020). Major role of surrounding environment in shaping biofilm community composition on marine plastic debris. *Frontiers in Marine Science*, *7* (May), 1–12. DOI: 10.3389/fmars.2020.00262.

Bastaraud, A. *et al.* (2020). Urbanization and waterborne pathogen emergence in low-income countries: Where and how to conduct surveys? *International Journal of Environmental Research and Public Health*, *17*(2), 1–19. DOI: 10.3390/ijerph17020480.

Bergmann, M. *et al.* (2019). White and wonderful? Microplastics prevail in snow from the Alps to the Arctic. *Science Advances*, *5*(8), 1–11. DOI: 10.1126/sciadv.aax1157.

Bhagwat, G., Zhu, Q., O'Connor, W., Subashchandrabose, S., Grainge, I., Knight, R., & Palanisami, T. (2021). Exploring the composition and functions of plastic microbiome using whole-genome sequencing. *Environmental Science & Technology*, *55*(8), 4899–4913.

Birarda, G. *et al.* (2021a). Plastics, (bio)polymers and their apparent biogeochemical cycle: An infrared spectroscopy study on foraminifera. *Environmental Pollution*, *279*, 116912. Elsevier Ltd. DOI: 10.1016/j.envpol.2021.116912.

Boldrini, A. *et al.* (2021b). Microplastics contamination versus inorganic particles: Effects on the dynamics of marine dissolved organic matter. *Environments*, *8*(3), 21. DOI: 10.3390/environments8030021.

Borrelle, S. B. *et al.* (2020). Predicted growth in plastic waste exceeds efforts to mitigate plastic pollution. *Science*, *369* (September), 1515–1518.

Bowley, J., Baker-Austin, C., Porter, A., Hartnell, R., & Lewis, C. (2021). Oceanic hitchhikers — Assessing pathogen risks from marine microplastic. *Trends in microbiology*, *29*(2), 107–116.

Brahney, J. *et al.* (2020). Plastic rain in protected areas of the United States. *Science*, *368*, 1257–1260.

Briand, J. F. *et al.* (2012). Pioneer marine biofilms on artificial surfaces including antifouling coatings immersed in two contrasting French Mediterranean coast sites. *Biofouling*, *28*(5), 453–463. DOI: 10.1080/08927014.2012.688957.

Bryant, J. A. *et al.* (2016). Diversity and activity of communities inhabiting plastic debris in the North Pacific Gyre. *mSystems*, *1*(3), e00024-16. DOI: 10.1128/mSystems.00024-16.

Capolupo, M. *et al.* (2020). Chemical composition and ecotoxicity of plastic and car tire rubber leachates to aquatic organisms. *Water Research*, *169*, 115270. Elsevier Ltd. DOI: 10.1016/j.watres.2019.115270.

Carpenter, E. J. *et al.* (1972). Polystyrene spherules in coastal waters. *Science*, *178*(4062), 749–750. Retrieved from https://www.jstor.org/stable/1735843.

Carpenter, E. J., & Smith, K. L. J. (1972). Plastics on the Sargasso Sea surface. *Science*, 175(4027), 1240–1241.

Carson, H. S. *et al.* (2013). The plastic-associated microorganisms of the North Pacific Gyre. *Marine Pollution Bulletin*, *75*, 126–132. DOI: 10.1016/j.marpolbul.2013.07.054.

Casabianca, S. *et al.* (2019). Plastic-associated harmful microalgal assemblages in marine environment. *Environmental Pollution*, *244*, 617–626. Elsevier Ltd. DOI: 10.1016/j.envpol.2018.09.110.

Casabianca, S. *et al.* (2020). Physical interactions between marine phytoplankton and PET plastics in seawater. *Chemosphere, 238*, 124560. Elsevier Ltd. DOI: 10.1016/j.chemosphere.2019.124560.

Casadevall, A., & Pirofski, L. (2018). What is a host? Attributes of individual susceptibility. *Infection and Immunity, 86*(2), 1–12.

Catao, E. C. P. *et al.* (2019). Shear stress as a major driver of marine biofilm communities in the NW Mediterranean Sea. *Frontiers in Microbiology, 10* (July), 1–16. DOI: 10.3389/fmicb.2019.01768.

Catao, E. C. P. *et al.* (2021). Metal resistance genes enrichment in marine biofilm communities selected by biocide-containing surfaces in temperate and tropical coastal environments. *Environmental Pollution, 268*, 115835. Elsevier Ltd. DOI: 10.1016/j.envpol.2020.115835.

Cavicchioli, R. *et al.* (2019). Scientists' warning to humanity: Microorganisms and climate change. *Nature Reviews Microbiology, 17*(9), 569–586. Springer US. DOI: 10.1038/s41579-019-0222-5.

CDC. (2019). Antibiotic resistance threats in the United States. Retrieved from https://www.cdc.gov/drugresistance/pdf/threats-report/2019-ar-threats-report-508.pdf. U. S. Centers for Disease Control and Prevention. Accessed on 21/07/2022.

Chamas, A. *et al.* (2020). Degradation rates of plastics in the environment. *ACS Sustainable Chemistry and Engineering, 8*(9), 3494–3511. DOI: 10.1021/acssuschemeng.9b06635.

Cluzard, M. *et al.* (2015). Intertidal concentrations of microplastics and their influence on ammonium cycling as related to the shellfish industry. *Archives of Environmental Contamination and Toxicology, 69*(3), 310–319. Springer US. DOI: 10.1007/s00244-015-0156-5.

Cornejo-D'Ottone, M., Molina, V., Pavez, J., & Silva, N. (2020). Greenhouse gas cycling by the plastisphere: The sleeper issue of plastic pollution. *Chemosphere, 246*, 125709.

Cózar, A. *et al.* (2014). Plastic debris in the open ocean. *Proceedings of the National Academy of Sciences, 111*(28), 10239–10244. DOI: 10.1073/pnas.1314705111.

Crawford, C. B., & Quinn, B. (2017). Physiochemical properties and degradation. In *Microplastic Pollutants*. Elsevier Science, Amsterdam. pp. 57–100. DOI: 10.1016/B978-0-12-809406-8.00004-9.

Crawford, R. L. (2011). Biodegradation: Principles, scope, and technologies. In *Comprehensive Biotechnology*, 2nd ed., pp. 3–13. Amsterdam: Elsevier B.V. DOI: 10.1016/B978-0-08-088504-9.00368-8.

Cunha, C. *et al.* (2019). Marine vs freshwater microalgae exopolymers as biosolutions to microplastics pollution. *Environmental Pollution, 249*, 372–380. DOI: 10.1016/j.envpol.2019.03.046.

Curren, E., & Leong, S. C. Y. (2019). Profiles of bacterial assemblages from microplastics of tropical coastal environments. *Science of the Total Environment, 655*, 313–320. Elsevier B.V. DOI: 10.1016/j.scitotenv.2018.11.250.

Dang, H. *et al.* (2008). Cross-ocean distribution of Rhodobacterales bacteria as primary surface colonizers in temperate coastal marine waters. *Applied and Environmental Microbiology, 74*(1), 52–60. DOI: 10.1128/AEM.01400-07.

Dang, H., & Lovell, C. R. (2016). Microbial surface colonization and biofilm development in marine environments. *Microbiology and Molecular Biology Reviews, 80*(1), 91–138. DOI: 10.1128/MMBR.00037-15.

Danso, D. *et al.* (2018). New insights into the function and global distribution of polyethylene terephthalate (PET) degrading bacteria and enzymes in marine and terrestrial metagenomes. *Applied and Environmental Microbiology, 84*(8), e02773-17. DOI: 10.1128/AEM.02773-17.

Danso, D., Chow, J., & Streita, W. R. (2019). Plastics: Environmental and biotechnological perspectives on microbial degradation. *Applied and Environmental Microbiology, 85*(19), 1–14. DOI: 10.1128/AEM.01095-19.

Datta, M. S. *et al.* (2016). Microbial interactions lead to rapid micro-scale successions on model marine particles. *Nature Communications, 7*(11965), 1–7. DOI: 10.1038/ncomms11965.

Davidov, K. *et al.* (2020). Identification of plastic-associated species in the Mediterranean Sea using DNA metabarcoding with nanopore MinION. *Scientific Reports, 10*(17533), 1–11. Nature Publishing Group UK.

de Haan, W. P., Sanchez-Vidal, A., & Canals, M. (2019). Floating microplastics and aggregate formation in the Western Mediterranean Sea. *Marine Pollution Bulletin, 140* (January), 523–535. DOI: 10.1016/j.marpolbul.2019.01.053.

de Smit, J. C. *et al.* (2021). Habitat-forming species trap microplastics into coastal sediment sinks. *Science of the Total Environment, 772*, 145520. The Authors. DOI: 10.1016/j.scitotenv.2021.145520.

De Tender, C. A. *et al.* (2015). Bacterial community profiling of plastic litter in the Belgian part of the North Sea. *Environmental Science & Technology, 49*, 9629–9638. DOI: 10.1021/acs.est.5b01093.

De Tender, C. A., Schlundt, C., Devriese, L. I. *et al.* (2017a). A review of microscopy and comparative molecular-based methods to characterize "Plastisphere" communities. *Analytical Methods, 9*(14), 2132–2143. Royal Society of Chemistry. DOI: 10.1039/c7ay00260b.

De Tender, C. A., Devriese, L. I., Haegeman, A. *et al.* (2017b). Temporal dynamics of bacterial and fungal colonization on plastic debris in the North Sea. *Environmental Science & Technology, 51*(13), 7350–7360. DOI: 10.1021/acs.est.7b00697.

Debroas, D., Mone, A., & Ter Halle, A. (2017). Plastics in the North Atlantic garbage patch: A boat-microbe for hitchhikers and plastic degraders. *Science of The Total Environment, 599–600*, 1222–1232. DOI: 10.1016/j.scitotenv.2017.05.059.

Delacuvellerie, A. *et al.* (2019). The plastisphere in marine ecosystem hosts potential specific microbial degraders including *Alcanivorax borkumensis* as a key player for the low-density polyethylene degradation. *Journal of Hazardous Materials, 380* (July), 120899. Elsevier. DOI: 10.1016/j.jhazmat.2019.120899.

Denaro, R. *et al.* (2020). Marine hydrocarbon-degrading bacteria breakdown poly(ethylene terephthalate) (PET). *Science of the Total Environment, 749*. 141608. DOI: 10.1016/j.scitotenv.2020.141608.

Devi, R. S. *et al.* (2015). Biodegradation of HDPE by Aspergillus spp. from marine ecosystem of Gulf of Mannar, India. *Marine Pollution Bulletin, 96*, 32–40. Elsevier Ltd. DOI: 10.1016/j.marpolbul.2015.05.050.

Devi, R. S. *et al.* (2019). Investigation of biodegradation potentials of high density polyethylene degrading marine bacteria isolated from the coastal regions of Tamil Nadu, India. *Marine Pollution Bulletin, 138*, 549–560. Elsevier. DOI: 10.1016/j.marpolbul.2018.12.001.

Dudek, K. L. *et al.* (2020). Microbial colonization of microplastics in the Caribbean Sea. *Limnology and Oceanography Letters, 5*(1), 5–17. DOI: 10.1002/lol2.10141.

Dussud, C., Hudec, C. *et al.* (2018a). Colonization of non-biodegradable and biodegradable plastics by marine microorganisms. *Frontiers in Microbiology, 9*, 1–13. DOI: 10.3389/fmicb.2018.01571.

Dussud, C., Meistertzheim, A. L. *et al.* (2018b). Evidence of niche partitioning among bacteria living on plastics, organic particles and surrounding seawaters. *Environmental Pollution, 236*, 807–816. DOI: 10.1016/j.envpol.2017.12.027.

Egger, M., Sulu-Gambari, F., & Lebreton, L. (2020). First evidence of plastic fallout from the North Pacific Garbage Patch. *Scientific Reports, 10*(1), 1–10. DOI: 10.1038/s41598-020-64465-8.

Eich, A. *et al.* (2015). Biofilm and diatom succession on polyethylene (PE) and biodegradable plastic bags in two marine habitats: Early signs of degradation in the pelagic and benthic zone? *PLoS One*, 10(9), 1–16. DOI: 10.1371/journal.pone.0137201.

Erni-Cassola, G. *et al.* (2019). Early colonization of weathered polyethylene by distinct bacteria in marine coastal seawater. *Microbial Ecology, 79*(3), 517–526. DOI: 10.1007/s00248-019-01424-5.

Esensoy, F. B., Senturk, Y., & Aytan, U. (2020). Microbial biofilm on plastics in the southeastern Black Sea. In *Marine Litter in the Black Sea*, Turkish Marine Research Foundation (TUDAV) Publication No: 56, Istanbul, Turkey. pp. 268–286.

Falah, W. *et al.* (2020). Polyethylene terephthalate degradation by Microalga *Chlorella vulgaris* along with pretreatment. *Materiale Plastice, 57*(3), 260–270. DOI: 10.37358/MP.20.3.5398.

Fazey, F. M. C., & Ryan, P. G. (2016). Biofouling on buoyant marine plastics: An experimental study into the effect of size on surface longevity. *Environmental Pollution, 210*, 354–360. Elsevier Ltd. DOI: 10.1016/j.envpol.2016.01.026.

Foster, T. J. (2017). Antibiotic resistance in *Staphylococcus aureus*. Current status and future prospects. *FEMS Microbiology Reviews*, *41* (April), 430–449. DOI: 10.1093/femsre/fux007.

Foulon, V. *et al.* (2016). Colonization of polystyrene microparticles by *Vibrio crassostreae*: Light and electron microscopic investigation. *Environmental Science and Technology*, *50*(20), 10988–10996. DOI: 10.1021/acs.est.6b02720.

Frere, L. *et al.* (2018). Microplastic bacterial communities in the Bay of Brest: Influence of polymer type and size. *Environmental Pollution*, *242*, 614–625. DOI: 10.1016/j.envpol.2018.07.023.

Galgani, L. *et al.* (2018). Polystyrene microplastics increase microbial release of marine chromophoric dissolved organic matter in microcosm experiments. *Scientific Reports*, *8*(1), 1–11. Springer US. DOI: 10.1038/s41598-018-32805-4.

Galgani, L. *et al.* (2019). Microplastics increase the marine production of particulate forms of organic matter. *Environmental Research Letters*, *14*(12). IOP Publishing. DOI: 10.1088/1748-9326/ab59ca.

Galgani, L. & Loiselle, S. A. (2021). Plastic pollution impacts on marine carbon biogeochemistry. *Environmental Pollution*, *268*, 115598. Elsevier Ltd. DOI: 10.1016/j.envpol.2020.115598.

Galloway, T. S., Cole, M., & Lewis, C. (2017). Interactions of microplastic debris throughout the marine ecosystem. *Nature Ecology and Evolution*, *1*(5), 1–8. Macmillan Publishers Limited. DOI: 10.1038/s41559-017-0116.

Gambarini, V. *et al.* (2021). Phylogenetic distribution of plastic-degrading microorganisms. *mSystems*, *6*(1), 1–13. DOI: 10.1128/msystems.01112-20.

Gan, Z. & Zhang, H. (2019). PMBD: A comprehensive plastics microbial biodegradation database. *Database: The Journal of Biological Databases and Curation*, *2019*, 1–11. DOI: 10.1093/database/baz119.

Gao, R. & Sun, C. (2021). A marine bacterial community capable of degrading poly(ethylene terephthalate) and polyethylene. *Journal of Hazardous Materials*, 416, 125928.

Gardette, M. *et al.* (2013). Photo- and thermal-oxidation of polyethylene: Comparison of mechanisms and influence of unsaturation content. *Polymer Degradation and Stability*, *98*, 2383–2390. DOI: 10.1016/j.polymdegradstab.2013.07.017.

Gerritse, J. *et al.* (2020). Fragmentation of plastic objects in a laboratory seawater microcosm. *Scientific Reports*, *10*(1), 1–16. DOI: 10.1038/s41598-020-67927-1.

Gewert, B. *et al.* (2018). Identification of chain scission products released to water by plastic exposed to ultraviolet light. *Environmental Science & Technology Letters*, *5*, 272–276. American Chemical Society. DOI: 10.1021/acs.estlett.8b00119.

Geyer, R., Jambeck, J. R., & Law, K. L. (2017). Production, use, and fate of all plastics ever made. *Science Advances*, *3*(7), 19–24. DOI: 10.1126/sciadv.1700782.

Giacomucci, L. *et al.* (2020). Biodegradation of polyvinyl chloride plastic films by enriched anaerobic marine consortia. *Marine Environmental Research, 158* (March), 104949. Elsevier Ltd. DOI: 10.1016/j.marenvres.2020.104949.

Girard, E. B. *et al.* (2020). Biodegradation of textile waste by marine bacterial communities enhanced by light. *Environmental Microbiology Reports, 12*(4), 406–418. DOI: 10.1111/1758-2229.12856.

Goldstein, M. C., Carson, H. S., & Eriksen, M. (2014). Relationship of diversity and habitat area in North Pacific plastic-associated rafting communities. *Marine Biology, 161*(6), 1441–1453. DOI: 10.1007/s00227-014-2432-8.

Gómez-Ramírez, A. L., Enriquez-Ocaña, L. F., Miranda-Baeza, A., Cordero Esquivel, B., López-Elías, J. A., & Martínez-Córdova, L. R. (2019). Biofilm-forming capacity of two benthic microalgae, Navicula incerta and Navicula sp., on three substrates (Naviculales: Naviculaceae). *Revista de Biología Tropical, 67*(3), 599–607.

Gregory, M. R. (2009). Environmental implications of plastic debris in marine settings- entanglement, ingestion, smothering, hangers-on, hitch-hiking and alien invasions. *Philosophical Transactions of the Royal Society B: Biological Sciences, 364*(1526), 2013–2025. DOI: 10.1098/rstb.2008.0265.

Guo, X. P. *et al.* (2020). Antibiotic resistance genes in biofilms on plastic wastes in an estuarine environment. *Science of the Total Environment, 745*, 140916. Elsevier B.V. DOI: 10.1016/j.scitotenv.2020.140916.

Gutierrez, T. (2018). Occurrence and roles of the obligate hydrocarbonoclastic bacteria in the ocean when there is no obvious hydrocarbon contamination. In McGenity, T. (Ed.), *Taxonomy, Genomics and Ecophysiology of Hydrocarbon-Degrading Microbes*. Cham: Springer. DOI: 10.1007/978-3-030-14796-9.

Haider, T. P. *et al.* (2019). Plastics of the future? The impact of biodegradable polymers on the environment and on society. *Angewandte Chemie — International Edition, 58*, 50–62. DOI: 10.1002/anie.201805766.

Hakkarainen, M. & Albertsson, A. (2004). Environmental degradation of polyethylene. In *Advances in Polymer Science*, vol. 169, pp. 177–199. DOI: 10.1007/b13523.

Hale, R. C. *et al.* (2020). A global perspective on microplastics. *Journal of Geophysical Research: Oceans, 125*(1), 1–40. DOI: 10.1029/2018JC014719.

Harrison, J. P. *et al.* (2011). Interactions between microorganisms and marine microplastics: A call for research. *Marine Technology Society Journal, 45*(2), 12–20. DOI: 10.4031/MTSJ.45.2.2.

Harrison, J. P. *et al.* (2014). Rapid bacterial colonization of low-density polyethylene microplastics in coastal sediment microcosms. *BMC Microbiology, 14*(232), 1–15.

Harshvardhan, K. & Jha, B. (2013). Biodegradation of low-density polyethylene by marine bacteria from pelagic waters, Arabian Sea, India. *Marine Pollution Bulletin*, *77*(1–2), 100–106. Elsevier Ltd. DOI: 10.1016/j.marpolbul.2013.10.025.

Harvey, B. P. *et al.* (2020). Ocean acidification alters bacterial communities on marine plastic debris. *Marine Pollution Bulletin*, *161*(PB), 111749. Elsevier Ltd. DOI: 10.1016/j.marpolbul.2020.111749.

Hawkey, P. M. (2015). Multidrug-resistant Gram-negative bacteria, *89*(4), 241–247. *Journal of Hospital Infection*. Elsevier Ltd. DOI: 10.1016/j.jhin.2015.01.008.

Hoffmann, L., Eggers, S. L., Allhusen, E., Katlein, C., & Peeken, I. (2020). Interactions between the ice algae Fragillariopsis cylindrus and microplastics in sea ice. *Environment International*, *139*, 105697.

Holmström, A. (1975). Plastic films on the bottom of the Skagerack. *Nature*, *255*, 622–623.

IACG. (2019). No Time to Wait: Securing the Future from Drug-Resistant Infections. Report to the Secretary-General of the United Nations. Retrieved from https://www.who.int/antimicrobial-resistance/interagency-coordination-group/IACG_final_report_EN.pdf?ua=1. United Nations Interagency Coordination Group on Antimicrobial Resistance. Accessed on 21/07/2022.

Jeon, H. J. & Kim, M. N. (2015). Functional analysis of alkane hydroxylase system derived from Pseudomonas aeruginosa E7 for low molecular weight polyethylene biodegradation. *International Biodeterioration & Biodegradation*, *103*, 141–146. Elsevier Ltd. DOI: 10.1016/j.ibiod.2015.04.024.

Jiang, P. *et al.* (2018). Microplastic-associated bacterial assemblages in the intertidal zone of the Yangtze Estuary. *Science of the Total Environment*, *624*, 48–54. Elsevier B.V. DOI: 10.1016/j.scitotenv.2017.12.105.

Jin, M. *et al.* (2020). How biofilms affect the uptake and fate of hydrophobic organic compounds (HOCs) in microplastic: Insights from an in situ study of Xiangshan Bay, China. *Water Research*, *184*, 116118. Elsevier Ltd. DOI: 10.1016/j.watres.2020.116118.

Johansen, M. P. *et al.* (2019). Biofilm-enhanced adsorption of strong and weak cations onto different microplastic sample types: Use of spectroscopy, microscopy and radiotracer methods. *Water Research*, *158*, 392–400. Elsevier Ltd. DOI: 10.1016/j.watres.2019.04.029.

Jones, K. L. *et al.* (2020). Microplastic accumulation in a Zostera marina L. bed at Deerness Sound, Orkney, Scotland. *Marine Pollution Bulletin*, *152* (September 2019), 110883. Elsevier. DOI: 10.1016/j.marpolbul.2020.110883.

Jones, P. R. *et al.* (2007). Bacterial community structure of biofilms on artificial surfaces in an estuary. *Microbial Ecology*, *53*(1), 153–162. DOI: 10.1007/s00248-006-9154-5.

Julie, D. *et al.* (2010). Ecology of pathogenic and non-pathogenic Vibrio parahaemolyticus on the French Atlantic coast. Effects of temperature, salinity, turbidity and chlorophyll *a*. *Environmental Microbiology*, *12*(4), 929–937. DOI: 10.1111/j.1462-2920.2009.02136.x.

Kaiser, D., Kowalski, N., & Waniek, J. J. (2017). Effects of biofouling on the sinking behavior of microplastics. *Environmental Research Letters*, *12*(12), 124003. DOI: 10.1088/1748-9326/aa8e8b.

Kaper, J. B., Nataro, J. P., & Mobley, H. L. T. (2004). Pathogenic Escherichia coli. *Nature Reviews Microbiology*, *2* (February), 123–140. DOI: 10.1038/nrmicro818.

Kathiresan, K. (2003). Polythene and plastics-degrading microbes from the mangrove soil. *Revista de Biología Tropical*, *51*(3), 629–634.

Keswani, A. *et al.* (2016). Microbial hitchhikers on marine plastic debris: Human exposure risks at bathing waters and beach environments. *Marine Environmental Research*, *118*, 10–19. Elsevier Ltd. DOI: 10.1016/j.marenvres.2016.04.006.

Kesy, K. *et al.* (2016). Polystyrene influences bacterial assemblages in Arenicola marina-populated aquatic environments in vitro. *Environmental Pollution*, *219*, 219–227. Elsevier Ltd. DOI: 10.1016/j.envpol.2016.10.032.

Kesy, K. *et al.* (2019). Spatial environmental heterogeneity determines young biofilm assemblages on microplastics in Baltic Sea mesocosms. *Frontiers in Microbiology*, *10* (August), 1–18. DOI: 10.3389/fmicb.2019.01665.

Kesy, K., Labrenz, M., Scales, B. S., Kreikemeyer, B., & Oberbeckmann, S. (2020). Vibrio colonization is highly dynamic in early microplastic-associated biofilms as well as on field-collected microplastics. *Microorganisms*, *9*(1), 76.

Kettner, M. T. *et al.* (2017). Microplastics alter composition of fungal communities in aquatic ecosystems. *Environmental Microbiology*, *19*, 4447–4459. DOI: 10.1111/1462-2920.13891.

Kettner, M. T. *et al.* (2019). The eukaryotic life on microplastics in brackish ecosystems. *Frontiers in Microbiology*, *10*, 538. DOI: 10.3389/fmicb.2019.00538.

Kirstein, I. V. *et al.* (2016). Dangerous hitchhikers? Evidence for potentially pathogenic Vibrio spp. on microplastic particles. *Marine Environmental Research*, *120*, 1–8. DOI: 10.1016/j.marenvres.2016.07.004.

Kirstein, I. V. *et al.* (2018). Mature biofilm communities on synthetic polymers in seawater — Specific or general? *Marine Environmental Research*, *142*, 147–154. Elsevier Ltd. DOI: 10.1016/j.marenvres.2018.09.028.

Kirstein, I. V. *et al.* (2019). The plastisphere – uncovering tightly attached plastic "specific" microorganisms. *PLoS One*, *14*(4), 1–17. DOI: 10.1371/journal.pone.0215859.

Knight, R. *et al.* (2018). Best practices for analysing microbiomes. *Nature Reviews Microbiology*, *16*(7), 410–422. Springer US. DOI: 10.1038/s41579-018-0029-9.

Kooi, M. *et al.* (2017). Ups and downs in the ocean: Effects of biofouling on vertical transport of microplastics. *Environmental Science and Technology*, *51*(14), 7963–7971. DOI: 10.1021/acs.est.6b04702.

Kooi, M. *et al.* (2018). Modeling the fate and transport of plastic debris in freshwaters: Review and guidance. In *Freshwater Microplastics, The Handbook of Environmental Chemistry* , Vol. 58, pp. 125–152. Cham: Springer Open. DOI: 10.1007/978-3-319-61615-5.

Krause, S. *et al.* (2020). Persistence of plastic debris and its colonization by bacterial communities after two decades on the abyssal seafloor. *Scientific Reports*, *10*(1), 1–15. DOI: 10.1038/s41598-020-66361-7.

Kumar, M., Xie, A., & Curley, J. (2016). Determining the potential secondary impacts associated with microorganismal biodegradation of microplastics in the marine environment. *The Journal of Experimental Secondary Science*, *3*(4), 1–11.

Kumar, S., Hatha, A. A. M., & Christi, K. S. (2007). Diversity and effectiveness of tropical mangrove soil microflora on the degradation of polythene carry bags. *International Journal of Tropical Biology*, *55*, 777–786.

Lacerda, A. L. F. *et al.* (2019). Plastics in sea surface waters around the Antarctic Peninsula. *Scientific Reports*, *9*(3977), 1–12. DOI: 10.1038/s41598-019-40311-4.

Lacerda, A. L. F. *et al.* (2020). Diverse groups of fungi are associated with plastics in the surface waters of the Western South Atlantic and the Antarctic Peninsula. *Molecular Ecology*, *29*(10), 1903–1918. DOI: 10.1111/mec.15444.

Laganà, P. *et al.* (2019). Do plastics serve as a possible vector for the spread of antibiotic resistance? First insights from bacteria associated to a polystyrene piece from King George Island (Antarctica). *International Journal of Hygiene and Environmental Health*, *222*(1), 89–100. Elsevier. DOI: 10.1016/j.ijheh.2018.08.009.

Lamb, J. B. *et al.* (2018). Plastic waste associated with disease on coral reefs. *Science*, *2010* (January), 26–29.

Laverty, A. L., Primpke, S., Lorenz, C., Gerdts, G., & Dobbs, F. C. (2020). Bacterial biofilms colonizing plastics in estuarine waters, with an emphasis on Vibrio spp. and their antibacterial resistance. *PloS One*, *15*(8), e0237704.

Lee, J.-W. *et al.* (2008). Bacterial communities in the initial stage of marine biofilm formation on artificial surfaces. *The Journal of Microbiology*, *46*(2), 174–182. DOI: 10.1007/s12275-008-0032-3.

Lee, Y. M. *et al.* (2016). Succession of bacterial community structure during the early stage of biofilm development in the Antarctic marine environment. *Korean Journal of Microbiology*, *52*(1), 49–58. DOI: 10.7845/kjm.2016.6005.

León, V. M. *et al.* (2018). Potential transfer of organic pollutants from littoral plastics debris to the marine environment. *Environmental Pollution, 236*, 442–453. DOI: 10.1016/j.envpol.2018.01.114.

Li, W., Zhang, Y., Wu, N., Zhao, Z., Xu, W. A., Ma, Y., & Niu, Z. (2019). Colonization characteristics of bacterial communities on plastic debris influenced by environmental factors and polymer types in the Haihe Estuary of Bohai Bay, China. *Environmental science & technology*, *53*(18), 10763–10773.

Lobelle, D. & Cunliffe, M. (2011). Early microbial biofilm formation on marine plastic debris. *Marine Pollution Bulletin*, *62*(1), 197–200. Elsevier Ltd. DOI: 10.1016/j.marpolbul.2010.10.013.

Lobelle, D., Kooi, M., Koelmans, A. A., Laufkötter, C., Jongedijk, C. E., Kehl, C., & van Sebille, E. (2021). Global modeled sinking characteristics of biofouled microplastic. *Journal of Geophysical Research: Oceans*, *126*(4), e2020JC017098.

Long, M. *et al.* (2015). Interactions between microplastics and phytoplankton aggregates: Impact on their respective fates. *Marine Chemistry*, *175*, 39–46. Elsevier B.V. DOI: 10.1016/j.marchem.2015.04.003.

Long, M. *et al.* (2017). Interactions between polystyrene microplastics and marine phytoplankton lead to species-specific hetero-aggregation. *Environmental Pollution*, *228*, 454–463. DOI: 10.1016/j.envpol.2017.05.047.

Lorenz, B. *et al.* (2020). Discrimination between pathogenic and non-pathogenic *E. coli* strains by means of Raman microspectroscopy. *Analytical and Bioanalytical Chemistry*, *412*, 8241–8247.

Lucas, N. *et al.* (2008). Polymer biodegradation: Mechanisms and estimation techniques. *Chemosphere*, *73*, 429–442. DOI: 10.1016/j.chemosphere.2008.06.064.

Machado, M. C. *et al.* (2020). The response of *Synechococcus* sp. PCC 7002 to micro-/nano polyethylene particles — Investigation of a key anthropogenic stressor. *PLoS One*, *15* (7 July), 1–14. DOI: 10.1371/journal.pone.0232745.

Masó, M. *et al.* (2003). Drifting plastic debris as a potential vector for dispersing harmful algal bloom (HAB) species. *Scientia Marina*, *67*(1), 107–111. DOI: 10.3989/scimar.2003.67n1107.

Masó, M. *et al.* (2016). Microfouling communities from pelagic and benthic marine plastic debris sampled across Mediterranean coastal waters. *Scientia Marina*, *80* (October), 117–127. DOI: 10.3989/scimar.04281.10A.

McGivney, E. *et al.* (2020). Rapid physicochemical changes in microplastic induced by biofilm formation. *Frontiers in Bioengineering and Biotechnology*, *8* (March), 1–14. DOI: 10.3389/fbioe.2020.00205.

Michels, J. *et al.* (2018). Rapid aggregation of biofilm-covered microplastics with marine biogenic particles. *Proceedings of the Royal Society B: Biological Sciences*, *285*(1885), 1–9. DOI: 10.1098/rspb.2018.1203.

Min, K., Cuiffi, J. D., & Mathers, R. T. (2020). Ranking environmental degradation trends of plastic marine debris based on physical properties and molecular structure. *Nature communications*, *11*(1), 1–11.

Mincer, T. J., Zettler, E. R., & Amaral-Zettler, L. A. (2016). Biofilms on plastic debris and their influence on marine nutrient cycling, productivity, and hazardous chemical mobility. Handbook of Environmental Chemistry, vol. 78, pp. 221–233. DOI: 10.1007/698_2016_12.

Misic, C. & Covazzi Harriague, A. (2019). Development of marine biofilm on plastic: Ecological features in different seasons, temperatures, and light regimes. *Hydrobiologia*, *835*(1), 129–145. Springer International Publishing. DOI: 10.1007/s10750-019-3934-7.

Mohanrasu, K. *et al.* (2018). Exploring multi potential uses of marine bacteria; An integrated approach for PHB production, PAHs and polyethylene biodegradation. *Journal of Photochemistry & Photobiology, B: Biology*, *185* (May), 55–65. Elsevier. DOI: 10.1016/j.jphotobiol.2018.05.014.

Möhlenkamp, P., Purser, A., & Thomsen, L. (2018). Plastic microbeads from cosmetic products: An experimental study of their hydrodynamic behaviour, vertical transport and resuspension in phytoplankton and sediment aggregates. *Elementa: Science of the Anthropocene* (2018) *6*(61), 1–16.

Moog, D. *et al.* (2019). Using a marine microalga as a chassis for polyethylene terephthalate (PET) degradation. *Microbial Cell Factories*, *18*(1), 1–15. BioMed Central. DOI: 10.1186/s12934-019-1220-z.

Morét-Ferguson, S. *et al.* (2010). The size, mass, and composition of plastic debris in the western North Atlantic Ocean. *Marine Pollution Bulletin*, *60*(10), 1873–1878. DOI: 10.1016/j.marpolbul.2010.07.020.

Mühlenbruch, M. *et al.* (2018). Mini-review: Phytoplankton-derived polysaccharides in the marine environment and their interactions with heterotrophic bacteria. *Environmental Microbiology*, *20*(8), 2671–2685. DOI: 10.1111/1462-2920.14302.

Muniz Silva, M. *et al.* (2019). Dispersal of potentially pathogenic bacteria by plastic debris in Guanabara Bay, RJ, Brazil. *Marine Pollution Bulletin*, *141* (March), 561–568. Elsevier. DOI: 10.1016/j.marpolbul.2019.02.064.

Murray, S. *et al.* (2015). 4 Alexandrium spp.: Genetic and ecological factors influencing saxitoxin production and proliferation. In: *Climate Change and Marine and Freshwater Toxins 2nd ed.*, De Gruyter Berlin/Boston pp. 133–166. DOI: 10.1515/978311-0625738-004.

Muthukrishnan, T., Al Khaburi, M., & Abed, R. M. (2019). Fouling microbial communities on plastics compared with wood and steel: Are they substrate- or location-specific? *Microbial ecology*, *78*(2), 361–374.

Näkki, P. *et al.* (2020). Polycyclic aromatic hydrocarbon sorption and bacterial community composition of biodegradable and conventional plastics incubated in coastal sediments. *Science of the Total Environment*, *755*, 143088. The Author(s). DOI: 10.1016/j.scitotenv.2020.143088.

Nauendorf, A. *et al.* (2016). Microbial colonization and degradation of polyethylene and biodegradable plastic bags in temperate fine-grained organic-rich marine sediments. *Marine Pollution Bulletin, 103*(1–2), 168–178. Elsevier Ltd. DOI: 10.1016/j.marpolbul.2015.12.024.

Ng, J. C. Y. & Chiu, J. M. Y. (2020). Changes in biofilm bacterial communities in response to combined effects of hypoxia, ocean acidification and nutrients from aquaculture activity in Three Fathoms Cove. *Marine Pollution Bulletin, 156* (October 2019), 111256. Elsevier. DOI: 10.1016/j.marpolbul.2020.111256.

Nguyen, T. H., Tang, F. H. M., & Maggi, F. (2020). Sinking of microbial-associated microplastics in natural waters. *PLoS One, 15*(2), 1–20. DOI: 10.1371/journal.pone.0228209.

Oberbeckmann, S. *et al.* (2014). Spatial and seasonal variation in diversity and structure of microbial biofilms on marine plastics in Northern European waters. *FEMS Microbiology Ecology, 90*, 478–492. DOI: 10.1111/1574-6941.12409.

Oberbeckmann, S., Kreikemeyer, B., & Labrenz, M. (2018). Environmental factors support the formation of specific bacterial assemblages on microplastics. *Frontiers in Microbiology, 8*, 1–12. DOI: 10.3389/FMICB.2017.02709.

Oberbeckmann, S. & Labrenz, M. (2020). Marine microbial assemblages on microplastics: Diversity, adaptation, and role in degradation. *Annual Review of Marine Science, 12*, 1–24.

Oberbeckmann, S., Osborn, A. M., & Duhaime, M. B. (2016). Microbes on a bottle: Substrate, season and geography influence community composition of microbes colonizing marine plastic debris. *PLoS One, 11*(8), 1–24. DOI: 10.1371/journal.pone.0159289.

Ogonowski, M. *et al.* (2018). Evidence for selective bacterial community structuring on microplastics. *Environmental Microbiology, 20*(8), 2796–2808. DOI: 10.1111/1462-2920.14120.

Paço, A. *et al.* (2017). Biodegradation of polyethylene microplastics by the marine fungus Zalerion maritimum. *Science of The Total Environment, 586*, 10–15. Elsevier B.V. DOI: 10.1016/j.scitotenv.2017.02.017.

Peeken, I., Primpke, S., Beyer, B., Gütermann, J., Katlein, C., Krumpen, T., Bergmann, M., Hehemann, L. & Gerdts, G. (2018). Arctic sea ice is an important temporal sink and means of transport for microplastic. *Nature communications, 9*(1), 1–12.

Peterson, J. (1996). Bacterial pathogenesis. In Baron, S. (Ed.), *Medical Microbiology*, 4th ed. University of Texas Medical Branch, Galveston.

Pham, T. H. *et al.* (2021). Global challenges in microplastics: From fundamental understanding to advanced degradations toward sustainable strategies. *Chemosphere, 267*, 129275. Elsevier Ltd. DOI: 10.1016/j.chemosphere.2020.129275.

Pinnell, L. J. & Turner, J. W. (2019). Shotgun metagenomics reveals the benthic microbial community response to plastic and bioplastic in a coastal marine environment. *Frontiers in Microbiology, 10*(June), 1–13. DOI: 10.3389/fmicb.2019.01252.

Pinnell, L. J. & Turner, J. W. (2020). Temporal changes in water temperature and salinity drive the formation of a reversible plastic-specific microbial community. *FEMS Microbiology Ecology, 96*(12), 1–10. Oxford University Press. DOI: 10.1093/femsec/fiaa230.

Pinto, M. *et al.* (2019). The composition of bacterial communities associated with plastic biofilms differs between different polymers and stages of biofilm succession. *PLoS One, 14*(6), 1–20. DOI: 10.1371/journal.pone.0217165.

Pinto, M. *et al.* (2020). Putative degraders of low-density polyethylene-derived compounds are ubiquitous members of plastic-associated bacterial communities in the marine environment. *Environmental Microbiology, 22*(11), 4779–4793. DOI: 10.1111/1462-2920.15232.

Pollet, T. *et al.* (2018). Prokaryotic community successions and interactions in marine biofilms: The key role of Flavobacteriia. *FEMS Microbial Ecology, 94*(6), 1–13. DOI: 10.1093/femsec/fiy083/4992301.

Porter, A. *et al.* (2018). Role of marine snows in microplastic fate and bioavailability. *Environmental Science and Technology, 52*(12), 7111–7119. American Chemical Society. DOI: 10.1021/acs.est.8b01000.

Quilliam, R. S., Jamieson, J., & Oliver, D. M. (2014). Seaweeds and plastic debris can influence the survival of faecal indicator organisms in beach environments. *Marine Pollution Bulletin, 84*(1–2), 201–207. DOI: 10.1016/j.marpolbul.2014.05.011.

Raghul, S. S. *et al.* (2014). Biodegradation of polyvinyl alcohol-low linear density polyethylene-blended plastic film by consortium of marine benthic vibrios. *International Journal of Environmental Science and Technology, 11*, 1827–1834. DOI: 10.1007/s13762-013-0335-8.

Rahman, M. & Brazel, C. S. (2004). The plasticizer market: An assessment of traditional plasticizers and research trends to meet new challenges. *Progress in Polymer Science, 29*, 1223–1248. DOI: 10.1016/j.progpolymsci.2004.10.001.

Rajeev, M. *et al.* (2019). Culture dependent and independent analysis and appraisal of early stage biofilm-forming bacterial community composition in the Southern coastal seawater of India. *Science of the Total Environment, 666*, 308–320. Elsevier B.V. DOI: 10.1016/j.scitotenv.2019.02.171.

Ramsperger, A. F. R. M. *et al.* (2020). Structural diversity in early-stage biofilm formation on microplastics depends on environmental medium and polymer properties. *Water, 12*(11), Article No. 3216, 1–21. DOI:10.3390/w12113216.

Reisser, J. *et al.* (2014). Millimeter-sized marine plastics: A new pelagic habitat for microorganisms and invertebrates. *PLoS One, 9*(6), 1–11. DOI: 10.1371/journal.pone.0100289.

Restrepo-Flórez, J.-M., Bassi, A., & Thompson, M. R. (2014). Microbial degradation and deterioration of polyethylene — A review. *International Biodeterioration and Biodegradation*, *88*, 83–90. Elsevier Ltd. DOI: 10.1016/j.ibiod.2013.12.014.

Roager, L. & Sonnenschein, E. C. (2019). Bacterial candidates for colonization and degradation of marine plastic debris. *Environmental Science and Technology*, *53*(20), 11636–11643. American Chemical Society. DOI: 10.1021/acs.est.9b02212.

Rodrigues, A. *et al.* (2019). Colonisation of plastic pellets (nurdles) by E. coli at public bathing beaches. *Marine Pollution Bulletin*, *139*, 376–380. Elsevier. DOI: 10.1016/j.marpolbul.2019.01.011.

Rogers, K. L. *et al.* (2020). Micro-by-micro interactions: How microorganisms influence the fate of marine microplastics. *Limnology and Oceanography Letters*, *5*(1), 18–36. DOI: 10.1002/lol2.10136.

Romera-Castillo, C. *et al.* (2018). Dissolved organic carbon leaching from plastics stimulates microbial activity in the ocean. *Nature Communications*, *9*(1), 1–7. Springer US. DOI: 10.1038/s41467-018-03798-5.

Rosato, A. *et al.* (2020). Microbial colonization of different microplastic types and biotransformation of sorbed PCBs by a marine anaerobic bacterial community. *Science of the Total Environment*, *705*, 135790. Elsevier B.V. DOI: 10.1016/j.scitotenv.2019.135790.

Royer, S.-J. *et al.* (2018). Production of methane and ethylene from plastic in the environment. *PLoS One*, *13*(8), 1–13. DOI: 10.1371/journal. pone.0200574.

Sánchez, C. (2020). Fungal potential for the degradation of petroleum-based polymers: An overview of macro- and microplastics biodegradation. *Biotechnology Advances*, *40* (December 2019), 107501. Elsevier. DOI: 10.1016/j.biotechadv.2019.107501.

Sapozhnikov, P. V., Snigirova, A., & Kalinina, O. (2020). Microphytes assemblages on the neustoplastics from the North Black Sea. In Aytan, Ü., Pogojeva, M., Simeonova, A. (Eds.), *Marine Litter in the Black Sea*, pp. 287–302. Istanbul, Turkey: Turkish Marine Research Foundation (TUDAV), No. 56. (November).

Sarker, I. *et al.* (2020). Assessing the toxicity of leachates from weathered plastics on photosynthetic marine bacteria prochlorococcus. *Frontiers in Marine Science*, *7* (September), 1–14. DOI: 10.3389/fmars.2020.571929.

Sarkhel, R., Sengupta, S., Das, P., & Bhowal, A. (2020). Comparative biodegradation study of polymer from plastic bottle waste using novel isolated bacteria and fungi from marine source. *Journal of Polymer Research*, *27*(1), 1–8.

Schlundt, C. *et al.* (2020). Spatial structure in the "Plastisphere": Molecular resources for imaging microscopic communities on plastic marine debris. *Molecular Ecology Resources*, *20* (November), 1–15. DOI: 10.1111/1755-0998.13119.

Schmidt, H. & Hensel, M. (2004). Pathogenicity islands in bacterial pathogenesis. *Clinical Microbiology Reviews*, *17* (1), 14–56. DOI: 10.1128/CMR.17.1.14.

Schwarz, A. E. *et al.* (2019). Sources, transport, and accumulation of different types of plastic litter in aquatic environments: A review study. *Marine Pollution Bulletin*, *143* (March), 92–100. Elsevier. DOI: 10.1016/j.marpolbul.2019.04.029.

Seeley, M. E. *et al.* (2020). Microplastics affect sedimentary microbial communities and nitrogen cycling. *Nature Communications*, *11*(1), 1–10. Springer US. DOI: 10.1038/s41467-020-16235-3.

Seymour, J. R., Amin, S. A., Raina, J. B., & Stocker, R. (2017). Zooming in on the phycosphere: the ecological interface for phytoplankton–bacteria relationships. *Nature microbiology*, *2*(7), 1–12.

Shah, A. A. *et al.* (2008). Biological degradation of plastics: A comprehensive review. *Biotechnology Advances*, *26*(3), 246–265. DOI: 10.1016/j.biotechadv.2007.12.005.

Shiu, R. F. *et al.* (2020). Nano- and microplastics trigger secretion of protein-rich extracellular polymeric substances from phytoplankton. *Science of the Total Environment*, *748*, 141469. Elsevier B.V. DOI: 10.1016/j.scitotenv.2020.141469.

Siboni, N. *et al.* (2007). Conditioning film and initial biofilm formation on ceramics tiles in the marine environment. *FEMS Microbiology Letters*, *274*, 24–29. DOI: 10.1111/j.1574-6968.2007.00809.x.

Song, J. *et al.* (2020). The travelling particles: Investigating microplastics as possible transport vectors for multidrug resistant *E. coli* in the Weser estuary (Germany). *Science of the Total Environment*, *720*, 137603. Elsevier B.V. DOI: 10.1016/j.scitotenv.2020.137603.

Stocker, R. & Seymour, J. R. (2012). Ecology and physics of bacterial chemotaxis in the ocean. *Microbiology and Molecular Biology Reviews*, *76*(4), 792–812. DOI: 10.1128/MMBR.00029-12.

Sudhakar, M. *et al.* (2008). Marine microbe-mediated biodegradation of low- and high-density polyethylenes. *International Biodeterioration & Biodegradation*, *61*(3), 203–213. DOI: 10.1016/j.ibiod.2007.07.011.

Summers, S., Henry, T., & Gutierrez, T. (2018). Agglomeration of nano- and microplastic particles in seawater by autochthonous and de novo-produced sources of exopolymeric substances. *Marine Pollution Bulletin*, *130* (March), 258–267. Elsevier. DOI: S0025326X18302042.

Syranidou, E., Karkanorachaki, K., Amorotti, F., Franchini, M. *et al.* (2017a). Biodegradation of weathered polystyrene films in seawater microcosms. *Scientific Reports*, *7*(1), 1–12. DOI: 10.1038/s41598-017-18366-y.

Syranidou, E., Karkanorachaki, K., Amorotti, F., Repouskou, E. *et al.* (2017b). Development of tailored indigenous marine consortia for the degradation of naturally weathered polyethylene films. *PLoS One*, *12*(8), 1–21.

Syranidou, E. *et al.* (2019). Biodegradation of mixture of plastic films by tailored marine consortia. *Journal of Hazardous Materials*, *375*, 33–42. DOI: 10.1016/j.jhazmat.2019.04.078.

Tagg, A. S. *et al.* (2019). Paint particles are a distinct and variable substrate for marine bacteria. *Marine Pollution Bulletin*, *146* (June), 117–124. Elsevier. DOI: 10.1016/j.marpolbul.2019.06.013.

Tetu, S. G. *et al.* (2019). Plastic leachates impair growth and oxygen production in Prochlorococcus, the ocean's most abundant photosynthetic bacteria. *Communications Biology*, *2*(1), 1–9. Springer US. DOI: 10.1038/s42003-019-0410-x.

Thiel, M. & Gutow, L. (2005). The ecology of rafting in the marine environment. I. The floating substrata. In: *Oceanography and Marine Biology: an annual review*, Vol. 42, pp. 181–264. Boca Raton, London, New York, Washington: CRC Press.

Tobias-Hünefeldt, S. P., Wenley, J., Baltar, F., & Morales, S. E. (2021). Ecological drivers switch from bottom–up to top–down during model microbial community successions. *The ISME Journal*, *15*(4), 1085–1097.

Tokiwa, Y. *et al.* (2009). Biodegradability of plastics. *International Journal of Molecular Sciences*. Multidisciplinary Digital Publishing Institute (MDPI), *10*(9), 3722–3742. DOI: 10.3390/ijms10093722.

Tourova, T. *et al.* (2020). Biodiversity of microorganisms colonizing the surface of polystyrene samples exposed to different aqueous environments. *Sustainability (Switzerland)*, *12*(9), 1–23. DOI: 10.3390/su12093624.

Tsiota, P. *et al.* (2018). Microbial degradation of HDPE secondary microplastics: Preliminary results. In *Proceedings of the International Conference on Microplastic Pollution in the Mediterranean Sea*, pp. 181–188. DOI: 10.1007/978-3-319-71279-6.

Tu, C., Chen, T., Zhou, Q., Liu, Y., Wei, J., Waniek, J. J., & Luo, Y. (2020). Biofilm formation and its influences on the properties of microplastics as affected by exposure time and depth in the seawater. *Science of The Total Environment*, 734, 139237.

Wang, G. X. *et al.* (2021a). Seawater-degradable polymers — Fighting the marine plastic pollution. *Advanced Science*, *8*(1), 1–26. DOI: 10.1002/advs.202001121.

Wang, J., Lu, J., Zhang, Y., Wu, J., & Luo, Y. (2021). Unique bacterial community of the biofilm on microplastics in coastal water. *Bulletin of Environmental Contamination and Toxicology*, *107*(4), 597–601.

Wang, J. *et al.* (2021b). The impact of microplastic-microbe interactions on animal health and biogeochemical cycles: A mini-review. *Science of the Total Environment*, *773*, 145697. Elsevier B.V. DOI: 10.1016/j.scitotenv.2021.145697.

Ward, C. P. & Reddy, C. M. (2020). Opinion: We need better data about the environmental persistence of plastic goods. *Proceedings of the National Academy of Sciences*, *117*(26), 202008009. DOI: 10.1073/pnas.2008009117.

Webb, H. K. *et al.* (2009). Poly(ethylene terephthalate) polymer surfaces as a substrate for bacterial attachment and biofilm formation. *Microbes and Environments*, *24*(1), 39–42. DOI: 10.1264/jsme2.ME08538.

Woodall, L. C., Jungblut, A. D., Hopkins, K., Hall, A., Robinson, L. F., Gwinnett, C., & Paterson, G. L. (2018). Deep-sea anthropogenic macrodebris harbours rich and diverse communities of bacteria and archaea. *PLoS One*, *13*(11), e0206220.

Wright, R. J. *et al.* (2021a). A multi-OMIC characterisation of biodegradation and microbial community succession within the PET plastisphere. *Microbiome*, *9*(141), 1–22. http://doi.wiley.com/10.4319/lom.2004.2.365.

Wright, R. J., Bosch, R. *et al.* (2020b). Plasticizer degradation by marine bacterial isolates: A proteogenomic and metabolomic characterization. *Environmental Science & Technology*, *54*(4), 2244–2256. DOI: 10.1021/acs.est.9b05228.

Wright, R. J., Erni-Cassola, G. *et al.* (2020a). Marine plastic debris — A new surface for microbial colonization. *Environmental Science and Technology*, *54*(19), 11657–11672. DOI: 10.1021/acs.est.0c02305.

Wright, R. J., Gibson, M. I., & Christie-Oleza, J. A. (2019). Understanding microbial community dynamics to improve optimal microbiome selection. *Microbiome*, *7*(85), 1–14. DOI: 10.1186/s40168-019-0702-x.

Wright, R. J., Langille, M. G. I., & Walker, T. R. (2021b). Food or just a free ride? A meta-analysis reveals the global diversity of the plastisphere. *The ISME Journal*, *15*(3), 789–806. Springer US. DOI: 10.1038/s41396-020-00814-9.

Wu, N., Zhang, Y., Li, W. *et al.* (2020a). Co-effects of biofouling and inorganic matters increased the density of environmental microplastics in the sediments of Bohai Bay coast. *Science of the Total Environment*, *717*. Elsevier B.V. DOI: 10.1016/j.scitotenv.2019.134431.

Wu, N., Zhang, Y., Zhao, Z., He, J., Li, W., Li, J., Ma, Y. & Niu, Z. (2020). Colonization characteristics of bacterial communities on microplastics compared with ambient environments (water and sediment) in Haihe Estuary. *Science of The Total Environment*, *708*, 134876.

Wu, X. *et al.* (2019). Selective enrichment of bacterial pathogens by microplastic biofilm. *Water Research*, *165*, 114979. Elsevier Ltd. DOI: 10.1016/j.watres.2019.114979.

Xu, X. *et al.* (2019). Marine microplastic-associated bacterial community succession in response to geography, exposure time, and plastic type in China's coastal seawaters. *Marine Pollution Bulletin*, *145* (May), 278–286. Elsevier. DOI: 10.1016/j.marpolbul.2019.05.036.

Yang, Y. *et al.* (2019). Plastics in the marine environment are reservoirs for antibiotic and metal resistance genes. *Environment International*, *123*(1), 79–86. Elsevier. DOI: 10.1016/j.envint.2018.11.061.

Yang, K. *et al.* (2020). Temporal dynamics of antibiotic resistome in the plastisphere during microbial colonization. *Environmental Science and Technology*, *54*(18), 11322–11332. DOI: 10.1021/acs.est.0c04292.

Ye, S., & Andrady, A. L. (1991). Fouling of floating plastic debris under Biscayne Bay exposure conditions. *Marine Pollution Bulletin*, *22*(12), 608–613. DOI: 10.1016/0025-326X(91)90249-R.

Yokota, K. *et al.* (2017). Finding the missing piece of the aquatic plastic pollution puzzle: Interaction between primary producers and microplastics. *Limnology and Oceanography Letters*, *2*(4), 91–104. DOI: 10.1002/lol2.10040.

Yoshida, S. *et al.* (2016). A bacterium that degrades and assimilates poly(ethylene terephthalate). *Science*, *351*, 1196–1199. DOI: 10.1126/science.aad6359.

Zaiko, A. *et al.* (2016). Metabarcoding improves detection of eukaryotes from early biofouling communities: Implications for pest monitoring and pathway management. *Biofouling*, *32*(6), 671–684.

Zettler, E. R., Mincer, T. J., & Amaral-Zettler, L. A. (2013). Life in the "plastisphere": Microbial communities on plastic marine debris. *Environmental Science and Technology*, *47*(13), 7137–7146. DOI: 10.1021/es401288x.

Zhang, Y. *et al.* (2020). Potential risks of microplastics combined with superbugs: Enrichment of antibiotic resistant bacteria on the surface of microplastics in mariculture system. *Ecotoxicology and Environmental Safety*, *187* (September 2019), 109852. Elsevier Inc. DOI: 10.1016/j.ecoenv.2019.109852.

Zhang, Y. *et al.* (2021). Does microplastic ingestion dramatically decrease the biomass of protozoa grazers? A case study on the marine ciliate Uronema marinum. *Chemosphere*, *267*, 129308. Elsevier Ltd. DOI: 10.1016/j.chemosphere.2020.129308.

Zhao, S. *et al.* (2018). Field-based evidence for microplastic in marine aggregates and mussels: Implications for trophic transfer. *Environmental Science and Technology*, *52*(19), 11038–11048. American Chemical Society. DOI: 10.1021/acs.est.8b03467.

Zhao, S. *et al.* (2020). Microbial carrying capacity and carbon biomass of plastic marine debris. *ISME Journal*, *15*(1), 67–77. Springer US. DOI: 10.1038/s41396-020-00756-2.

Zheng, Y., Yanful, E. K., & Bassi, A. S. (2005). A review of plastic waste biodegradation. *Critical Reviews in Biotechnology*, *25*, 243–250. DOI: 10.1080/07388550500346359.

Zhu, L. *et al.* (2020). Photochemical dissolution of buoyant microplastics to dissolved organic carbon: Rates and microbial impacts. *Journal of Hazardous Materials*, *383* (August 2019), 121065. Elsevier. DOI: 10.1016/j.jhazmat.2019.121065.

Zobell, C. E. (1943). The effect of solid surfaces upon bacterial activity. *Journal of Bacteriology*, *46*(1), 39–56. DOI: 10.1128/jb.46.1.39-56.1943.

Zrimec, J. *et al.* (no date). Plastic-degrading potential across the global microbiome correlates with recent pollution trends. *bioRxiv*, 1–23.

Zsolnay, Á. (2003). Dissolved organic matter: Artefacts, definitions, and functions. *Geoderma, 113*(3–4), 187–209. DOI: 10.1016/S0016-7061(02)00361-0.

Zumstein, M. T. *et al.* (2019). Dos and do nots when assessing the biodegradation of plastics. *Environmental Science & Technology*, *53*(17), 9967–9969. American Chemical Society. DOI: 10.1021/acs.est.9b04513.

https://doi.org/10.1142/9789811259111_0012

Chapter 11

A Review of Biodegradable Plastics from Multidisciplinary Perspectives

Winnie Courtene-Jones*, **Francesca De Falco and Imogen E. Napper**

International Marine Litter Research Unit, University of Plymouth, Plymouth, UK

**winnie.courtene-jones@plymouth.ac.uk*

1. Introduction

Plastics are lightweight, versatile, affordable and highly durable materials that have become ubiquitous in our everyday lives. Plastics have brought numerous societal benefits and are used for wide-ranging applications across sectors including transportation, medicine, agriculture and within domestic households (Thompson *et al.*, 2009; Napper & Thompson, 2020). However, many plastic objects have applications as single-use, disposable items such as packaging. Due to the short usage of these items and the increasing global production of plastics, which has seen a seven-fold increase since the mid-1970s (PlasticsEurope, 2019), plastics are a major component of waste which can pose challenges for solid waste management infrastructure (Kaza *et al.*, 2018). A global analysis of the production, use and end-of-life fate of plastics concluded that out of the 8300 Mt of plastics produced between 1905 and 2015, 2500 Mt are in use, 4,900 Mt have been discarded either in landfills or the environment, 800 Mt have been incinerated and only 600 Mt have been recycled

(Geyer *et al.*, 2017). Numerous studies report the presence of plastics in aquatic and terrestrial environments (Courtene-Jones *et al.*, 2017; Napper *et al.*, 2020; Lusher *et al.*, 2015; Hurley *et al.*, 2020) and their accumulation in marine ecosystems over the last 70 years (Ostle *et al.*, 2019; Courtene-Jones *et al.*, 2019; Thompson *et al.*, 2004; Law *et al.*, 2014). Indeed, the accumulation of end-of-life plastics, both in managed waste systems and in the environment, is seen as a global environmental issue (GESAMP, 2016).

Similar to hydrocarbon-based polymers, biopolymers have also been available for over a century (Morris, 1986) but have remained largely undeveloped (Gross & Kalra, 2002). However, over recent decades biodegradable polymers have received increased attention (Babu *et al.*, 2013; Philp *et al.*, 2013). This is due to a myriad of reasons including increased economic opportunities and technological advances, attempts to move away from reliance on fossil fuels and growing societal concern to the widespread environmental accumulation of conventional plastics and their associated impacts on wildlife, ecosystem services and human well-being (Beaumont *et al.*, 2019; Lettner *et al.*, 2017) together with the assumption that bio-based materials may be less harmful.

Biodegradable and bio-based plastics currently only make up a small volume of the total plastic market (see Section 2 for definitions). In 2020, the global production capacity for biodegradable polymers was 1.227 Mt, with a further 0.884 Mt of bio-based polymers. This corresponds to a total of ~0.5% (0.3% biodegradable plastics and 0.2% bio-based polymers) of the 360 Mt of plastics produced (PlasticsEurope, 2019; European Bioplastics, 2021). Demand for biodegradable and bio-based plastics has increased over the last two decades largely due to the reasons stated earlier (Philp *et al.*, 2013; Jayanth *et al.*, 2017), and future projections expect production capacity to increase further still to an estimated 2.87 Mt by 2025 (European Bioplastics, 2021). Bio-based and biodegradable plastics are used for an increasing number of applications, including flexible and rigid packaging materials (e.g., wrapping, trash bags and food containers), textiles and hygiene products (e.g., fabrics, diapers and sanitary products), consumer goods (e.g., egg cartons, toys and tableware) and agricultural/horticulture items (e.g., mulch films, planters and plant ties) (Yin & Yang, 2020).

In 2018, the European Commission established its European Plastic Strategy in a Circular Economy (European Commission, 2018). Based on the principles of resource efficiency and the waste hierarchy, the circular economy is designed to keep materials in the loop as long as possible and prevent the depletion of natural resources (Prieto-Sandoval *et al.*, 2018). A recent report highlighted the role of biodegradable plastics within a circular economy and recognises that in some applications where it is challenging to remove or collect a plastic product or its fragments from the environment after use (e.g., agricultural mulch films); or where it is difficult to separate plastic from organic material that is destined for a composting waste stream (e.g., organic waste bags (European Parliament, 2018)) or wastewater treatment, biodegradable plastics have the potential to bring advantages, compared with conventional plastics (SAPEA, 2020). As such, biodegradable polymers have been proposed as part of the solution to the environmental accumulation of plastics, with several European strategies and directives, such as the Waste Framework Directive and the EU Directive on packaging and packaging waste, recognise the role of biodegradable plastics (European Commission, 2018; European Parliament, 1994; European Union, 2019; European Parliament, 2018). In addition, the development of biodegradable plastics is also an attempt to achieve some of the United Nations Sustainable Development Goals (SDGs), laid out in the 2030 Agenda for Sustainable Development, for example, SDG 12: responsible consumption and production, SDG13: climate action, SDG 14: life below water and SDG 15: life on land (UNEP, 2016). However, it is important to emphasise that simply replacing conventional plastics with biodegradable alternatives is not a viable solution to address plastic pollution, poor waste management or the inappropriate disposal of waste, i.e., littering (SAPEA, 2020; UNEP, 2015; Viera *et al.*, 2020; Haider *et al.*, 2019).

Most plastics labelled as biodegradable require specific conditions (temperature, oxygen availability, microorganisms, humidity and light (see Sections 2–4)) in order to degrade and are usually required to undergo biological decomposition within an industrial facility. The specific conditions required for complete biodegradation may not be present within natural environments and as such, there is a potential that this group of plastic may still cause environmental harm in the same way as conventional plastics. Currently, there is limited information regarding the mechanisms and timescales for biodegradation of polymers and products in natural environments, and a lack of assessment regarding the ecological, societal and economic risks associated. Such information is required to advance the development of biodegradable materials and inform holistic impact assessments.

2. Classifications of Biodegradable Plastics

In an attempt to discuss plastics classified as biodegradable, bio-based or compostable, it is first necessary to clarify and specify the correct terminology to address such materials. A term that is often misused and which creates confusion is "bioplastic". According to European Bioplastics, "a plastic material is defined as a bioplastic if it is either bio-based, biodegradable, or features both properties", so the term is used for general purposes (European Bioplastics, 2021). However, the International Union of Pure and Applied Chemistry (IUPAC) gives a different meaning, defining bioplastic as a "bio-based (composed or derived in whole or in part of biological products issued from the biomass) polymer that can be shaped by flow at some stage in its processing into finished products" (Vert *et al.*, 2012). Since a bio-based polymer is not necessarily biodegradable or "environmentally friendly", the IUPAC discourages the use of the term "bioplastic" and it will not be used for the purposes of the current chapter, preferring the term "bio-based" (Vert *et al.*, 2012).

The term "biodegradable" has also been debated, depending on the definition of "biodegradation". Both the IUPAC and the American Society for Testing and Materials (ASTM) provide similar definitions: "degradation of a polymeric item due to cell-mediated phenomena" — IUPAC (Vert *et al.*, 2012); "degradation resulting from the action of naturally occurring microorganisms such as bacteria, fungi, and algae" — ASTM (Lambert & Wagner, 2017). Additional terms were also suggested by the Organisation for Economic Co-operation and Development (OECD) which distinguish between "primary" and "ultimate" biodegradation (OECD, 1992). "Primary" biodegradation is a chemical alteration of a substance due to biological activity and causes the loss of one of the specific properties of the substance. Instead, "ultimate" biodegradation refers to a substance that, under aerobic conditions, is completely utilised by microorganisms, resulting in the production of carbon dioxide (CO_2) water, mineral salts and biomass.

Recently, a working group of the Science Advice for Policy by European Academies (SAPEA) produced a report on "Biodegradability of plastics in the open environment" and proposed a more comprehensive definition of biodegradation. In detail, they introduced the concept of plastic biodegradation not as a material property but as a system one that includes not only the specific material properties that make it potentially biodegradable but also the specific conditions of the environment that receives such material (SAPEA, 2020). This is a more complete way to consider and define biodegradation that takes into account the fact that a specific plastic material cannot biodegrade the same in all the different types of environments (i.e., marine, riverine and terrestrial) and even in the same environment as conditions may be very different (i.e., humidity, pH and microbial community). The SAPEA definition of "plastic biodegradation" is reported in Table 1, along with the main terminology related to the topic.

It is important to highlight that a bio-based polymer can be either biodegradable or not; a biodegradable polymer can be bio-based but also fossil-based; a compostable polymer is also biodegradable but not vice versa. Figure 1 summarises the distinctions and interconnections among the different categories. A term that is not discussed within this chapter but is worth considering briefly is "oxo-degradable", which has been used for a category of plastics that have been at the centre of different debates. These materials are conventional fossil-based polymers (i.e., LDPE) that contain additives to accelerate their oxidation and fragmentation by the action of ultraviolet (UV) light or heat and oxygen. In 2019, they were banned by the European Union due to concerns about their actual impact on the environment (Manfra *et al.*, 2021).

The lack of consensus on the classification of biodegradable plastics adds more confusion and challenges to a scientific field that still requires more research. Further investigations are needed to fully understand the behaviour of such materials in the environment and if they can effectively be part of the solution for plastic pollution. Therefore, a common agreement and harmonisation on the accepted definition of "biodegradation" are of striking importance.

3. Chemical Composition of Biodegradable Polymers

Biodegradable polymers can be found in nature or synthesised from natural or fossil sources. Natural polymers are created during the growth cycles of all organisms and they comprise polysaccharides (i.e., cellulose, starch and chitosan), polypeptides (i.e., gelatin) and bacterial polyesters (i.e., polyhydroxyalkanoates) (Chandra & Rustgi, 1998). Among natural polysaccharides, one of the most used is starch which is easily biosynthesised by plants (i.e., corn, rice and potato) and composed of linear amylose polymers and highly branched amylopectin polymers (Fink, 2020). Starch can be easily modified to obtain a thermoplastic polymer, but its fast degradation via hydrolysis limits its applications. For this reason, it is

Table 1. The main terminology related to biodegradable polymers and those used within this chapter.[a]

Term	Definition
Bio-based polymer	A polymer composed or derived in whole, or in part of biological products issued from the biomass (including plants, animals and marine or forestry materials).
Biodegradable plastic	Plastic is considered biodegradable when it breaks down to basic elemental components (water, biomass and gas) with the aid of microorganisms.
Biodegradation	Microbial conversion of all the organic constituents of a material to carbon dioxide, new microbial biomass and mineral salts under oxic conditions or to carbon dioxide, methane, new microbial biomass and mineral salts under anoxic conditions.
Compostable plastic	Plastic capable of undergoing biological decomposition in a compost site as part of an available program, such that the material is not visually distinguishable and breaks down into carbon dioxide, water, inorganic compounds and biomass, at a rate consistent with known compostable materials.
Degradable polymer	Polymer in which macromolecules are able to undergo chain scissions, resulting in a decrease of molar mass.
Degradation	Degradation that results in desired changes in the values of in-use properties of the material because of macromolecule cleavage and molar mass decrease.
Fossil-based/ conventional polymer	A polymer derived from petrochemicals.
Macromolecule	Molecule of high relative molecular mass, the structure of which essentially comprises the multiple repetition units derived, actually or conceptually, from molecules of low relative molecular mass.
Plastic	A material that contains, as an essential ingredient, one or more organic polymeric substances of large molecular weight (i.e., polymers). A plastic is solid in its finished form but can be shaped by flow during manufacturing or finishing into finished articles.
Polymer	Substance composed of macromolecules.

Note: [a]Based on Kyrikou and Briassoulis (2007), Vert *et al.* (2012), Lambert *et al.* (2017), and SAPEA (2020).

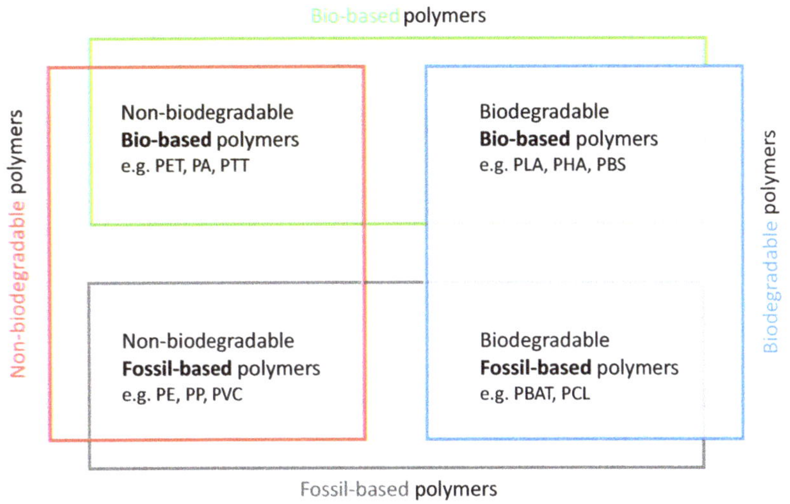

Figure 1. Categories of biodegradable and non-biodegradable polymers.

Source: This schematic was adapted from European Bioplastics (2021).

often blended with other synthetic polymers for longevity, creating materials like Mater-bi® produced by Novamont, which is now widely used in applications such as packaging, agricultural mulch films, disposable cutlery or consumer goods (Moeini *et al*., 2020).

Polyhydroxyalkanoates (PHAs) are a class of polymers first discovered in 1926, synthesised and intracellularly accumulated in most bacteria under unfavourable growth conditions. In general, PHAs are composed of R(-)-3-hydroxyalkanoic acid monomers (carbon atoms in the range C3–C14), with a variety of saturated/unsaturated and straight/branched chains containing aliphatic or aromatic side groups (Ojumu *et al*., 2004). The most common PHAs are polyhydroxybutyrate (PHB) that is synthesised by the polymerisation of 3-hydroxybutyrate monomers and the copolymer polyhydroxybutyrate-valerate (PHBV) used in packaging applications (Siracusa *et al*., 2008).

Beyond natural occurring polymers, biodegradable plastics can be also artificially synthesised using conventional fossil-based resources or renewable biological ones. Among the polymers considered biodegradable and bio-based, the one that has received great attention and is used in numerous applications is poly(lactic acid) (PLA), an aliphatic polyester considered compostable. The building block of PLA is lactic acid that can be derived from renewable resources like starch and sugar through fermentation. PLA has been mainly used in biomedical applications like tissue scaffolds and implant devices. However, recent advancements in its manufacturing have broadened its applications that now range from disposable food service ware to packaging and textiles (Lim *et al.*, 2008).

Another interesting bio-based biodegradable polyester is represented by poly(butylene succinate) (PBS) and its copolymers that are used as food packaging, bags, mulch films, flushable hygiene products, etc. PBS is synthesised by polycondensation of succinic acid and 1,4-butanediol that can be both derived from fossil or renewable resources (Xu & Guo, 2010). One of its most promising copolymers is poly(butylene succinate-co-butylene adipate) (PBSA), characterised by a lower crystallinity than PBS and greater flexibility of the polymer chains, features that make PBSA more susceptible to biodegradation as we further discuss in the following (Ray *et al.*, 2007).

Finally, biodegradable polymers can also be obtained from petrochemical resources, for instance, poly(butylene adipate-co-butylene terephthalate) (PBAT) and poly(ε-caprolactone) (PCL, Figure 1). The first is a copolymer of terephthalic acid, butanediol and adipic acid, mainly used in agricultural applications also in blends with PLA and starch (Agarwal, 2020). PCL is an aliphatic polyester synthesised through ring-opening polymerisation of the cyclic ester ε-caprolactone, applied in medicine and sustainable packaging (Bartnikowski *et al.*, 2019).

Abbreviations of the polymers referred to within this chapter are summarised in Box 1.

Box 1. Abbreviations and definitions of polymers.

HDPE	High-density polyethylene
LDPE	Low-density polyethylene
PA	Polyamide
PBAT	Poly(butylene adipate-co-butylene terephthalate)
PBS	Poly(butylene succinate)
PBSA	Poly(butylene succinate-co-butylene adipate)
PBSe	Poly(butylene sebacate)
PBSeT	Poly(butylene sebacate-co-terephtalate)

PCL	Poly(ε-caprolactone)
PE	Polyethylene
PET	Polyethylene terephthalate
PHA	Polyhydroxyalkanoates
PHB	Polyhydroxybutyrate
PHBV	Polyhydroxybutyrate-valerate
PLA	Poly(lactide acid)
PLGA	Poly(lactic-co-glycolic acid)
PLLA	Poly(l-lactide acid)
PP	Polypropylene
PTT	Polytrimethylene terephthalate
PVC	Polyvinyl chloride

4. Biodegradation of Biodegradable Polymers

In recent years, several reviews have summarised the data available on the degradation of some of the most widely used biodegradable polymers (i.e., see Lambert *et al.*, 2017; Haider *et al.*, 2019; Agarwal, 2020; Manfra *et al.*, 2021). Different aspects of polymer chemistry affect biodegradation. In general, biodegradation comprises two steps (Agarwal, 2020). The first involves the breakdown of macromolecular chains into molecules of lower molecular weight like oligomers and monomers. This fragmentation can be caused by hydrolysis (in the presence or absence of enzymes), oxidation or other mechanisms mainly depending on both the polymer chemical structure and the receiving environment characteristics. The second step consists of the microbial uptake of these low molar mass molecules, resulting in the formation of CO_2, CH_4, biomass and water. The biodegradability of synthetic polymers can depend on the presence in their polymer chains of characteristics similar to those of naturally occurring polymers (Chandra & Rustgi, 1998; Kijchavengkul & Auras, 2008). As natural polymers like starch generally biodegrade through hydrolysis followed by oxidation (Chandra & Rustgi, 1998), aliphatic polyesters like PLA, PHB and PCL can biodegrade especially thanks to the hydrolysable ester bond in the main chain that is sensitive to microbial attack (Tserki *et al.*, 2006).

The main factors affecting biodegradability can be distinguished in two categories: polymer characteristics and exposure conditions (Kijchavengkul & Auras, 2008). Considering the first one, factors like mechanical properties, glass transition temperature, morphology, hydrophilicity and crystallinity, all have an influence on biodegradation (Eubeler *et al.*, 2010). The nature of the chemical bonds present in the polymer chain is very important to determine biodegradation. For example, conventional polyolefins (i.e., PE and PP) have carbon–carbon single bonds that make them hydrophobic and resistant to degradation (Kyrikou & Briassoulis, 2007). To be susceptible to mechanisms like hydrolysis, polymer chains need to contain hydrolysable covalent bonds as in ester, anhydride and amide groups; whereas, other characteristics of the polymer structure (i.e., branched chains and unsaturated bonds) may influence oxidation (Lucas *et al.*, 2008). The presence of the functional groups previously mentioned has an additional positive effect on biodegradation, since they increase the flexibility of the chain, an aspect discussed in detail in the following (Kijchavengkul & Auras, 2008). The molecular weight of the polymer also influences the biodegradation rate, since a high molecular weight implies longer chains with more bonds to cleave to break down the polymer in oligomers and monomers (Kyrikou & Briassoulis, 2007). Moreover, by increasing the molecular weight, the glass transition temperature (temperature at which the polymer passes from the glassy to the rubbery state) increases and the polymer becomes less flexible (Kijchavengkul & Auras, 2008). The accessibility of the polymer chains to degradation agents like microorganisms and water is an important aspect that is linked to characteristics like polymer flexibility and its morphology. Regarding flexibility, for instance, the presence of bulky side groups can limit the movements of the polymer chains, whereas the presence of carbon double bonds increases flexibility by facilitating the rotation around contiguous single bonds (Kijchavengkul & Auras, 2008). For the morphology, many of the widely used biodegradable polymers are semi-crystalline (i.e., PLA and PBS); the crystalline regions not only reduce the flexibility of the polymer increasing its stiffness and density but are also less accessible to water, oxygen and microbes than the amorphous ones (Kijchavengkul & Auras, 2008; Kyrikou & Briassoulis, 2007; Laycock *et al.*, 2017; Lucas *et al.*, 2008). Since they are impermeable to water, the crystalline regions slow hydrolytic degradation and, in addition, reduce transport processes like solvent and gas diffusion due to their well-organised cross-linked crystalline domains (Laycock *et al.*,

2017; Lucas *et al.*, 2008). The disorganised amorphous domains are more flexible and accessible to degradation agents and water can penetrate these regions more easily causing faster hydrolysis (Laycock *et al.*, 2017).

Passing to the exposure conditions, they can be further discriminated in abiotic and biotic factors (Kijchavengkul & Auras, 2008). Temperature, pH and moisture are among the abiotic factors that can affect the hydrolysis reaction and also the microbial activity (Haider *et al.*, 2019). By increasing temperature and moisture, both hydrolysis and microbial action increase, especially if the temperature is above the glass transition of the polymer and so it is more flexible and accessible to water and microbes (Kijchavengkul & Auras, 2008). In biotic conditions, enzymes (proteins with a complex 3D structure) are the most important contributors and their degradation activity can be influenced by oxygen levels, pH, nutrients and microbial population (Laycock *et al.*, 2017).

Another important aspect to consider is that a plastic item is not only composed of polymers but also included are additives that have their specific influence on biodegradation and must be taken into account. Additives are used because they are chemically and physically active, modifying the polymer chemical structure, mechanical properties and optical characteristics (Kyrikou & Briassoulis, 2007). They can be divided into functional additives (e.g., plasticisers, stabilisers and flame retardants), colorants (e.g., dyes and pigments), fillers (e.g., calcium carbonate and clay) and reinforcements (e.g., glass and carbon fibres) (Hahladakis *et al.*, 2018). Depending on the types of additives, their presence can accelerate or hinder biodegradation (Kijchavengkul & Auras, 2008). In the framework of biodegradable plastics, additives should be considered to have their dedicated regulatory assessment and be included in the testing of the material as a whole, and those certified as biodegradable should consider the polymer as well as the additives that are incorporated (SAPEA, 2020).

5. Standardisation and Certification of Biodegradable Plastics

Most of the research performed on the biodegradation of polymeric materials has been focused on soil or compost environments, with little attention to the aquatic systems (Eubeler *et al.*, 2009; Kijchavengkul & Auras, 2008; Kyrikou & Briassoulis, 2007).

Since the 1980s, different efforts have been made to develop standardised tests for biodegradation, starting with guidelines developed by the Organization for Economic Cooperation and Development (OECD) (OECD, 1992). The American Society for Testing and Materials (ASTM), the European Committee for Standardization (CEN) and International Organization for Standardization (ISO) have also been working on standards that are available by paying a fee, but difficult to compare due to the different definitions of biodegradability adopted (Kyrikou & Briassuoilis 2007; Harrison *et al*., 2018; Filiciotto & Rothenberg, 2021).

In general, all these methods refer to specific environments with specific characteristics and microorganism population, and are often a balance between real-time conditions and feasible testing time (Eubeler *et al*., 2009; Filiciotto & Rothenberg, 2021). Their main criticisms are related to the lack of representability of real environmental conditions and on the analytical procedure applied to assess and quantify biodegradation (Haider *et al*., 2019). Currently, the main procedures applied involve measurements of oxygen consumption or CO_2 evolution, polymer weight loss, molecular weight evolution, tensile properties, enzyme assays, extent of fragmentation and ecotoxicity (Kyrikou & Briassoulis, 2007; Lott *et al*., 2020). It is important not to use a single analytical procedure to assess biodegradation, since it can cause misleading interpretations (Albertsson & Hakkarainen, 2017).

The recent SAPEA report has tried to provide some clarity and new guidelines for a realistic assessment of plastic biodegradation (SAPEA, 2020). Considering regulatory aspects, the definition of plastic biodegradation reported in Table 1 needs to be accompanied by some specifications that include the type of open environment in which the biodegradation occurs and the extent of biodegradation reached in a pre-defined timeframe. The report introduces a new approach for biodegradation standard testing, based on a three-tier testing scheme that is composed of the following steps: (1) the plastic material is tested at laboratory scale possibly using real inoculums, not prepared *ad hoc*; (2) the material undergoes field testing in real natural conditions; (3) the plastic material in tested in tanks (mesocosms), simulating real environmental conditions in larger volumes than compared to laboratory tests. By combining the data obtained from these three types of tests, it is possible to obtain a full assessment of the biodegradation of the tested material (SAPEA, 2020).

Standard test methods give the means to compare the biodegradability of a material worldwide but do not provide certifications that are instead given by accredited bodies following payment (Filiciotto & Rothenberg, 2021). Usually, independent accredited testing facilities test the material following standard procedures and pass the results to the certifier that review them and eventually give the certification to the client (SAPEA, 2020). Currently, the main institutions that provide certifications are TÜV Austria, DIN CERTCO (Germany), Biodegradable Products Institute (BPI; USA), Japan BioPlastics Association (JBPA), Australasian Bioplastics Association (ABA) and the European Union (Filiciotto & Rothenberg, 2021; SAPEA, 2020). Further efforts and harmonisation actions are needed to create effective certification and labelling systems that are based on environmentally realistic test methods to prevent false claims by the industry and misleading consumers.

6. Environmental Considerations of Biodegradable Plastics

Biodegradable plastics have a wide range of applications (European Bioplastics, 2021) which can lead to their introduction or leakage into the natural environment. For example, biodegradable plastics are used directly in the environment as agricultural mulch films (Kasirajan & Ngouajio, 2012). Other uses are as fibres used in wet wipes and sanitary products which when disposed of via the sewage system can shed fibres; these can be discharged into aquatic environments or be introduced to land via the application of sewage sludge as a fertiliser (O Briain *et al.*, 2020; Zubris and Richards, 2005; Daria *et al.*, 2020). Research to understand the environmental risks associated with biodegradable plastics is currently in the early stages, and further work is, therefore, required to evaluate their impacts and consider these in relation to conventional plastics.

Ecotoxicological data for biodegradable polymers are scarce. In contrast to many other chemical substances, the environmental impact assessment of biodegradable polymers is not covered by the European legislation on chemicals (REACH). A recent proposal calls for more stringent criteria for biodegradable polymers (ECHA, 2020), which may increase future testing of these materials. When considering the potential ecological impact of plastic polymers either petrochemical or biodegradable, the exposure time and concentration can influence their ecotoxicity. Currently, biodegradable plastics only constitute ~0.3% of the overall plastic production, and consequently, the emissions to the environment are considerably lower than for conventional plastics. As the quantities of biodegradable plastics are expected to increase (European Bioplastics, 2021) so too may the quantities in the environment (i.e., the concentration to which an organism is exposed). The exposure time is linked to persistence, i.e., the biodegradability of the polymer in a particular environment. Studies indicate that biodegradation rates can vary widely between environments (terrestrial or aquatic), exposure conditions and seasons (e.g., temperature, oxygen availability and pH) and the microbial communities within the environment (Sintim *et al.*, 2020; Napper & Thompson, 2019; Dilkes-Hoffman *et al.*, 2019a; Haider *et al.*, 2019; Narancic *et al.*, 2018), in addition to the properties of the polymer itself (chemical composition, crystallinity, thickness and shape) (Volova *et al.*, 2010). Where conditions for biodegradation are not favourable, biodegradable plastics may persist in the environment for substantial periods of time (see Sections 6.1 and 6.2).

A further concern is related to additives contained within biodegradable plastics (SAPEA, 2020). As with conventional plastics, bio-based plastics contain additives used to enhance functionality or convey certain desirable properties. While bio-polymers are often promoted as a more sustainable alternative to petrochemical plastics, it is not known whether they represent a safer alternative with regard to the chemicals they contain. Several chemicals, such as phthalates and per- and polyfluoroalkyl substances (PFAS) which are used in conventional plastic products, have also been detected in biodegradable or bio-based items (Anderson & Shenkar, 2021; Timshina *et al.*, 2021). Additionally, the behaviour of chemical additives, such as their potential to leach into the environment and their persistence after the polymer has completely biodegraded, has not been evaluated.

Recent studies indicate that products made from biodegradable polymers may have similar toxicity to conventional plastics (Zimmermann *et al.*, 2019, 2020). Non-targeted chemical screening of 42 samples of biodegradable and bio-based plastic products, many of which were used as food contact material, indicated that those composed of starch and cellulose contained the greatest number of chemical features (typically >10,000) (Zimmermann *et al.*, 2020). Of the 42 products tested, 29 contained chemicals which induced baseline toxicity, 18 of which induced oxidative stress, 10 antiandrogenicity and one estrogenicity in the bacterium *Aliivibrio fischeri.* While the authors did not identify every chemical signature, they found that certain chemicals were present in all of the plastics tested (bio-based and conventional plastics), and others were specific to polymers/products. The study also indicated that across all of these end-points, more commercial products induced toxicity than raw materials (Zimmermann *et al.*, 2020) emphasising the need for the testing of final products.

6.1. *Biodegradable plastics in terrestrial environments*

To date, the majority of research into the degradation of biodegradable plastics has been focussed on terrestrial ecosystems, and, in particular, studies have considered their use in agricultural applications. Modern agriculture relies heavily on the use of plastic materials. In some cases, these are applied directly to the open environment, for example, mulch films, which are challenging and costly to recover and separate from soil (Kasirajan & Ngouajio, 2012). Plastic mulch films are used globally to increase crop yields while lowering the consumption of water, fertilisers and herbicides, and minimising soil erosion (Kasirajan & Ngouajio, 2012) and thus convey clear societal and economic advantages in terms of food production and security.

Most mulch films are composed of polyethylene and residual fragments in soils can accumulate reducing water infiltration, altering soil ecosystems and ultimately plant germination and growth (Liu *et al.*, 2014; Gao *et al.*, 2019). Biodegradable mulch films offer similar physical barrier properties to conventional plastic films (Bandopadhyay *et al.*, 2018) but are designed to be ploughed into the soil over which they were used, where it is anticipated they will biodegrade (Hayes & Flury, 2018). As such they are proposed to offer a promising advantage over conventional polymers (SAPEA, 2020). Biodegradable mulch films (PLA, PHA, PBAT and PBS) have been used since the 1980s (Sintim & Flury, 2017) and presently account for ~5% of the market share (APE Europe, 2020).

The standards for soil degradation (ISO 17556:2019 or ASTM D5988: 2018) specify that at least 90% of the organic carbon must be converted to carbon dioxide within two years. The deterioration of mulch film varies with soil type and climate (temperature, moisture and solar radiation), meaning biodegradation rates can vary geographically as well as seasonally (Miles *et al.*, 2017; Li *et al.*, 2014b; Sintim *et al.*, 2020). Field studies conducted in two regions of Vietnam reported that under natural soil conditions PHA films lost 98% and 47% of their weight (Boyandin *et al.*, 2013). At three test sites in the USA, commercial PCL mulch films buried at a depth of 8–12 cm in natural soil for 24 months lost between 11–98 % of their area (Li *et al.*, 2014b). In many agricultural systems, mulch films are used for a single growing cycle and may be applied to the same field year after year. Hence there is the potential for localised accumulation (Miles *et al.*, 2017; Bandopadhyay *et al.*, 2018).

When ploughed into soil, biodegradable mulch films provide an input of carbon, albeit a relatively small one. The growth of soil microbes in agricultural soils is usually carbon limited and studies have shown that microbes can respond to these small inputs of carbon, causing increases in biomass and enzyme activity (Li *et al.*, 2014a), and changes in microbial community structure, for example, enrichment of fungi (Li *et al.*, 2014b; Moreno & Moreno, 2008; Ma *et al.*, 2016; Muroi *et al.*, 2016). Contrary to this, Bandopadhyay *et al.* (2020) found only minor site-specific effects on soil microbial communities and their function between biodegradable and conventional plastic mulch film treatments. It is not altogether clear what overall effect the changes in microbiota may have on the soil ecosystem and functioning as a whole, or how long-lasting these effects may persist.

Further studies have indicated that in the short term (<3 years), soil-biodegradable plastic mulch films may cause minor effects and hence pose only minimal risk to soil physical, chemical and biological properties (Sintim *et al.*, 2019; Li *et al.*, 2014a). For instance, the soil quality index (based on the assessment of pH, electrical conductivity, total organic carbon, microbial biomass and β-glucosidase activity which measure metabolic capability) was shown to marginally reduce after an 18-month exposure to biodegradable mulch films compared to the control (no-mulch) treatment (Li *et al.*, 2014a). Similar to polyethylene mulches, studies have reported that biodegradable mulch films have no measurable impact on the nitrification potential of soils (Ardisson *et al.*, 2014; Kapanen *et al.*, 2008) which is a further measure of soil health. A four-year study concluded that biodegradable mulches increased soil aggregate stability and water infiltration compared to conventional mulches, which were deemed as beneficial impacts to the soil (Sintim *et al.*, 2021). Further work is required to build upon these initial studies and examine whether the same observations occur in different geographic regions.

Tests where seeds were sown in soils which had previously been exposed to 1% (w/w) biodegradable plastic fragments for 6–7 months did not find significant impacts on germination or the growth of cress, barley or sorghum (Rychter *et al.*, 2006; Sforzini *et al.*, 2016; Muroi *et al.*, 2016). However, studies which have incorporated biodegradable fragments into soils have detected impacts on plant species. Following a 2-month exposure, a significant reduction in shoot biomass, total plant biomass and fruit biomass was observed for wheat plants grown in soils containing 1% (w/w) starch-based biodegradable plastic, compared to control groups (Qi *et al.*, 2018). Ryegrass germination and growth were also significantly reduced when grown in soils containing PLA fragments (0.1% w/w) (Boots *et al.*, 2019). While residues of conventional plastic-film mulches can also impact plant growth (Liu *et al.*, 2021), biodegradable polymers have been demonstrated to have a greater negative effect on crop growth and yield than polyethylene (Qi *et al.*, 2018, 2020; Boots *et al.*, 2019).

Among terrestrial organisms, earthworms are key ecosystem engineers and important for soil health. Conventional plastic particles can reduce the growth rate, increase mortality rate and alter the metabolic activity and burrowing behaviour of earthworms (*Lumbricus terrestris, Eisenia foetida* and *Eisenia andrei*) (Cao *et al.*, 2017; Huerta Lwanga *et al.*, 2016; Rodriguez-Seijo *et al.*, 2017; Prendergast-Miller *et al.*, 2019). Studies on the impacts of biodegradable plastics on earthworms are limited. Boots *et al.* (2019) report that PLA fragments in soil (0.1% w/w) caused a reduction in worm biomass. Anecic species of earthworms such as *L. terrestris* construct permanent, long vertical burrows. They feed at the soil surface and drag organic matter into the burrows; this process has also been shown to redistribute conventional and biodegradable plastics within soils (Zhang *et al.*, 2018; Rillig *et al.*, 2017) and may enhance the degradation of biodegradable plastics (Sanchez-Hernandez *et al.*, 2020).

As discussed earlier, studies show differing results when considering the effects of biodegradable plastics on (i) microbial communities, (ii) soil health and (iii) terrestrial species. Evidence has indicated that biodegradable plastics may have little effect on the physical and chemical properties of the soil (Sintim *et al.*, 2019; Li *et al.*, 2014a), while the presence of fragments can cause detrimental effects on crop growth (Qi *et al.*, 2018, 2020; Boots *et al.*, 2019). International standards (ISO 17556:2019 or ASTM D5988) specify that 90% biodegradation must be achieved in two years, thus there is potential for localised accumulation if there is year-upon-year application of film; the effect of this remains currently unexplored.

6.2. *Biodegradable plastics in aquatic environments*

Aquatic environments span a diversity of habitats including rivers, lakes, estuaries and coastal ecosystems, the open ocean and the deep sea, and consequently, there is substantial variability in temperature, oxygen availability, light and microorganism assemblages. The TÜV AUSTRIA Biodegradable Water conformity mark (TÜV AUSTRIA, 2013) states that there should be 90% biodegradation in aquatic environments in 56 days; however, this assumes temperatures are between 20 and 25°C which are not representative of the majority of natural aquatic environments.

The biodegradation rates of polymers differ widely between aquatic environments and polymer types (Green *et al.*, 2015; Volova *et al.*, 2007; Nauendorf *et al.*, 2016; Briassoulis *et al.*, 2019; Bagheri *et al.*, 2017), as also found for terrestrial studies. Based on available data, it was calculated that a PHA bottle may take between 1.5 to 3.5 years to biodegrade completely in the marine environment (Dilkes-Hoffman *et al.*, 2019b). Bagheri *et al.* (2017) compared the biodegradation of a PLA film in artificial seawater and freshwater with PLGA, PCL, PHB and PBAT. PLGA had completely degraded in ≈270 days while PHB had lost only ~9% of its initial weight after 365 days and all other polymers showed no significant degradability in either treatment. These studies suggest that either intact items or fragments may remain in the aquatic environment for prolonged periods with the potential for ecological impacts.

Laboratory and field experiments comparing the deterioration of PHB at the sediment–water interface with PBS and polybutylene sebacate-co-terephtalate (PBSeT) report all materials had biodegraded after 200 days (Briassoulis *et al.*, 2020). Evidence suggests that the biodegradation rate, or the degree of disintegration, was greater for polymers (PHA, PHB, PBSe and PBSeT) in contact with the benthos than when in the water column (Briassoulis *et al.*, 2019; Mayer, 1990). This may be due to the presence of diverse microorganisms within the benthic zone which could increase biodegradation, or due to bio-fouling the samples suspended in the photic zone may have reduced biodegradation rates, both of which require further investigation (Briassoulis *et al.*, 2019; Lott *et al.*, 2020).

Research comparing the biogeochemical effects of conventional and biodegradable plastic bags found that the presence of both types of bags caused anoxic conditions within underlying sediment and the reduction of primary productivity and significantly reduced the abundance of infaunal invertebrates (Green *et al.*, 2015). Sediment containing PLA had approximately half the concentration of ammonium in porewater as sediments containing conventional polymers (PVC and HDPE), which the authors hypothesise may be due to the chemical structure of PLA which can absorb cations from the environment and thus reduce concentrations in the porewater (Green *et al.*, 2016, 2017). A study investigating the effects of biodegradable plastic bags on seagrass meadows in the Mediterranean Sea demonstrated that the reduction in sediment oxygen and pH levels caused by the presence of the biodegradable bags also influenced the growth of seagrass leading to increased inter- and intraspecific competition between plants (Balestri *et al.*, 2017). These studies indicate that the presence of biodegradable plastics has the potential to rapidly, within 1–2 months, alter marine assemblages and the ecosystem services they provide.

Research to examine the acute toxicity of PLLA particles (10–100 μg/L) on mussels (*Mytilus edulis*) demonstrated there was a dose-dependent downregulation of lipids, particularly glycerophospholipids which are important for the structure of biological membranes (Khalid *et al.*, 2021). A reduction in the *in vitro* fertilisation rates of solitary ascidians was found to correlate with higher exposure concentrations of PLA and PET control, but no statistical significance was identified between material types (Anderson & Shenkar, 2021). The authors attribute the effects to the presence of additives, such as phthalates, which were present in both materials.

Chronic exposures (30–60 days) to PLA induced stress responses in benthic invertebrates including oysters and lugworms (Green *et al.*, 2016, 2017; Green, 2016). Lugworms were also shown to exhibit behavioural responses, producing fewer burrows/casts when exposed to PLA, PE and PVC than in control treatments (Green *et al.*, 2016). The process of bioturbation through the action of burrowing organisms is important for the redistribution of oxygen and nutrients in sediments. The findings of this study, therefore, suggest potential ecological consequences through the reduction of bioturbation by lugworms. Following a 52-day exposure to PLA microplastics (25 mgL^{-1}), mussels were shown to produce fewer byssal threads which also had a lower strength of attachment, compared to control groups (Green *et al.*, 2019). Mussels are important ecosystem engineers, modifying the structural complexity of benthic habitats due to their aggregation into beds. The authors highlight that both conventional and biodegradable microplastics could impact the functioning and structure of sedimentary habitats.

Biodegradable plastics have been shown to deteriorate into microplastic fragments under artificial freshwater and seawater experiments (Wei *et al.*, 2021; González-Pleiter *et al.*, 2019). PBAT fragmented more readily than polyethylene under aquatic conditions and the rate of PBAT fragmentation was faster in seawater than in freshwater, attributed to faster hydrolysis under the basic pH conditions in seawater (Wei *et al.*,

2021). The effect of nano-sized PHB particles derived from the fragmentation of larger particles under simulated natural environmental conditions on three freshwater species was investigated. PHB nanoplastics induced an increase in reactive oxygen species in all organisms and damage to cell membranes which caused a significant decrease in the growth of cyanobacteria (*Anabaena* sp.) and green algae (*Chlamydomonas reinhardtii*) and a significant immobilisation of *Daphnia magna* (González-Pleiter *et al.*, 2019). While this study shows a potent acute toxicological effect, it is not clear what specific concentrations of nano-PHB were used and whether these were environmentally relevant.

In summary, the rate of biodegradation varies considerably between different marine and freshwater environments. Exposure to biodegradable plastics has been shown to induce a range of deleterious effects on aquatic organisms including oxidative stress and behavioural responses such as reduced burrowing activity in lugworms (Green, 2016; Green *et al.*, 2016, 2017; González-Pleiter *et al.*, 2019). Further work is required to elucidate the role of additives compared to the polymer itself on toxicity and observe a range of endpoints across a diversity of aquatic organisms. Further research is also needed to consider the wider biogeochemical effects of biodegradable plastics on aquatic ecosystems which is currently lacking.

7. Social Considerations of Biodegradable Plastics

Widespread concern about plastic pollution in the environment has encouraged consumers to opt for alternatives to conventional plastic products. Over the last decade, partly driven by this shift in consumer demand, there has been substantial development of alternative plastics with higher biodegradable properties. Manufacturers have marketed these products using ecological strategies, known as green marketing, to encourage consumers to buy or use them (Koenig-Lewis *et al.*, 2014).

In green marketing, ecological claims have been criticised for confusing or misleading buyers (Koenig-Lewis *et al.*, 2014). This leaves consumers vulnerable to "greenwashing", where some companies seek to make environmental claims about their products based on partial analysis of the underlying science in the expectation that a majority of consumers believe such claims (Belz & Peattie, 2012; Polonsky *et al.*, 1997). This has led to calls for more rationality in promoting and evaluating "green" products.

The UK's Competition and Markets Authority (CMA), a non-ministerial government department in the United Kingdom, has been concerned that the surge in demand for green products and services could incentivise some businesses to make vague, false or misleading claims about the sustainability or environmental impact of the products or services they sell (CMA, 2020). Examples of misleading behaviour could include the following (CMA, 2020):

- Exaggerating the positive environmental impact of a product or service
- Using complex or jargon-heavy language
- Implying that items are eco-friendly through packaging and logos when this is not true

Labels, logos and material types are the most important features for consumers in identifying environmentally friendly products and packaging (Lindh *et al.*, 2016; Scott & Vigar-Ellis, 2014). For example, a study conducted in South Africa (Scott & Vigar-Ellis, 2014) investigated how consumers identify environmentally friendly packaging of food products. They reported that 45% of the participants looked for labels, while 30% looked for images or logos such as the recycling logo. The packaging material itself was used by 18% of the participants to judge if packaging was sustainable, and only 12% of the participants admitted to not knowing the difference between sustainable and other packaging.

The disposal of a product or packaging is another feature that consumers use to identify sustainability. A study of consumers in Italy showed that the aspect of disposal (i.e., whether the packaging is recyclable, non-recyclable or biodegradable) of food packaging was most important compared to other packaging attributes. The participants regarded biodegradable packaging as having many advantages over recyclable and non-recyclable packaging and consumers relied heavily on labelling and logos to help them determine whether they viewed a product as sustainable (Arboretti Giancristofaro & Bordignon, 2016).

Consumers are also shown to be unfamiliar with and, therefore, confused by the term 'bio-based' plastic (Koutsimanis *et al.*, 2012; Sijtsema *et al.*, 2016). For example, research by Koutsimanis *et al.* (2012) found that only 55% of the participating US consumers answered correctly to the question: "Which raw materials are used to produce bio-based containers?" While the term bio-based plastic may imply that the material is fully biodegradable, this is not the case (see Section 4).

It is necessary to have clear definitions and product labelling to indicate appropriate usage and disposal of products and packaging (Thompson *et al.*, 2009). With the development of new materials, such as biodegradable plastics, it may lead to confusion over which is the correct waste stream for disposal. For example, in Germany, only 10% of consumers dispose of fossil fuel-based plastic packaging incorrectly, compared to 63% of consumers that dispose of compostable, bio-based plastic incorrectly (Taufik *et al.*, 2020).

There needs to be the availability of a dedicated waste stream for the consumer to use and the appropriate infrastructure; such as an industrial composting facility. Plus, the consumer needs sufficient understanding to correctly separate their waste accordingly (Napper & Thompson, 2019). A lack of clarity and understanding from the consumer may lead to them disposing of an item incorrectly, e.g., via the recycling waste stream which could cause contamination of the recyclate. Other potential barriers for biodegradable plastics are higher prices, lack of availability, unfamiliarity and perceived lower quality (see Section 8) (de Jonge & van Trijp, 2013; Hughner *et al.*, 2007; Magnier & Crié, 2015; Stern, 2000).

We should not consider biodegradable plastics as a complete technical solution, thereby excusing our environmental responsibility, such as littering. It has been reported that the general public find it more acceptable to drop items composed of biodegradable plastics than non-biodegradable plastics because it is thought they are less harmful to the environment (Dilkes-Hoffman *et al.*, 2019a). The importance of reducing or reusing plastic and biodegradable plastic items as much as possible throughout their lifespan and reducing reliance on single-use products needs to be communicated.

8. Economic Considerations of Biodegradable Plastics

Bio-based plastics are derived from biomass, the majority of which originates from first-generation feedstock, defined as crops suitable for human or animal consumption (Brizga *et al.*, 2020). With the projected expansion of the bio-based plastics market over the next decade (European Bioplastics, 2021) and proposed substitutions of conventional polymers with biodegradable alternatives in some applications, it is predicted that there may be increased competition for land or resources to meet this demand (Posen *et al*., 2017; Brizga *et al*., 2020). A study by Brizga *et al.* (2020) estimated that at least 7.4 million hectares of land would be required to satisfy the land requirements to replace all petrochemical-based plastic packaging with bio-based alternatives in Europe. On a global scale, they estimate that 61 million hectares would be required to produce enough bio-based feedstock (Brizga *et al*., 2020). This may result in changes to land use and present challenges in terms of natural resources and food security (Escobar *et al*., 2018) and as such demands for raw materials from bio-based plastics must be balanced against competing claims on land.

A number of economic barriers are frequently cited in relation to the slow adoption and development of bio-based and biodegradable polymers. Compared with conventional petrochemical-based plastics, the material price for bio-based polymers is high and the technical production processes are complex, making the final product more expensive than conventional plastics (Shen *et al*., 2020; Fahim *et al*., 2019). The expansion of the bio-based market may lead to the wider commercialisation of biodegradable plastics and thereby attain economies of scale (Ren, 2003).

The global interest in 'zero-waste' (which has principles based on the circular economy and the waste hierarchy (Franco-García *et al*., 2019)) has offered new business opportunities for biodegradable plastics (Brizga *et al*., 2020; Zaman, 2015). Biodegradable plastics may provide opportunities to move towards a more circular economy, through higher resource efficiency, but only if waste products can re-enter the economy through a biological cycle (Confente *et al*., 2020; SAPEA, 2020). Appropriate management and disposal practices are required for this to be achieved. In the absence of clear labelling (Sections 5 and 7), there is confusion over how to appropriately dispose of biodegradable plastic products (Sijtsema *et al*., 2016) and as such they may enter inappropriate waste streams.

Contamination of recycling streams presents a problem. Biodegradable plastics may contaminate existing plastic recycling streams leading to low-quality recyclate, which may not be usable in the production of new products (Alaerts *et al*., 2018; Rujnić-Sokele & Pilipović, 2017) and, therefore, interrupt the circular flow of materials. For example, a study found that contamination of 0.1% PLA can cause structural problems to PET, rendering the material unusable for many applications (Alaerts *et al*., 2018). Alternatively, biodegradable plastics may enter organic recycling waste streams, where any residual material from incomplete degradation may compromise the quality of the resulting compost. As the compost derived from this process is largely intended to be returned to the natural environment, any plastic residues may also be introduced to the environment where there may be unintended environmental consequences (see Section 6) (Hann *et al.*, 2020). As such, evaluating any unintended consequences arising from end-of-life disposal of biodegradable plastics is required.

9. Conclusion and Outlook

The degradability of a polymer largely depends on its chemical structure, rather than its carbon source (i.e., bio-based or petrochemical-based). The potential benefits in terms of biodegradability will only be realised if end-of-life items reach an appropriate receiving environment for biodegradation to occur. Humidity, temperature, oxygen availability and microorganisms vary across environments, resulting in different biodegradation rates. Additionally, manufacturers include additives at various concentrations and so materials made from the same polymer should not be considered chemically identical (Lambert & Wagner, 2017), placing emphasis on the need for testing (and certifying) biodegradability of the end product rather than on the pure polymer. This calls for an effective framework of standards and certifications of biodegradable materials that take into account the plastic material as composed of polymers plus additives and the type of open environment where the biodegradation should occur.

The absence of clear labelling has led to consumer confusion when trying to choose 'sustainable' products and the appropriate way to dispose of items once they have reached the end of their life in service. This confusion can lead to an increase in biodegradable items becoming littered in the environment (Dilkes-Hoffman *et al.*, 2019a) or disposed of via an uncorrected managed waste stream, e.g., with conventional plastic recycling, which may contaminate the resulting recyclate and rendering it unusable (Alaerts *et al.*, 2018; Rujnić-Sokele & Pilipović, 2017).

Currently, there are rather limited empirical data on the environmental effects of biodegradable plastics. The persistence of fragments from incomplete degradation along with residues of any chemical additives can induce a range of effects, including reduction of crop growth and yields, and in terrestrial and aquatic organisms, a range of stress and behavioural responses have been reported, along with wider implications on ecosystem functioning.

Looking ahead, as biodegradable plastic consumption is projected to increase, economic opportunities may be presented. In certain applications, biodegradable plastics can have a role as part of a circular economy (see Hann *et al.*, 2020; SAPEA, 2020). Biodegradable polymers may offer advantages over conventional plastics where degradability is part of the product's function, for example, as agricultural mulch films, bags to contain organic waste or drug delivery agents in biomedicine (SAPEA, 2020; Haider *et al.*, 2019). However, substituting all commodity plastics with biodegradable alternatives may have unintended environmental and economic repercussions and are not a solution for plastic pollution (Narancic *et al.*, 2018). Further product testing and research are required to holistically appraise their effect on society, economy and the environment.

References

Agarwal, S. (2020). Biodegradable polymers: Present opportunities and challenges in providing a microplastic-free environment. *Macromolecular Chemistry and Physics, 221*, 2000017.

Alaerts, L., Augustinus, M., & Van Acker, K. (2018). Impact of bio-based plastics on current recycling of plastics. *Sustainability, 10*(5), 1487.

Albertsson, A.-C. & Hakkarainen, M. (2017). Designed to degrade. *Science, 358*, 872–873.

Anderson, G. & Shenkar, N. (2021). Potential effects of biodegradable single-use items in the sea: Polylactic acid (PLA) and solitary ascidians. *Environmental Pollution, 268*, 115364.

APE Europe. (2020). Statistics, plasticulture in Europe [Online]. Retrieved from https://apeeurope.eu/statistics. Accessed on 29/05/2020.

Arboretti Giancristofaro, R., & Bordignon, P. (2016). Consumer preferences in food packaging: CUB models and conjoint analysis. *British Food Journal, 118*, 527–540.

Ardisson, G. B., Tosin, M., Barbale, M., & Degli-Innocenti, F. (2014). Biodegradation of plastics in soil and effects on nitrification activity. A laboratory approach. *Frontiers Microbiology, 5*, 710.

ASTM International. (2018). ASTM D5988-18, Standard test method for determining aerobic biodegradation of plastic materials in soil. ASTM International.

Babu, R. P., O'connor, K., & Seeram, R. (2013). Current progress on bio-based polymers and their future trends. *Progress in Biomaterials, 2*(8), 1–16.

Bagheri, A. R., Laforsch, C., Greiner, A., & Agarwal, S. (2017). Fate of so-called biodegradable polymers in seawater and freshwater. *Global Challenges, 1*, 1700048.

Balestri, E., Menicagli, V., Vallerini, F., & Lardicci, C. (2017). Biodegradable plastic bags on the seafloor: A future threat for seagrass meadows? *Science of the Total Environment*, *605–606*, 755–763.

Bandopadhyay, S., Martin-Closas, L., Pelacho, A. M., & Debruyn, J. M. (2018). Biodegradable plastic mulch films: Impacts on soil microbial communities and ecosystem functions. *Frontiers Microbiology*, *9*, 819.

Bandopadhyay, S., Sintim, H. Y., & Debruyn, J. M. (2020). Effects of biodegradable plastic film mulching on soil microbial communities in two agroecosystems. *PeerJ*, *8*, e9015.

Bartnikowski, M., Dargaville, T. R., Ivanovski, S., & Hutmacher, D. W. (2019). Degradation mechanisms of polycaprolactone in the context of chemistry, geometry and environment. *Progress in Polymer Science*, *96*, 1–20.

Beaumont, N. J., Aanesen, M., Austen, M. C., Borger, T., Clark, J. R., Cole, M., Hooper, T., Lindeque, P. K., Pascoe, C., & Wyles, K. J. (2019). Global ecological, social and economic impacts of marine plastic. *Marine Pollution Bulletin*, *142*, 189–195.

Belz, F.-M. & Peattie, K. (2012). *Sustainability Marketing: A Global Perspective. John Wiley & Sons Ltd, Chichester*, pp 1–352.

Boots, B., Russell, C. W., & Green, D. S. (2019). Effects of microplastics in soil ecosystems: Above and below ground. *Environmental Science and Technology*, *53*, 11496–11506.

Boyandin, A. N., Prudnikova, S. V., Karpov, V. A., Ivonin, V. N., Đỗ, N. L., Nguyễn, T. H., Lê, T. M. H., Filichev, N. L., Levin, A. L., Filipenko, M. L., Volova, T. G., & Gitelson, I. I. (2013). Microbial degradation of polyhydroxyalkanoates in tropical soils. *International Biodeterioration & Biodegradation*, *83*, 77–84.

Briassoulis, D., Pikasi, A., Briassoulis, C., & Mistriotis, A. (2019). Disintegration behaviour of bio-based plastics in coastal zone marine environments: A field experiment under natural conditions. *Science of the Total Environment*, *688*, 208–223.

Briassoulis, D., Pikasi, A., Papardaki, N. G., & Mistriotis, A. (2020). Aerobic biodegradation of bio-based plastics in the seawater/sediment interface (sublittoral) marine environment of the coastal zone — Test method under controlled laboratory conditions. *Science of the Total Environment*, *722*, 137748.

Brizga, J., Hubacek, K., & Feng, K. (2020). The unintended side effects of bioplastics: Carbon, land, and water footprints. *One Earth*, *3*, 45–53.

Cao, D., Wang, X., Luo, X., Liu, G., & Zheng, H. (2017). Effects of polystyrene microplastics on the fitness of earthworms in an agricultural soil. *IOP Conference Series: Earth and Environmental Science*, *61*, 012148.

Chandra, R. & Rustgi, R. (1998). Biodegradable polymers. *Progress in Polymer Science*, *23*, 1273–1335.

CMA. (2020). CMA to examine if 'eco-friendly' claims are misleading. Competition and Markets Authority [Online]. Retrieved from https://www.gov.uk/government/news/cma-to-examine-if-eco-friendly-claims-are-misleading. Accessed 16/04/2020.

Confente, I., Scarpi, D., & Russo, I. (2020). Marketing a new generation of bio-plastics products for a circular economy: The role of green self-identity, self-congruity, and perceived value. *Journal of Business Research, 112*, 431–439.

Courtene-Jones, W., Quinn, B., Gary, S. F., Mogg, A. O. M., & Narayanaswamy, B. E. (2017). Microplastic pollution identified in deep-sea water and ingested by benthic invertebrates in the Rockall Trough, North Atlantic Ocean. *Environmental Pollution, 231*, 271–280.

Courtene-Jones, W., Quinn, B., Ewins, C., Gary, S. F., & Narayanaswamy, B. E. (2019). Consistent microplastic ingestion by deep-sea invertebrates over the last four decades (1976-2015), a study from the North East Atlantic. *Environmental Pollution*, 244, 503–512.

Daria, M., Krzysztof, L., & Jakub, M. (2020). Characteristics of biodegradable textiles used in environmental engineering: A comprehensive review. *Journal of Cleaner Production, 268*, 122129.

de Jonge, J. & van Trijp, H. C. M. (2013). Meeting heterogeneity in consumer demand for animal welfare: A reflection on existing knowledge and implications for the meat sector. *Journal of Agricultural and Environmental Ethics, 26*, 629–661.

Dilkes-Hoffman, L. S., Ashworth, P., Laycock, B., Pratt, S., & Lant, P. (2019a). Public attitudes towards bioplastics — Knowledge, perception and end-of-life management. *Resources, Conservation and Recycling, 151*.

Dilkes-Hoffman, L. S., Lane, J. L., Grant, T., Pratt, S., Lant, P. A., & Laycock, B. (2018). Environmental impact of biodegradable food packaging when considering food waste. *Journal of Cleaner Production, 180*, 325–334.

Dilkes-Hoffman, L. S., Lant, P. A., Laycock, B., & Pratt, S. (2019b). The rate of biodegradation of PHA bioplastics in the marine environment: A meta-study. *Marine Pollution Bulletin, 142*, 15–24.

ECHA. (2020). Opinion on an Annex XV dossier proposing restrictions on intentionally-added microplastics. Committee for Risk Assessment (RAC), Committee for Socio-economic Analysis (SEAC).

Escobar, N., Haddad, S., Börner, J., & Britz, W. (2018). Land use mediated GHG emissions and spillovers from increased consumption of bioplastics. *Environmental Research Letters, 13*, 125005.

Eubeler, J. P., Zok, S., Bernhard, M., & Knepper, T. P. (2009). Environmental biodegradation of synthetic polymers I. Test methodologies and procedures. *TrAC Trends in Analytical Chemistry, 28*(9), 1057–1072.

European Bioplastics. (2021a). Bioplastics market data 2020 [Online]. Retrieved from https://www.european-bioplastics.org/market/. Accessed 12/05/ 2021.
European Bioplastics. (2021b). What are bioplastics? [Online]. Retrieved from https://www.european-bioplastics.org/bioplastics/. Accessed on 12/05/2021.
European Commission. (2018). *A European Strategy for Plastics in a Circular Economy*. European Commission.
European Parliament. (1994). Directive 94/62/EC on packaging and packaging waste.
European Parliament. (2018). Directive (EU) 2018/851 of the European Parliament and of the Council of 30 May 2018 amending Directive 2008/98/EC on waste.
European Union. (2019). Directive (EU) 2019/904 on the reduction of the impact of certain plastic products on the environment.
Fahim, I. S., Chbib, H., & Mahmoud, H. M. (2019). The synthesis, production & economic feasibility of manufacturing PLA from agricultural waste. *Sustainable Chemistry and Pharmacy*, *12*, 100142.
Filiciotto, L. & Rothenberg, G. (2021). Biodegradable plastics: Standards, policies, and impacts. *ChemSusChem, 14*(1), 56–72.
Fink, J. K. (2020). Carbohydrate-related polymers. In: *The Chemistry of Bio-based Polymers*, pp. 213–296. Hoboken: Wiley.
Franco-García, M. L., Carpio-Aguilar, J. C., & Bressers, H. (2019). Towards zero waste, circular economy boost: Waste to resources. In Franco-García, M. L., Carpio-Aguilar, J. C., & Bressers, H. (Eds.), *Towards Zero Waste*. Greening of Industry Networks Studies, pp. 1–8. Cham: Springer Nature.
Gao, H., Yan, C., Liu, Q., Ding, W., Chen, B., & Li, Z. (2019). Effects of plastic mulching and plastic residue on agricultural production: A meta-analysis. *Science of the Total Environment*, *651*, 484–492.
GESAMP. (2016). Sources, fate and effects of microplastics in the marine environment: Part two of a global assessment. In: Kershaw, P.J., Rochman, C.M (eds.), IMO/FAO/UNESCO-IOC/UNIDO/WMO/IAEA/UN/UNEP/UNDP Joint Group of Experts on the Scientific Aspects of Marine Environmental Protection, GESAMP No. 93, pp. 1–220.
Geyer, R., Jambeck, J. R., & Law, K. L. (2017). Production, use, and fate of all plastics ever made. *Science Advances*, *3*, e1700782.
González-Pleiter, M., Tamayo-Belda, M., Pulido-Reyes, G., Amariei, G., Leganés, F., Rosal, R., & Fernández-Piñas, F. (2019). Secondary nanoplastics released from a biodegradable microplastic severely impact freshwater environments. *Environmental Science: Nano*, *6*, 1382–1392.
Green, D. S. (2016). Effects of microplastics on European flat oysters, *Ostrea edulis* and their associated benthic communities. *Environmental Pollution, 216*, 95–103.

Green, D. S., Boots, B., Blockley, D. J., Rocha, C., & Thompson, R. (2015). Impacts of discarded plastic bags on marine assemblages and ecosystem functioning. *Environmental Science and Technology*, *49*(9), 5380–5389.

Green, D. S., Boots, B., O'Connor, N. E., & Thompson, R. (2017). Microplastics affect the ecological functioning of an important biogenic habitat. *Environmental Science and Technology*, *51*(1), 68–77.

Green, D. S., Boots, B., Sigwart, J., Jiang, S., & Rocha, C. (2016). Effects of conventional and biodegradable microplastics on a marine ecosystem engineer (*Arenicola marina*) and sediment nutrient cycling. *Environmental Pollution*, *208*, 426–434.

Green, D. S., Colgan, T. J., Thompson, R. C., & Carolan, J. C. (2019). Exposure to microplastics reduces attachment strength and alters the haemolymph proteome of blue mussels (*Mytilus edulis*). *Environmental Pollution*, *246*, 423–434.

Gross, R. A. & Kalra, B. (2002). Biodegradable polymers for the environment. *Science*, *297*(5582), 803–807.

Hahladakis, J. N., Velis, C. A., Weber, R., Iacovidou, E., & Purnell, P. (2018). An overview of chemical additives present in plastics: Migration, release, fate and environmental impact during their use, disposal and recycling. *Journal of Hazardous Materials*, *344*, 179–199.

Haider, T. P., Volker, C., Kramm, J., Landfester, K., & Wurm, F. R. (2019). Plastics of the future? The impact of biodegradable polymers on the environment and on society. *Angewandte Chemie International Edition*, *58*, 50–62.

Hann, S., Scholes, R., Molteno, S., Hilton, M., Favoino, E., & Jakobsen, L. G. (2020). *Relevance of Biodegradable and Compostable Consumer Plastic Products and Packaging in a Circular Economy*. Report for European Commission, DG Environment. Luxembourg.

Harrison, J. P., Boardman, C., O'Callaghan, K., Delort, A. M., & Song, J. (2018). Biodegradability standards for carrier bags and plastic films in aquatic environments: A critical review. *Royal Society Open Science*, *5*, 171792.

Hayes, D. G. & Flury, M. (2018). Summary and assessment of EN 17033:2018, a new standard for biodegradable plastic mulch film. Retrieved from https://ag.tennessee.edu/biodegradablemulch/Documents/EU%20regs%20factsheet.pdf. Accessed on 18/07/2022.

Huerta Lwanga, E., Gertsen, H., Gooren, H., Peters, P., Salanki, T., van Der Ploeg, M., Besseling, E., Koelmans, A. A., & Geissen, V. (2016). Microplastics in the terrestrial ecosystem: Implications for *Lumbricus terrestris* (*Oligochaeta L.*). *Environmental Science and Technology*, *50*, 2685–2691.

Hughner, R. S., Mcdonagh, P., Prothero, A., Shultz, C. J., & Stanton, J. (2007). Who are organic food consumers? A compilation and review of why people purchase organic food. *Journal of Consumer Behaviour*, *6*, 94–110.

Hurley, R., Horton, A., Lusher, A., & Nizzetto, L. (2020). Plastic waste in the terrestrial environment. In Letcher, T. M. (Ed.), *Plastic Waste and Recycling*, pp. 163–193. London, San Diego, Cambridge, Oxford: Academic Press.

ISO 17556. (2019). Plastics- determination of the ultimate aerobic biodegradability of plastic materials in soil by measuring the oxygen demand in a respirometer or the amount of carbon dioxide evolved. International Organization for Standardization.

Jayanth, D., Kumar, P. S., Nayak, G. C., Kumar, J. S., Pal, S. K., & Rajasekar, R. (2017). A review on biodegradable polymeric materials striving towards the attainment of green environment. *Journal of Polymers and the Environment*, *26*, 838–865.

Kapanen, A., Schettini, E., Vox, G., & Itävaara, M. (2008). Performance and environmental impact of biodegradable films in agriculture: A field study on protected cultivation. *Journal of Polymers and the Environment*, *16*, 109–122.

Kasirajan, S. & Ngouajio, M. (2012). Polyethylene and biodegradable mulches for agricultural applications: A review. *Agronomy for Sustainable Development*, *32*, 501–529.

Kaza, S., Yao, L., Bhada-Tata, P., & Van Woerden, F. (2018). *What a Waste 2.0: A Global Snapshot of Solid Waste Management to 2050*. Washington, DC: World Bank.

Khalid, A., Zalouk-Vergnoux, A., Benali, S., Mincheva, R., Raquez, J. M., Bertrand, S., & Poirier, L. (2021). Are bio-based and biodegradable microplastics impacting for blue mussel (*Mytilus edulis*)? *Marine Pollution Bulletin*, *167*, 112295.

Kijchavengkul, T. & Auras, R. (2008). Compostability of polymers. *Polymer International*, *57*, 793–804.

Koenig-Lewis, N., Palmer, A., Dermody, J., & Urbye, A. (2014). Consumers' evaluations of ecological packaging — rational and emotional approaches. *Journal of Environmental Psychology*, *37*, 94–105.

Koutsimanis, G., Getter, K., Behe, B., Harte, J., & Almenar, E. (2012). Influences of packaging attributes on consumer purchase decisions for fresh produce. *Appetite*, *59*, 270–280.

Kyrikou, I. & Briassoulis, D. (2007). Biodegradation of agricultural plastic films: A critical review. *Journal of Polymers and the Environment*, *15*, 125–150.

Lambert, S. & Wagner, M. (2017). Environmental performance of bio-based and biodegradable plastics: The road ahead. *Chemical Society Reviews*, *46*, 6855–6871.

Law, K. L., Moret-Ferguson, S. E., Goodwin, D. S., Zettler, E. R., Deforce, E., Kukulka, T., & Proskurowski, G. (2014). Distribution of surface plastic debris in the eastern Pacific Ocean from an 11-year data set. *Environmental Science and Technology*, *48*, 4732–4738.

Laycock, B., Nikolić, M., Colwell, J. M., Gauthier, E., Halley, P., Bottle, S., & George, G. (2017). Lifetime prediction of biodegradable polymers. *Progress in Polymer Science*, *71*, 144–189.

Lettner, M., Schöggl, J.-P., & Stern, T. (2017). Factors influencing the market diffusion of bio-based plastics: Results of four comparative scenario analyses. *Journal of Cleaner Production*, *157*, 289–298.

Li, C., Moore-Kucera, J., Lee, J., Corbin, A., Brodhagen, M., Miles, C., & Inglis, D. (2014a). Effects of biodegradable mulch on soil quality. *Applied Soil Ecology*, *79*, 59–69.

Li, C., Moore-Kucera, J., Miles, C., Leonas, K., Lee, J., Corbin, A., & Inglis, D. (2014b). Degradation of potentially biodegradable plastic mulch films at three diverse U.S. locations. *Agroecology and Sustainable Food Systems*, *38*, 861–889.

Lim, L. T., Auras, R., & Rubino, M. (2008). Processing technologies for poly(lactic acid). *Progress in Polymer Science*, *33*, 820–852.

Lindh, H., Olsson, A., & Williams, H. (2016). Consumer perceptions of food packaging: Contributing to or counteracting environmentally sustainable development? *Packaging Technology and Science*. *29*(1), 3–23.

Liu, E. K., He, W. Q., & Yan, C. R. (2014). 'White revolution' to 'white pollution' — Agricultural plastic film mulch in China. *Environmental Research Letters*, *9*, 091001.

Liu, Y., Huang, Q., Hu, W., Qin, J., Zheng, Y., Wang, J., Wang, Q., Xu, Y., Guo, G., Hu, S., & Xu, L. (2021). Effects of plastic mulch film residues on soil-microbe-plant systems under different soil pH conditions. *Chemosphere*, *267*, 128901.

Lott, C., Eich, A., Unger, B., Makarow, D., Battagliarin, G., Schlegel, K., Lasut, M. T., & Weber, M. (2020). Field and mesocosm methods to test biodegradable plastic film under marine conditions. *PLoS One*, *15*, e0236579.

Lucas, N., Bienaime, C., Belloy, C., Queneudec, M., Silvestre, F., & Nava-Saucedo, J. E. (2008). Polymer biodegradation: Mechanisms and estimation techniques. *Chemosphere*, *73*, 429–442.

Lusher, A. L., Tirelli, V., O'connor, I., & Officer, R. (2015). Microplastics in Arctic polar waters: The first reported values of particles in surface and subsurface samples. *Scientific Reports*, *5*, 14947.

Ma, Z., Ma, Y., Qin, L., Liu, J., & Su, H. (2016). Preparation and characteristics of biodegradable mulching films based on fermentation industry wastes. *International Biodeterioration & Biodegradation*, 111, 54–61.

Magnier, L. & Crié, D. (2015). Communicating packaging eco-friendliness: An exploration of consumers' perceptions of eco-designed packaging. *International Journal of Retail and Distribution Management*, *43*, 350–366.

Manfra, L., Marengo, V., Libralato, G., Costantini, M., De Falco, F., & Cocca, M. (2021). Biodegradable polymers: A real opportunity to solve marine plastic pollution? *Journal of Hazardous Materials*, *416*, 125763.

Mayer, J. M. (1990). Degradation kinetics of polymer films in marine and soil systems under accelerated conditions. In: DAWES, E.A. (ed.) *Novel Biodegradable Microbial Polymers*. Springer, Dordrecht, pp. 469–471.

Miles, C., Devetter, L., Ghimire, S., & Hayes, D. G. (2017). Suitability of biodegradable plastic mulches for organic and sustainable agricultural production systems. *HortScience*, *52*, 10–15.

Moeini, A., Mallardo, S., Cimmino, A., Dal Poggetto, G., Masi, M., Di Biase, M., van Reenen, A., Lavermicocca, P., Valerio, F., Evidente, A., Malinconico, M., & Santagata, G. (2020). Thermoplastic starch and bioactive chitosan sub-microparticle biocomposites: Antifungal and chemico-physical properties of the films. *Carbohydrate Polymers*, *230*, 115627.

Moreno, M. M. & Moreno, A. (2008). Effect of different biodegradable and polyethylene mulches on soil properties and production in a tomato crop. *Scientia Horticulturae*, 116, 256–263.

Morris, P. J. T. (1986). *Polymer Pioneers: A Popular History of the Science and Technology of Large Molecules*. Center for History of Chemistry, p. 88.

Muroi, F., Tachibana, Y., Kobayashi, Y., Sakurai, T., & Kasuya, K.-I. (2016). Influences of poly(butylene adipate-co-terephthalate) on soil microbiota and plant growth. *Polymer Degradation and Stability*, *129*, 338–346.

Napper, I. E., Davies, B. F. R., Clifford, H., Elvin, S., Koldewey, H. J., Mayewski, P. A., Miner, K. R., Potocki, M., Elmore, A. C., Gajurel, A. P., & Thompson, R. C. (2020). Reaching new heights in plastic pollution — Preliminary findings of microplastics on Mount Everest. *One Earth*, *3*, 621–630.

Napper, I. E. & Thompson, R. C. (2019). Environmental deterioration of biodegradable, oxo-biodegradable, compostable, and conventional plastic carrier bags in the sea, soil, and open-air over a 3-year period. *Environmental Science and Technology*, *53*, 4775–4783.

Napper, I. E. & Thompson, R. C. (2020). Plastic debris in the marine environment: History and future challenges. *Global Challenges*, *4*, 1900081.

Narancic, T., Verstichel, S., Reddy Chaganti, S., Morales-Gamez, L., Kenny, S. T., De Wilde, B., Babu Padamati, R., & O'connor, K. E. (2018). Biodegradable plastic blends create new possibilities for end-of-life management of plastics but they are not a panacea for plastic pollution. *Environmental Science and Technology*, *52*, 10441–10452.

Nauendorf, A., Krause, S., Bigalke, N. K., Gorb, E. V., Gorb, S. N., Haeckel, M., Wahl, M., & Treude, T. (2016). Microbial colonization and degradation of polyethylene and biodegradable plastic bags in temperate fine-grained organic-rich marine sediments. *Marine Pollution Bulletin*, *103*, 168–178.

O Briain, O., Marques Mendes, A. R., Mccarron, S., Healy, M. G., & Morrison, L. (2020). The role of wet wipes and sanitary towels as a source of white microplastic fibres in the marine environment. *Water Research*, *182*, 116021.

OECD. (1992). OECD 203, *Guidelines for the Testing of Chemicals*. OECD Publishing, Paris.

Ojumu, T. V., Yu, J., & Solomon, B. O. (2004). Production of polyhydroxyalkanoates, a bacterial biodegradable polymer. *African Journal of Biotechnology*, *3*, 18–24.

Ostle, C., Thompson, R. C., Broughton, D., Gregory, L., Wootton, M., & Johns, D. G. (2019). The rise in ocean plastics evidenced from a 60-year time series. *Nature Communications*, *10*, 1622.

Philp, J. C., Bartsev, A., Ritchie, R. J., Baucher, M. A., & Guy, K. (2013). Bioplastics science from a policy vantage point. *New Biotechnology*, *30*, 635–646.

PlasticsEurope. (2019). *Plastics — The Facts 2019. An Analysis of European Plastics Production, Demand and Waste Data*. Belgium: PlasticsEurope Brussels.

Polonsky, M. J., Carlson, L., Grove, S., & Kangun, N. (1997). International environmental marketing claims: Real changes or simple posturing? *International Marketing Review*, *14*, 218–232.

Posen, I. D., Jaramillo, P., Landis, A. E., & Griffin, W. M. (2017). Greenhouse gas mitigation for U.S. plastics production: Energy first, feedstocks later. *Environmental Research Letters*, *12*, 034024.

Prendergast-Miller, M. T., Katsiamides, A., Abbass, M., Sturzenbaum, S. R., Thorpe, K. L., & Hodson, M. E. (2019). Polyester-derived microfibre impacts on the soil-dwelling earthworm *Lumbricus terrestris*. *Environmental Pollution*, *251*, 453–459.

Prieto-Sandoval, V., Jaca, C., & Ormazabal, M. (2018). Towards a consensus on the circular economy. *Journal of Cleaner Production*, *179*, 605–615.

Qi, Y., Beriot, N., Gort, G., Huerta Lwanga, E., Gooren, H., Yang, X., & Geissen, V. (2020). Impact of plastic mulch film debris on soil physicochemical and hydrological properties. *Environmental Pollution*, *266*, 115097.

Qi, Y., Yang, X., Pelaez, A. M., Huerta Lwanga, E., Beriot, N., Gertsen, H., Garbeva, P., & Geissen, V. (2018). Macro- and micro- plastics in soil-plant system: Effects of plastic mulch film residues on wheat (*Triticum aestivum*) growth. *Science of the Total Environment*, *645*, 1048–1056.

Ray, S. S., Bandyopadhyay, J., & Bousmina, M. (2007). Thermal and thermomechanical properties of poly[(butylene succinate)-co-adipate] nanocomposite. *Polymer Degradation and Stability*, *92*, 802–812.

Ren, X. (2003). Biodegradable plastics: A solution or a challenge? *Journal of Cleaner Production*, *11*, 27–40.

Rillig, M. C., Ziersch, L., & Hempel, S. (2017). Microplastic transport in soil by earthworms. *Scientific Reports*, *7*, 1362.

Rodriguez-Seijo, A., Lourenço, J., Rocha-Santos, T. A. P., Da Costa, J., Duarte, A. C., Vala, H., & Pereira, R. (2017). Histopathological and molecular effects of microplastics in *Eisenia andrei Bouché*. *Environmental Pollution, 220*, 495–503.

Rujnić-Sokele, M. & Pilipović, A. (2017). Challenges and opportunities of biodegradable plastics: A mini review. *Waste Management & Research, 35*, 132–140.

Rychter, P., Biczak, R., Herman, B., Smyłła, A., Kurcok, P., Adamus, G., & Kowalczuk, M. (2006). Environmental degradation of polyester blends containing Atactic poly(3-hydroxybutyrate). Biodegradation in soil and ecotoxicological impact. *Biomacromolecules*, *7*, 3125–3131.

Sanchez-Hernandez, J. C., Capowiez, Y., & Ro, K. S. (2020). Potential use of earthworms to enhance decaying of biodegradable plastics. *ACS Sustainable Chemistry & Engineering*, *8*, 4292–4316.

SAPEA. (2020). *Biodegradability of Plastics in the Open Environment*. Berlin: Science Advice for Policy by European Academies.

Scott, L. & Vigar-Ellis, D. (2014). Consumer understanding, perceptions and behaviours with regard to environmentally friendly packaging in a developing nation. *International Journal of Consumer Studies*, *38*, 642–649

Sforzini, S., Oliveri, L., Chinaglia, S., & Viarengo, A. (2016). Application of biotests for the determination of soil ecotoxicity after exposure to biodegradable plastics. *Frontiers in Environmental Science*, *4*, 68.

Shen, M., Song, B., Zeng, G., Zhang, Y., Huang, W., Wen, X., & Tang, W. (2020). Are biodegradable plastics a promising solution to solve the global plastic pollution? *Environmental Pollution*, *263*, 114469.

Sijtsema, S. J., Onwezen, M. C., Reinders, M. J., Dagevos, H., Partanen, A., & Meeusen, M. (2016). Consumer perception of bio-based products-an exploratory study in 5 European countries. *NJAS — Wageningen Journal of Life Sciences*, *77*, 61–69.

Sintim, H. Y., Bandopadhyay, S., English, M. E., Bary, A. I., Debruyn, J. M., Schaeffer, S. M., Miles, C. A., Reganold, J. P., & Flury, M. (2019). Impacts of biodegradable plastic mulches on soil health. *Agriculture, Ecosystems & Environment*, *273*, 36–49.

Sintim, H. Y., Bandopadhyay, S., English, M. E., Bary, A., Liquet y González, J. E., Debruyn, J. M., Schaeffer, S. M., Miles, C. A., & Flury, M. (2021). Four years of continuous use of soil-biodegradable plastic mulch: Impact on soil and groundwater quality. *Geoderma, 381*, 114665.

Sintim, H. Y., Bary, A. I., Hayes, D. G., Wadsworth, L. C., Anunciado, M. B., English, M. E., Bandopadhyay, S., Schaeffer, S. M., Debruyn, J. M., Miles,

C. A., Reganold, J. P., & Flury, M. (2020). In situ degradation of biodegradable plastic mulch films in compost and agricultural soils. *Science of the Total Environment*, *727*, 138668.

Sintim, H. Y. & Flury, M. (2017). Is biodegradable plastic mulch the solution to agriculture's plastic problem? *Environmental Science and Technology*, *51*, 1068–1069.

Siracusa, V., Rocculi, P., Romani, S., & Rosa, M. D. (2008). Biodegradable polymers for food packaging: A review. *Trends in Food Science & Technology*, *19*, 634–643.

Stern, P. C. (2000). New environmental theories: Toward a coherent theory of environmentally significant behavior. *Journal of Social Issues*, *56*, 407–424.

Taufik, D., Reinders, M. J., Molenveld, K., & Onwezen, M. C. (2020). The paradox between the environmental appeal of bio-based plastic packaging for consumers and their disposal behaviour. *Science of the Total Environment*, *705*, 135820.

Thompson, R. C., Olsen, Y., Mitchell, R. P., Davis, A., Rowland, S. J., John, A. W. G., Mcgonigle, D., & Russell, A. E. (2004). Lost at sea: Where is all the plastic. *Science*, *304*, 838.

Thompson, R. C., Swan, S. H., Moore, C. J., & Vom Saal, F. S. (2009). Our plastic age. *Philosophical Transactions of the Royal Society B*, *364*, 1973–1976.

Timshina, A., Aristizabal-Henao, J. J., Da Silva, B. F., & Bowden, J. A. (2021). The last straw: Characterization of per- and polyfluoroalkyl substances in commercially-available plant based drinking straws. *Chemosphere*, *277*, 130238.

Tserki, V., Matzinos, P., Pavlidou, E., Vachliotis, D., & Panayiotou, C. (2006). Biodegradable aliphatic polyesters. Part I. Properties and biodegradation of poly(butylene succinate-co-butylene adipate). *Polymer Degradation and Stability*, *91*, 367–376.

TÜV AUSTRIA. (2013). OK biodegradable Water. *Bio products — Degradation in water*. TÜV AUSTRIA, Belgium.

UNEP. (2015). Biodegradable plastic & marine litter, misconceptions, concerns and impacts on marine environments. United Nations Environment Programme (UNEP), Nairobi.

UNEP. (2016). Marine plastic debris and microplastics — Global lessons and research to inspire action and guide policy change. United Nations Environment Programme (UNEP), Nairobi.

Vert, M., Doi, Y., Hellwich, K.-H., Hess, M., Hodge, P., Kubisa, P., Rinaudo, M., & Schué, F. (2012). Terminology for biorelated polymers and applications (IUPAC recommendations 2012). *Pure and Applied Chemistry*, *84*, 377–410.

Viera, J. S. C., Marques, M. R. C., Nazareth, M. C., Jimenez, P. C., & Castro, Í. B. (2020). On replacing single-use plastic with so-called biodegradable ones: The case with straws. *Environmental Science & Policy*, *106*, 177–181.

Volova, T. G., Boyandin, A. N., Vasiliev, A. D., Karpov, V. A., Prudnikova, S. V., Mishukova, O. V., Boyarskikh, U. A., Filipenko, M. L., Rudnev, V. P., Bá Xuân, B., Việt Dũng, V., & Gitelson, I. I. (2010). Biodegradation of polyhydroxyalkanoates (PHAs) in tropical coastal waters and identification of PHA-degrading bacteria. *Polymer Degradation and Stability*, *95*, 2350–2359.

Volova, T. G., Gladyshev, M. I., Trusova, M. Y., & Zhila, N. O. (2007). Degradation of polyhydroxyalkanoates in eutrophic reservoir. *Polymer Degradation and Stability*, *92*, 580–586.

Wei, X.-F., Bohlén, M., Lindblad, C., Hedenqvist, M., & Hakonen, A. (2021). Microplastics generated from a biodegradable plastic in freshwater and seawater. *Water Research*, *198*, 117123.

Xu, J. & Guo, B. H. (2010). Poly(butylene succinate) and its copolymers: research, development and industrialization. *Biotechnology Journal*, *5*, 1149–1163.

Yin, G.-Z. & Yang, X.-M. (2020). Biodegradable polymers: a cure for the planet, but a long way to go. *Journal of Polymer Research*, *27*, 38, 1-14.

Zaman, A. U. (2015). A comprehensive review of the development of zero waste management: Lessons learned and guidelines. *Journal of Cleaner Production*, *91*, 12–25.

Zhang, L., Sintim, H. Y., Bary, A. I., Hayes, D. G., Wadsworth, L. C., Anunciado, M. B., & Flury, M. (2018). Interaction of Lumbricus terrestris with macroscopic polyethylene and biodegradable plastic mulch. *Science of the Total Environment*, *635*, 1600–1608.

Zimmermann, L., Dierkes, G., Ternes, T. A., Volker, C., & Wagner, M. (2019). Benchmarking the in vitro toxicity and chemical composition of plastic consumer products. *Environmental Science and Technology*, *53*, 11467–11477.

Zimmermann, L., Dombrowski, A., Volker, C., & Wagner, M. (2020). Are bioplastics and plant-based materials safer than conventional plastics? In vitro toxicity and chemical composition. *Environmental International*, *145*, 106066.

Zubris, K. A. & Richards, B. K. (2005). Synthetic fibers as an indicator of land application of sludge. *Environmental Pollution*, *138*, 201–211.

https://doi.org/10.1142/9789811259111_0013

Chapter 12

Solutions, Approaches and Mitigation Strategies for Plastic Waste Management

Eamonn McGuinty and Tony R. Walker*

School for Resource and Environmental Studies, Dalhousie University, Halifax, Canada

**trwalker@dal.ca*

1. Introduction

Plastic pollution is pervasively affecting all ecosystems globally, particularly marine ecosystems (Borrelle *et al.*, 2020). As one of the most universally used materials in modern society, plastic dominates the global waste problem. The convenience of plastics, especially single-use plastics, has resulted in increased global production and consumption leading to increased plastic waste generation. Since 1950, production and use of plastics have increased faster than any other material, partly due to the durability and low costs of production (Geyer *et al.*, 2017). Furthermore, since the creation of the first commercial plastic polymers, plastic production and distribution have become one of the leading industries around the world in terms of employment, GDP contributions and overall economic impacts (Geyer *et al.*, 2017). In 2018, global plastic production reached 359 million metric tonnes (MT) (PlasticsEurope, 2019) but is estimated to double within 20 years (Lebreton & Andrady, 2019). Due to low degradation of plastics coupled with unsustainable production, use and disposal, plastic pollution has become a severe transboundary threat to marine ecosystems (Prata *et al.*, 2019; Silva *et al.*, 2021).

As most consumer plastics are designed for single use with finite utility, it has resulted in increasing quantities of plastic waste (Schnurr *et al.*, 2018). Global plastic waste generation has been further exacerbated during the COVID-19 pandemic which has resulted in huge increases in the use and disposal of single-use plastics and personal protective equipment (PPE) (Prata *et al.*, 2020; Silva *et al.*, 2020, 2021; Kitz *et al.*, 2022; Molloy *et al.*, 2022). Prata *et al.* (2020) estimated that mismanagement of PPE during the COVID-19 pandemic resulted in the monthly use of 129 billion face masks and 65 billion gloves globally. Even before the COVID-19 pandemic, Borrelle *et al.* (2020) estimated that 19–23 Mt of plastic waste generated globally in 2016 entered aquatic and marine ecosystems. Furthermore, plastic waste is predicted to reach up to 53 Mt annually by 2030 (Borrelle *et al.*, 2020). Therefore, widespread plastic mismanagement threatens the ability of the global community to sustainably manage plastic production, use and disposal (Walker, 2021). In response to the plastic pollution problem faced globally, multilevel mitigation strategies are urgently needed, accompanied by real-world solutions and innovative approaches to reduce and transform the global plastics economy (Walker, 2018).

Addressing the multifaceted issue of plastic waste and pollution at the global scale, specifically in our oceans, requires a robust framework encompassing international agreements, national and regional policies, industry action and participation, technologies and innovative processes (BRS Conventions, 2019; Prata *et al.*, 2019; Borrelle *et al.*, 2020; Lau *et al.*, 2020). These elements coupled with strong consumer, government and industry-level support can lead to a large paradigm shift which many stakeholders and multilateral institutions have long been advocating (Xanthos & Walker, 2017; Schnurr *et al.*, 2018; UNEP, 2018; Walker *et al.*, 2021). To achieve a concerted global waste management strategy and a zero plastic waste future in the world's oceans, unprecedented solutions, approaches and mitigation strategies must be widely adopted and implemented immediately (Prata *et al.*, 2019; NIEPS, 2020; Borrelle *et al.*, 2020; Silva *et al.*, 2020). Furthermore, a large systemic change in plastic waste management needs to include cutting-edge technologies and innovations to enable enhanced industry action and efficient policymaking (Adam *et al.*, 2020; Clayton *et al.*, 2020; Schmaltz *et al.*, 2020; Bezerra *et al.*, 2021). By increasing and scaling technology adoption rates across multiple sectors, process changes and efficiencies can be uncovered through waste reduction, waste management and environmental recovery. Furthermore, to avoid a massive build-up of plastic around the natural world, further action is required to accelerate the reduction of consumption and innovation in the plastic value chain (Lau *et al.*, 2020).

Marine plastic pollution has been a growing concern globally for decades (Barnes *et al.*, 2009). Because plastic pollution recognises no international boundaries, it has become a global persistent pollution problem requiring many solutions to reduce sources of plastic pollution (i.e., leakage) at local, national and international levels (Xanthos & Walker, 2017; Walker, 2018; Schnurr *et al.*, 2018; Adam *et al.*, 2020; Clayton *et al.*, 2020; Bezerra *et al.*, 2021). Acknowledging the scale of the problem, there have been many initiatives at different levels to reduce ocean plastic waste, including mechanisms and policies, communication, and plastic collection and treatment, with extensive participation of international organisations, governmental and non-governmental organisations, as well as corporations in the private sector (UNEP, 2018; BRS Conventions, 2019; Prata *et al.*, 2019). Marine plastic pollution is being addressed internationally by the United Nations in the UN Sustainable Development Goals (SDGs) under Goal 14 which is to *Conserve and sustainably use the oceans, seas and marine resources for sustainable development.* Specifically, in target 14.1: *By 2025, prevent and significantly reduce marine pollution of all kinds, in particular from land-based activities, including marine debris and nutrient pollution.* This will be measured by indicator 14.1.1: *Index of coastal eutrophication and floating plastic debris density.* Thus, a reduction in the density of floating plastic debris is understood by the international community as being central to the sustainable use of the oceans and a leading factor in performance metrics. However, an internationally accepted index of floating plastic debris density does not currently exist (Walker, 2021). The United Nations Environment Program (UNEP) is working with the Joint Group of Experts on Scientific Aspects of Marine Environmental Protection (GESAMP) (GESAMP, 2019). GESAMP is playing an important role in working towards coordinated, science-based solutions to reduce marine plastic litter.

Solutions required from all stakeholders to reduce plastic waste are diverse and many cannot be properly addressed in this chapter. However, some current and future solutions, approaches and mitigation strategies to improve the management of plastic waste discussed in this chapter include the following:

(1) Extended producer responsibility (EPR) programs that include exclusive partnerships with cutting-edge packaging suppliers (Diggle & Walker, 2020).

(2) Prevention initiatives to reduce single-use plastics including behavioural economics tactics (Xanthos & Walker, 2017; Schnurr *et al.*, 2018; Adam *et al.*, 2020; Clayton *et al.*, 2020; Bezerra *et al.*, 2021).
(3) Circular economy systems' deployment, corporate social responsibility (CSR) including more robust financial and environmental reporting standards, increased consumer awareness and improved performance measures.
(4) Performance measures for industrial action include plastic and greenhouse gas emissions reduction to eliminate the anticipated unsustainable global plastic waste levels.
(5) While striving for a zero-waste closed-loop system, responses to plastic waste in the world's oceans have been insufficient (Walker & Xanthos, 2018). Solution responses include employment of technologies such as the following:
    a. robotics, sensors, artificial intelligence (AI);
    b. software-as-a-service (SaaS);
    c. geographic information systems (GIS); and
    d. open-source platforms which would serve as an integral part of the global commitment to create shared value amongst citizens, commercial players and government-driven plans.

Accelerating action to address plastic waste management challenges faced is imperative from all angles of our economic and ecological systems. The most effective solutions, given the scale of this wicked problem, lie in cohesive approaches to unify policy, industry action, technology and innovative solutions and each is discussed in turn in the following sections.

2. Policy Solutions

Although there are no international standards for monitoring marine debris to help inform policies to reduce plastic pollution, many countries around the world already have existing national marine plastic databases or are currently developing and implementing them (GESAMP, 2019; Walker, 2021). Understanding the extent and magnitude of the plastic pollution problem, including understanding the types and categories of problem plastics in the environment, helps inform government policies and strategies to reduce sources of plastic pollution (Ambrose *et al.*, 2019). As marine debris is a transboundary issue, solutions require integrated and robust regional and national action plans to address marine debris. Although action plans

Table 1. Policy instruments to minimise single-use plastics (modified from UNEP, 2018).

Policy instruments to reduce single-use plastics	**Strategies to implement policy instruments**	**Examples of policy instruments in action**	**Outcomes of policy instrument solutions**
Voluntary reduction instruments	• Requires agreements between all stakeholders (government, consumers, and industry), but industry usually adopts corporate strategies to achieve results • Voluntary consumer choices develop with increased awareness	• Voluntary agreements between all stakeholders • Agreements between industry and governments (e.g., extended producer responsibility, EPR) • Make reusable alternatives to single-use plastics widely available	• Voluntary actions allow all stakeholders to adopt single-use plastic reduction initiatives (e.g., switching to reusables or sustainable plastic alternatives)
Increase public education and awareness	• Interventions normally required to change towards pro-environmental consumer behaviour	• Prioritise environmental education in schools • Initiate social media education campaigns for the wider public	• Fosters pro-environmental consumer behaviour
Types of policy instruments	• Legislative or regulatory instruments • Market-based or economic instruments • Combination of legislative and • market-based instruments	• Policy instruments vary widely and vary according to jurisdiction or waste management and recycling infrastructure available (e.g., legislative bans, taxes, or levies on single-use plastics, EPR and independent corporate sustainability actions)	• Reduced use and disposal of single-use plastics by consumers and industry • Tax revenues from bans may be incorporated into government sustainability or waste reduction initiatives and education

vary from country to country and by region, many have common objectives and priorities which include (i) policy support and planning; (ii) research, innovation and capacity building; (iii) raising public awareness, education and outreach; and (iv) private sector engagement.

There are numerous policy instruments available for governments at national or sub-national levels to reduce use and waste and improve recycling of single-use plastics which are illustrated in Table 1. Policy instruments include regulatory or legislative bans on certain types of plastics. This has historically been focused on plastic grocery bags (Xanthos & Walker, 2017), but recently, legislative bans have been expanded to include bans on microbeads and other types of single-use plastics (e.g., food packaging and utensils) (Xanthos & Walker, 2017; Schnurr *et al.*, 2018; Adam *et al.*, 2020; Clayton *et al.*, 2020; Bezerra *et al.*, 2021). Other policy instruments to reduce single-use plastics include economic instruments such as taxes or levies (i.e., market-based instruments) (Adam *et al.*, 2020). Both taxes and levies involve applying nominal charges which are designed to change human behaviour and dissuade the use of certain types of plastics (i.e., problem or harmful plastics) found in the environment. Revenues from taxes or levies have been used in other jurisdictions to fund green initiatives, such as education, green bond issuances and fund recycling schemes among others. Another policy instrument for local or federal governments to consider implementing is EPR. EPR strategies leverage corporate resources and implement mandatory requirements to reduce plastic waste generated by consumers. Implementation of EPR strategies allows local jurisdictions to gain greater control over their waste streams and develop risk-sharing and waste management responsibilities between their operations and industrial participants including plastics (Diggle & Walker, 2020).

While policy instruments have had varying degrees of success around the world to reduce sources of marine plastic pollution, top-down approaches may sometimes be difficult to implement (Clayton *et al.*, 2020; Bezerra *et al.*, 2021), can be ill conceived by governments if consumers have not been consulted (Nwafor & Walker, 2020) and may be unpopular with the plastics industry, manufacturers, retailers and consumers lacking environmental awareness (Schnurr *et al.*, 2018). However, bottom-up approaches have been successful at reducing the use of single-use plastics and ultimately sources of plastic pollution. Bottom-up approaches include voluntary reduction strategies by consumers and businesses alike as well as raising public awareness about the environmental threats posed by plastic pollution. Voluntary reduction strategies by individuals such as the use of reusable water bottles, bags and straws have increasingly become popular following popular media campaigns such as the *Blue Planet II* documentary by the BBC and the National Geographic slogan *Planet or Plastic* to raise awareness about the global plastic waste crisis (Schnurr *et al.*, 2018). Industrial voluntary reduction strategies include business recognition programs and certifications for plastic waste reduction, in addition to the gamification and incentivisation for consumer rewards to enhance reuse (Varkey *et al.*, 2021). Public education and raising awareness can take many forms, but successful awareness campaigns include voluntary beach clean ups by citizen scientists. Beach clean ups by citizen scientists can help remove stranded beach debris, although removal rates of plastic pollution pale compared to plastic pollution inputs into the world's oceans. However, one of the most impactful outcomes of empowering citizen scientists is that it helps compile useful plastic debris characterisation data to help inform government policy and also helps trigger behaviour change and pro-environmental actions (Ambrose *et al.*, 2019; Walker *et al.*, 2021).

The list below represents a 10-step roadmap based on the UNEP (2018) report to guide governments at national or sub-national levels to implement effective policy approaches to reduce plastic pollution.

2.1. *Assess baseline conditions*

An assessment of baseline conditions can help to understand which plastic debris items are problem items. In other words, it provides policymakers with evidence of categories of plastics most commonly found in the environment. Assessment of baseline conditions will help determine what are the most problematic plastics that require government action (e.g., plastic bottles or plastic bags) (UNEP, 2018; Ambrose *et al.*, 2019). Baseline conditions will also help determine the extent of the problem; the impacts that mismanaged plastics are having on human health and wildlife, the environment, and the economy; and help identify the sources of plastic pollution (e.g., citizen negligence, poor collection systems and improper disposal sites). Estimating the consumers' willingness to pay for a certain good or service can provide a powerful tool as part of the baseline assessment process to ensure that the instrument chosen will eventually be dissuasive and influence consumer behaviour (Walker *et al.*, 2021). In the case of a levy, for instance, it would be important to set a tax that is high enough to discourage consumers from requesting plastic bags (Xanthos & Walker, 2017; Schnurr *et al.*, 2018).

2.2. *Evaluate appropriateness of possible policy actions*

Based on findings of baseline surveys, it will be important to evaluate what are the most appropriate instruments that will be beneficial in addressing specific issues that have been identified. Instruments such as levies or bans on plastic grocery bags require existence and proper integration of effective legal and fiscal systems (UNEP, 2018).

2.3. *Assess sustainable development impacts of preferred options*

Once appropriateness of possible actions has been assessed, governments may consider a shortlist of suitable instruments. Before selecting the most appropriate option that would address issues and needs identified in the baseline assessment, a key step is to study the sustainable development impacts of the shortlisted choices, taking into consideration all sectors (including food and/or other retail) and all segments of the population. For example, although the environmental benefits of introducing a ban might be highly positive, the social impacts on a large part of the population might be unsustainable, making bans not the most desirable option. For example, in many African countries, effectiveness of outright bans of plastic bags has not been proven due to lack of enforcement and lack of suitable sustainable alternatives to plastic bags (Adam *et al.*, 2020; Nwafor & Walker, 2020; Bezerra *et al.*, 2021).

2.4. *Stakeholder engagement*

Acceptance from the broadest range of stakeholders is of utmost importance and can be ensured through calls for early inputs, policy discussion meetings and wide-reaching awareness campaigns. Special attention should be paid to mapping the main stakeholder groups that will be affected by the new policy and their relative power. Opposition from industry can be expected, especially when using regulatory instruments. Being able to present evidence-based options can help support the policy chosen and ensure successful results. The most common stakeholder groups that might be engaged from the onset include national and local government, national waste management authorities, local waste management authorities, trade and industry associations, single-use plastic producers, retailers, individual citizens and environmental non-governmental organisations (NGOs).

2.5. *Raise awareness*

Resistance is likely to decrease if consumers are aware of the social, environmental and economic impacts of mismanaged single-use plastics (UNEP, 2018). These can be communicated through a variety of methods, ranging from educational programmes, workshops in schools, extensive multimedia awareness-raising campaigns (TV, radio, newspapers and social media), development and distribution of information material and demonstrating and/or distributing alternative options to single-use plastics (reusable bags and reusable bottles). Campaigns should have a clear and simple message, relevant to a wide range of stakeholders. For example, if a levy is to be introduced, it would be important for the public to fully understand the link between the fee that will have to be paid and the environmental benefits that will derive (Clayton *et al.*, 2020). The public should also be made aware of eventual punitive measures included in the law. Consumers should be aware that the easiest way to minimise environmental impact of plastic bags is to reuse them frequently and dispose of them properly at the end-of-life. In the case of bans or levies, awareness-raising activities should be launched before the imposition of the instrument and persist after the introduction, to facilitate acceptance of the policy over the long term (Schnurr *et al.*, 2018).

2.6. *Support uptake of eco-friendly alternatives*

Before banning plastic bags (or any single-use plastic), governments may wish to verify the presence of appropriate sustainable alternatives (Nwafor & Walker, 2020). If affordable and resistant alternatives to plastic bags are lacking, the ban may end up negatively impacting the poorest segments of the population. Eco-friendly and fit-for-purpose alternatives should provide the same or better properties of the items that are to be regulated (UNEP, 2018). When promoting the use of alternatives, governments may wish to consider their environmental and life cycle impacts, which in most cases are determined primarily in the production stage. For example, a common alternative, to replace single-use plastic bags, is paper bags. Although paper bags degrade much quicker in the environment, they require more energy to be produced, are more expensive and once discarded take more space in collection trucks and landfills.

2.7. *Provide incentives to industry*

When wanting to regulate production and consumption of single-use plastics, governments may face resistance from the plastic industry as well as from packaging importers and distributors. To limit resistance and gain as much support as possible, governments may wish to consider providing incentives to industry. It may be beneficial to introduce incentives long before new legislation is put into effect in order to guarantee enough time for plastic manufactures, distributors and retailers to adapt to the new stipulations. Such measures might include provisions to allow time to adapt to the transition (for instance, provide enough time for retailers to deplete existing plastic bags stocks, begin alerting consumers of the upcoming change and purchase new alternatives), and tax rebates and financial incentives to stimulate the production of cost-effective alternatives to plastic bags can also be considered (Clayton *et al.*, 2020; Bezerra *et al.*, 2021).

2.8. *Use revenues to support environmental programs and promote international cooperation and innovation*

When introducing a levy on single-use plastic products, consideration should be given to how revenues from that economic instrument will be used. To maximise public benefits, revenues from levies could be used to support specific environmental projects, boost local recycling industry and create job opportunities in plastic recycling (Xanthos & Walker, 2017). Finance awareness initiatives promote, for instance, waste minimisation. Given that the main objective of a levy is to be dissuasive, it is important that levy revenues are used for activities that are time-bound. For example, if the levy is successful, people will be dissuaded from continuing to use single-use plastic bags, and revenues should be expected to gradually decrease and eventually stop (Schnurr *et al.*, 2018). To ensure a transparent process and maximise public support, it is important to widely communicate the chosen purpose for which the revenues will be utilised (UNEP, 2018).

2.9. *Enforce the policy*

To guarantee good governance, enforcement and monitoring of policies, it is important to clearly distribute and define roles and responsibilities between local, national and sub-national authorities and organisations. For example, in the case of a levy on retailers, it should be made clear to the retailers how and when the levy should be collected or deposited. In the case of a levy on consumers, the public should be made aware of the amount they are expected to pay. When the law includes punitive measures, such as fines or even prison time in extreme cases, prosecution of offenders will help ensure compliance with the policy (UNEP, 2018).

2.10. *Continue to monitor marine debris nationally and adjust policies as necessary*

It is important to monitor progress and effectiveness of introduced policies and adjust accordingly. Progress could be monitored in several ways, including through audits, surveys, impact assessments and focus-group interviews. It would be advisable to review policy instruments on a regular basis (for instance, with a yearly or quarterly frequency for the first three to five years, and afterwards every five years or as deemed necessary). In Ireland, the consumption of plastic bags increased a few years after the levy was introduced as people became used to it. As a result, the levy was revised upward to address this *rebound effect* (Xanthos & Walker, 2017). In the case of total bans, law enforcement and monitoring of compliance are critical to ensure that prohibited items are not illegally produced or imported from neighbouring countries, as seen in parts of Africa where punitive legislation (e.g., outright bans) has been implemented (Adam *et al.*, 2020; Nwafor & Walker, 2020). To gather data on effectiveness, governments may consider including a reporting obligation (providing a standard template to allow for comparability across years) in legislation to estimate reduction in consumption. Once progress and effectiveness have been estimated, these would inform and form part of the new baseline scenario. The steps presented in this road map for policymakers could then be reviewed and revised to ensure that measures in place continue to be effective and appropriate (UNEP, 2018).

3. Industry Action

As it is still not clear which strategies, action plans and mitigation techniques are required to avoid further impacts from the global problem of plastic pollution, industrial action is imperative (Borrelle *et al.*, 2020). Harnessing corporate engagement, redesigning and engineering new products, alternative packaging solutions and a robust reduction framework is critical. Companies also need to take on more responsibilities for the full lifecycle of their plastic products. Furthermore, as industry is aware of the issues they face going forward, a global commitment and alliance such as the *New Plastics Economy Global Commitment* and the *Plastics Pact* with the Ellen MacArthur Foundation and the UN Environment Program (UNEP) is a positive step forward. In this commitment, over 450 signatories promise to increase plastic recycling as part of a broader commitment to the circular economy principles (Häußler *et al.*, 2021; Nature Editorial, 2021). Furthermore, the grouping of signatories include companies representing 20% of all plastic packaging produced globally, in addition to governments, NGOs, universities, industry associations, investors and other organisations (Ellen MacArthur Foundation, 2021). The *Plastics Pact* also harnesses a network of initiatives that bring together all key stakeholders at the national or regional level to implement solutions towards a circular economy for plastic (Ellen MacArthur Foundation, 2021). Examples of government commitments in the *Plastics Pact* include engagements with the *UK Plastics Pact*, *Pacte National sur les emballages plastiques* in France, *Circula El Plàstico* in Chile, *Plastic Pact NL* from the Netherlands and the *Canada Plastics Pact* (Ellen MacArthur Foundation, 2021). In line with business and industry engagement, the *New Plastics Economy Global Commitment* plans to engage circular economy principles defined as the following:

(1) Eliminate unnecessary and problematic packaging through redesign and innovation
(2) Move from single use to reusability
(3) Ensure plastic packaging is designed for reuse, recyclability and composting

(4) Increase the reuse, recyclability and composting of plastic packaging
(5) Increase recycled content in new plastic packaging

As the latest briefing from the global commitment shows an uneven level of progress among participants, companies need to be nudged and pressed harder to act (Ellen MacArthur Foundation, 2020). If major companies are pressed to take responsibility for the entire lifecycle of plastic production and consumption, they are generally less inclined to use materials that are difficult to reuse or recycle (Diggle & Walker, 2020). Product redesign needs to consider the end-of-life factor, targeting plastic resins that can be easily recycled. A proposed global treaty for plastic waste reduction is critical for long-term success alongside the basic principles of recyclable content in packaging, virgin content reduction in new products and investing in alternative solutions alongside policies being implemented.

The production and distribution of sustainable plastic alternatives have become one of the biggest and most urgent questions in industrial activity. As researchers, investors and executives seek to find ways to reduce plastic waste and improve plastic recyclability (Häußler *et al.*, 2021), simple solutions such as switching to alternative packaging and plastic use seem relatively straightforward. However, given the intensity of recycling and recycled plastics being less durable than virgin plastics, for alternatives to be completely adopted, they require to be more sustainable than regular plastics throughout the entire product lifecycle (Diggle & Walker, 2020; Häußler *et al.*, 2021; Nature Editorial, 2021). A reduction in consumption is also a major pain point for the industry to address a future where plastic never becomes waste. Finally, a reduction in consumption is a basic principle of the waste hierarchy and industry participants can adopt this along with the three pillars of plastic management: reduce, reuse and recycle. In addition, industry will also be regulated by EPR and will have to follow EPR regulations in respective jurisdictions. As the coalition towards a zero plastic waste future forms, current signatories of the *New Plastics Economy Global Commitment* represent a broad base of stakeholders which policymakers must continue to engage with. Major stakeholders involved include the following:

(1) Government agencies and departments including municipalities
(2) Major consumer packaged goods companies
(3) Investment banks and financial institutions
(4) Plastic producers and fossil fuel businesses

(5) Pharmaceutical and healthcare businesses
(6) Retailers and wholesalers, including food and other consumer goods companies
(7) Fashion and apparel businesses
(8) NGOs and non-profits
(9) Cosmetics and electronics companies
(10) Other important industrial organisations (manufacturing, agriculture, fishing and construction)

Successful outcomes will create a circular economy model, where a large paradigm shift can be achieved. As a unique multistakeholder, industry-led, cross-value chain collaboration platform which aims to tackle plastic waste and pollution, bringing together businesses, governments, NGOs and other key actors in the plastics value chain is essential to achieve measurable and long-term goals.

4. Technology Solutions

The current state of plastic waste in the world's oceans has guided many to believe the adoption of cutting-edge technologies can enhance the waste management process and inform decision-making on pollution at large. The notion of adoption, testing and enablement of technologies in practical terms is something governments, industry and consumers need to consider as a viable solution to plastic waste management.

While plastics can degrade, the rate at which particles break down is inconsistent and can be affected by multiple factors, such as polymer type, temperature and light exposure (Jambeck *et al.*, 2015). When plastics are leaked into water, the degradation rate is temperature dependent and the lack of sunlight exposure can also slow down decomposition (Jambeck *et al.*, 2015). As a result, the bulk of micro- and macro-sized plastics range from different products with different sources after human consumption. Macroplastics range from anywhere between 25 mm to over 1 m in size, whereas microplastics are widely acknowledged as smaller than 5 mm (UNEP, 2018). The majority of waste found in oceans and shorelines comes from discarded single-use plastics and fishing gear (NIEPS, 2020). With time, plastic pollution will degrade and form particles in fresh and marine waters, impacting ecological processes, organisms and long-term human health (Lebreton *et al.*, 2018). To manage, track, discard and develop a secondary market system for plastic waste in the world's oceans, it is critical that technological solutions be embraced.

Given the increasing amount of plastic pollution in our oceans and the demonstrated risks imposed on human health, biota and natural ecosystems, it is anticipated the frequency of occurrence of physical effects on individual environmental receptors will continue at a rapid pace if pollution trends are not appropriately mitigated (Prata *et al.*, 2019). Through technological change and transformation, significant improvements can be made in ocean health (Schmaltz *et al.*, 2020). Actions across the value chain or that promote technological breakthroughs and new models will most likely result in the systemic changes necessary to achieve a zero plastic waste future.

While many trends and important developments are underway, it is important to note the top-level themes and categories of the following technological categories readily available. As emerging technologies can be used to target hotspots of marine plastic pollution, they can also be leveraged to develop inventories to identify plastic pollution prevention and clean-up efficiencies (Schmaltz *et al.*, 2020). Better knowledge base and data collection, including plastic inventory development, are a combination that is imperative to be coupled with improved human behaviour and robust policymaking. The following includes some emerging solutions that can lead to enhanced prevention and collection of plastic waste:

(1) Robotics, Sensors, and Internet of Things (IoT)
(2) Artificial Intelligence (AI)
(3) Software-as-a-Service (SaaS)
(4) Platform-as-a-Service (PaaS)
(5) Open-source platforms
(6) GIS Technology
(7) Hardware systems, including filters, rakes, skimmers and traps
(8) Boating and large-scale ocean cleanup vessels
(9) Autonomous vehicles and drone technologies

As reported by the UN Secretary-General's Independent Expert Advisory Group on a Data Revolution for Sustainable Development, without high-quality geospatial data, the task of designing, monitoring and evaluating effective policies to achieve the UN SDGs is almost impossible. The implementation of databases and management systems cannot ignore the abilities and effectiveness of new data management technologies, artificial intelligence, cloud computing, quantum computing, cloud storage and algorithmic forecasts, coupled with accurate geospatial data. In particular, for ocean plastic waste, these new data management technologies together with increased volume of accessible geospatial data are making it possible to manage, share, process and analyse large volumes of data in near real time as well democratising access to the data itself. Furthermore, the development of mobile applications, across numerous mobile phone operating systems, including both basic software platforms (feature phones) and advanced software platforms (smartphones) is critical for more accurate management, sharing, processing and accessing large volumes of data in a decentralised manner. The use of mobile phones, and specifically smartphones, has increased significantly in specific countries where its correlation to high levels of plastic waste is positive (Lyons *et al.*, 2020). The use of image recognition and the application of artificial intelligence tools are also tremendous value-additions for better plastic waste management.

Use of artificial intelligence-based solutions provides fast, scalable and potentially cost-effective automatic methods for identifying and evaluating ocean plastic waste. As more remotely sensed observational data are available from autonomous underwater vehicles (AUV) and remote platform aerial systems (RPAS) in forms of videos and imagery, machine learning provides a solid, sustainable and trustworthy alternative to the standard visual-census approach, from both a time and classification perspective. While various governments, industry, innovators, scientists, civil society members and advocacy groups are working to move towards a more technologically driven and circular economy for plastic waste, a number of significant challenges stand in the way:

(1) The cannibalisation effect of primary and secondary markets for plastics
(2) Weakness in end markets for recycled plastics
(3) Recycling and collection rates below needed levels
(4) Insufficient options for recovery of wasted plastic
(5) Significant costs of plastic pollution borne by municipalities and individuals

Although technologies exist to address the problem, the infrastructures, policies, business processes and financing needs are not met to enable rapid technology deployment. A substantial shift of investment away from the production of new plastic to the development of reuse and refill systems and sustainable substitute materials is needed, in addition to expanded recycling facilities, collection infrastructure and new delivery models. By supporting and stimulating investments in new technologies around the world through incentives and regulatory easing, innovators and researchers can focus on value-addition and fast-tracking commercialisation. Furthermore, by unlocking barriers to growth and regulatory pressures, the focus on development, testing and scale-up of technologies can prevent plastic waste and pollution, increase the flow of alternatives and new forms of plastics, find new technologies for value recovery and secondary markets and other innovative business practices to improve the management of plastics throughout the value chain. From preventative solutions to collection technologies, both mobile and infrastructure-related, the global scale needed can be achieved with the right policies to drive technological progress.

The technologies available cannot tackle the plastic pollution problem alone (Schmaltz *et al*., 2020). The readily available solutions in the marketplace allow for a scan of opportunities and enable policymakers and industry leaders to reflect on the challenges at scale. In addition, the plastic pollution challenge forces the innovation agenda to focus on lifecycle stages, microplastics' identification, technology implementation costs and the deployment locations for specific technologies (Borrelle *et al*., 2020; Lau *et al*., 2020). The solutions for plastic pollution face hurdles and highlight the need for collaborative efforts across multiple scales of implementation and within the countries with the highest rates of plastic mismanagement and plastic leakage to combat the major battle faced in managing and mitigating plastic pollution (Schmaltz *et al*., 2020). Consistent reoccurring needs for technologies developed and solutions readily available can include the following:

(1) Localised solutions for global pollution for clean-up goals in plastic waste hotspots
(2) Lifecycle stages of collected plastics and focusing on leakage prevention
(3) Microplastic prevention technologies
(4) Capital requirements for successful implementation and support
(5) Location and mapping efficiencies
(6) A multifaceted global unified platform for data sharing and decision-making

5. Innovation Solutions

The global discourse on combatting plastic pollution emphasises the development of a circular economy to reduce the production of plastic that may reach the natural environment (Nordic Council of Ministers, 2020). In ocean plastic waste management, a plan advancing the circular economy is needed, which would reduce the use of virgin materials and integrate the formal and informal sectors in the plastic waste paradigm. Stressing further is the importance of more sustainable management innovation of plastics throughout their lifecycle to increase sustainable consumption and production patterns, not only limited to the circular economy principles and other sustainable economic models (Nordic Council of Ministers, 2020). In addition, plastic waste management in our oceans should consider innovative approaches to resource efficiency, embedding the fundamental three "Rs" (reduce, reuse and recycle) in materials management (NIEPS, 2020).

As plastic has become ubiquitous on store shelves and in homes, the convenience factor comes with a significant price: plastic waste enters the ocean at a rate of 23 Mt per year (Borelle *et al.*, 2020). Additionally, over 400 Mt of plastics are produced every year and the present growth rate presents a forecast of the doubling of plastic production by 2040 (Geyer *et al.*, 2017). A comprehensive and long-term governance strategy must address the prevention of plastic waste in addition to enhancing sustainable designs and production of primary materials. Specific levers, such as a global plastic protocol, secondary markets' development, technological innovations and enabling policies, can help capture the material value of up to US$80–120 billion in new plastics economies, as per the World Economic Forum, Ellen MacArthur Foundation, & McKinsey & Company (2016). Geyer *et al.* (2017) identified that 99% of plastics are derived from fossil-based feedstocks, including oil, natural gas and coal (2017). A need for alternative production processes, input and material sources and mass distribution is essential to meet ambitious targets and curb plastic pollution in our oceans. As innovative solutions are necessary to enable the transition to a new plastic economy, the following themes are critical to advance its standardisation:

(1) Circular economy transition
(2) Shift from petroleum to bio-based products
(3) Secondary markets' development

(4) Biodegradable product innovation
(5) Carbon capture and incineration
(6) Recovery and recycling infrastructure development and investments
(7) Value re-creation and reverse logistics

The value chain of plastics indicates the improvement potential for circular material flows between upstream, midstream and downstream producers and consumers (Figure 1). This figure embodies the potential for a circular economy transition towards a secondary and tertiary market, including reduced leakage during plastic production.

A coherent global strategy to solve the urgent need for plastic waste management remains elusive. Different agencies, groups, institutions and experts have voiced different responses ranging from plastic elimination to turning it into fuels, and from developing biodegradable substitutes to recycling plastic back to secondary markets (PEW & SYSTEMIQ, 2020). Understanding the effectiveness of different solutions and new innovative models and their associated economic, social and environmental

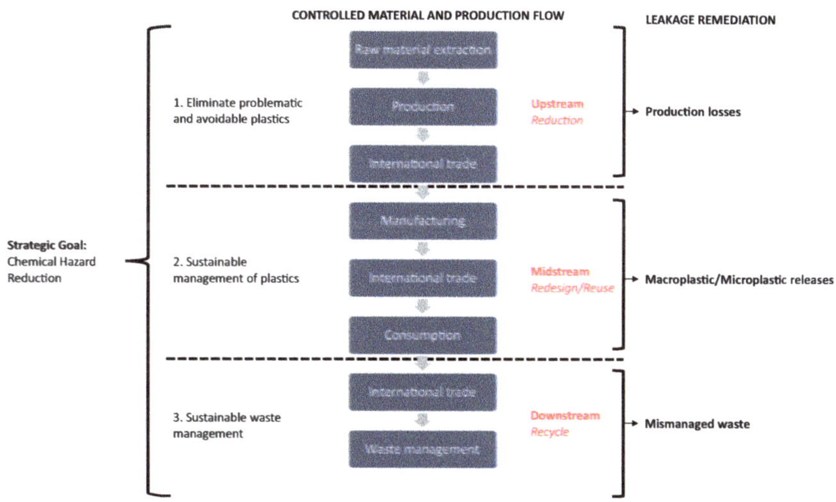

Figure 1. Value chain of plastics indicating circular materials' flows.

Note: This figure is an adaptation of an original work by the Nordic Council of Ministers. Responsibility for the views and opinions expressed in the adaptation rests solely with the authors with its author(s). The views and opinions in this adaptation have not been approved by the Nordic Council of Ministers (2020).

implications is crucial to making progress towards reducing and preventing ocean plastic pollution. Therefore, a concurrent, ambitious and global implementation of multiple complementary system interventions to enhance a system change in ocean plastic waste is critical (PEW & SYSTEMIQ, 2020). This can be achieved through the following steps:

(1) Reduce growth in plastic production and consumption through elimination, reuse and new delivery models
(2) Implement and tailor (based on jurisdiction) policy instruments such as taxes/levies, regulatory or legislative bans on problem or difficult to recycle single-use plastics
(3) Implement harmonised EPR programs to improve recycling rates and to reduce disposal of single-use plastics
(4) Substitute plastic with sustainable paper and compostable alternative materials
(5) Design products and packaging for improved end-of-life and recycling options
(6) Improve waste collection and recycling rates in middle- and low-income countries
(7) Increase mechanical recycling infrastructure
(8) Develop plastic-to-plastic conversion
(9) Invest and build better disposal facilities as a transitional measure
(10) Reduce plastic waste exports and enforce country-specific transparency and responsibility

6. Conclusions

This chapter provides a brief overview of some current and future solutions, approaches and mitigation strategies for the management of plastic waste, particularly leakage into the marine environment. The current state of plastic waste mismanagement also signals to society, policymakers and industry leaders that there is no silver bullet to reduce plastic pollution. Findings presented in the chapter indicate that there is an urgent and coordinated need for action combining both upstream and downstream solutions to curb the global plastic pollution problem. Through the use of appropriate and region-specific policies, technologies, industry action and innovation, plastic production and consumption can be addressed to decelerate and eliminate plastic accumulation in natural environments, including the world's oceans. Future innovations and resource-efficient business models, including those with low-carbon intensities, can be met with the matching of appropriate policies and industry actions to ensure successful outcomes. Without coordinated international, national, regional and local community action, plastic pollution into the world's oceans will continue.

References

Adam, I., Walker, T. R., Bezerra, J. C., & Clayton, A. (2020). Policies to reduce single-use plastic marine pollution in West Africa. *Marine Policy*, *116*, 103928.

Ambrose, K. K., Box, C., Boxall, J., Brooks, A., Eriksen, M., Fabres, J., Fylakis, G., & Walker, T. R. (2019). Spatial trends and drivers of marine debris accumulation on shorelines in South Eleuthera, The Bahamas using citizen science. *Marine Pollution Bulletin*, *142*, 145–154.

Barnes, D. K., Galgani, F., Thompson, R. C., & Barlaz, M. (2009). Accumulation and fragmentation of plastic debris in global environments. *Philosophical Transactions of the Royal Society B: Biological Sciences*, *364*(1526), 1985–1998.

Bezerra, J. C., Walker, T. R., Clayton, C. A., & Adam, I. (2021). Single-use plastic bag policies in the Southern African development community. *Environmental Challenges*, *3*, 100029

Borrelle, S. B., Ringma, J., Law, K. L., Monnahan, C. C., & Lebreton, L. (2020). Predicted growth in plastic waste exceeds efforts to mitigate plastic pollution. *Science*, *369*(6510), 1515–1518. https://doi.org/10.1126/science.aba3656.

Brooks, A. L., Wang, S., & Jambeck, J. R. (2018). The Chinese import ban and its impact on global plastic waste trade. *Science Advances*, *4*(6), eaat0131.

BRS Conventions. (2019). COPs 2019 press release: Governments agree landmark decisions to protect people and planet from hazardous chemicals and waste, including plastic waste, synergies among the basel, rotterdam and stockholm conventions. Retrieved from http://www.brsmeas.org/?tabid=8005. Geneva. Accessed 12/05/2022.

Clayton, C. A., Walker, T. R., Bezerra, J. C., & Adam, I. (2021). Policy responses to reduce single-use plastic marine pollution in the Caribbean. *Marine Pollution Bulletin*, *162*, 111833.

Diggle, A. & Walker, T. R. (2020). Implementation of harmonized extended producer responsibility strategies to incentivize recovery of single-use plastic packaging waste in Canada. *Waste Management*, *110*, 20–23.

Ellen MacArthur Foundation. (2021). Global commitment. Retrieved from https://ellenmacarthurfoundation.org/global-commitment/. Accessed 12/05/2022.

GESAMP. (2019). Guidelines or the monitoring and assessment of plastic litter and microplastics in the ocean (Kershaw P. J., Turra, A., & Galgani, F. (Eds.)), (IMO/FAO/UNESCO-IOC/UNIDO/WMO/IAEA/UN/UNEP/UNDP/ISA Joint Group of Experts on the Scientific Aspects of Marine Environmental Protection). Rep. Stud. GESAMP No. 99, p. 130.

Geyer, R., Jambeck, J. R., & Law, K. L. (2017). Production, use, and fate of all plastics ever made. *Science Advances*, *3*(7), e1700782.

Häußler, M., Eck, M., Rothauer, D. & Mecking, S. (2021). Closed-loop recycling of polyethylene-like materials. *Nature*, 590, 423–427.

Jambeck, J. R., Geyer, R., Wilcox, C., Siegler, T. R., Perryman, M., Andrady, A., Narayan, R., & Law, K. L. (2015). Plastic waste inputs from land into the ocean. *Science*, *347*(6223), 768–771.

Kitz, R., Walker, T. R., Charlebois, S., & Music, J. (2022). Food packaging during the COVID-19 pandemic: Consumer perceptions. *International Journal of Consumer Studies*, *46*(2), 434–448.

Lau, W. W., Shiran, Y., Bailey, R. M., Cook, E., Stuchtey, M. R., Koskella, J., Velis, C. A., Godfrey, L., Boucher, J., Murphy, M. B., Thompson, R. C., Jankowska, E., Castillo Castillo, A., Pilditch, T. D., Dixon, B., Koerselman, L., Kosior, E., Favoino, E., Gutberlet, J., Baulch, S., Atreya, M. E., Fischer, D., He, K. K., Petit, M. M., Sumaila, U. R., Neil, E., Bernhofen, M. V., Lawrence, K., & Palardy, J. E. (2020). Evaluating scenarios toward zero plastic pollution. *Science*, *369*(6510), 1455–1461.

Lebreton. L. & Andrady, A. (2019). Future scenarios of global plastic waste generation and disposal. *Palgrave Communications*, *5*(1), 1–11.

Lebreton, L., Slat, B., Ferrari, F., Sainte-Rose, B., Aitken, J., Marthouse, R., Hajbane, S., Cunsolo, S., Schwarz, A., Levivier, A., Noble, K., Debeljak, P., Maral, H., Schoeneich-Argent, R., Brambini, R., & Reisser, J. (2018). Evidence that the Great Pacific Garbage Patch is rapidly accumulating plastic. *Scientific Reports*, *8*(1), 4666.

Lyons, Y., Neo, M. L., Lim, A., & Tay, Y. L. (2020). Status of research, legal and policy efforts on marine plastics in ASEAN+3: A gap analysis at the interface of science, law and policy. Singapore, pp. 1–418. Coordinating Body on the Seas of East Asia (COBSEA) and National University of Singapore (NUS), https://wedocs.unep.org/bitstream/handle/20.500.11822/33383/1/NUS.pdf. Accessed on 21/07/2022.

Molloy, S., Varkey, P., & Walker, T. R. (2022). Opportunities for single-use plastic reduction in the food service sector during COVID-19. *Sustainable Production and Consumption*, *30*, 1082–1094.

Nature Editorial. (2021). Chemistry can help make plastics sustainable — But it isn't the whole solution. *Nature*, *590*, 363–364.

Nicholas Institute for Environmental Policy Solutions — Duke University (NIEPS), Karasik, R., Vegh, T., & Diana, Z. (2020). 20 years of government responses to the global plastic pollution problem (NI R 20-05). Nicholas Institute for Environmental Policy Solutions. Durham, NC, pp 1–311. Retrieved from https://nicholasinstitute.duke.edu/sites/default/files/publications/20-Years-of-Government-Responses-to-the-Global-Plastic-Pollution-Problem-New_1. Accessed on 21/07/2022.

Nordic Council of Ministers. (2020). Possible elements of a new global agreement to prevent plastic pollution. Retrieved from https://pub.norden.org/temanord2020-535/#. Accessed on 21/07/2022.

Nwafor, N. & Walker, T. R. (2020). Plastic bags prohibition bill: A developing story of crass legalism aiming to reduce plastic marine pollution in Nigeria. *Marine Policy*, *120*, 104160.

PEW & SYSTEMIQ. (2020). Breaking the plastic wave. Retrieved from https://www.pewtrusts.org/-/media/assets/2020/07/breakingtheplasticwave_report.pdf. Accessed on 21/07/2022.

PlasticsEurope. (2019). *Plastics — The Facts 2019. An Analysis of European Plastics Production, Demand and Waste Data.* Belgium: PlasticsEurope Brussels.

Prata, J. C., Silva, A. L. P., da Costa, J. P., Mouneyrac, C., Walker, T. R., Duarte, A. C., & Rocha-Santos, T. (2019). Solutions and integrated strategies for the control and mitigation of plastic and microplastic pollution. *International Journal of Environmental Research and Public Health*, *16*(13), 2411.

Prata, J. C., Silva, A. L., Walker, T. R., Duarte, A. C., & Rocha-Santos, T. (2020). COVID-19 pandemic repercussions on the use and management of plastics. *Environmental Science & Technology*, *54*(13), 7760–7765.

Raubenheimer, K., Urho, N. (2020). Possible elements of a new global agreement to prevent plastic pollution. *Nordic Council of Ministers*, Denmark, Copenhagen, pp 1–148.

Schmaltz, E., Melvin, E. C., Diana, Z., Gunady, E. F., Rittschof, D., Somarelli, J. A., Virdin, J., & Dunphy-Daly, M. M. (2020). Plastic pollution solutions: Emerging technologies to prevent and collect marine plastic pollution. *Environment International*, *144*, 106067.

Schnurr, R. E., Alboiu, V., Chaudhary, M., Corbett, R. A., Quanz, M. E., Sankar, K., Srain, H. S., Thavarajah, V., Xanthos, D., & Walker, T. R. (2018). Reducing marine pollution from single-use plastics (SUPs): A review. *Marine Pollution Bulletin*, *137*, 157–171.

Silva, A. L. P., Prata, J. C., Walker, T. R., Campos, D., Duarte, A. C., Soares, A. M., Barcelò, D., & Rocha-Santos, T. (2020). Rethinking and optimising plastic waste management under COVID-19 pandemic: Policy solutions based on redesign and reduction of single-use plastics and personal protective equipment. *Science of the Total Environment*, *742*, 140565.

Silva, A. L., Prata, J. C., Walker, T. R., Duarte, A. C., Ouyang, W., Barcelò, D., & Rocha-Santos, T. (2021). Increased plastic pollution due to COVID-19 pandemic: Challenges and recommendations. *Chemical Engineering Journal*, *405*, 126683.

UNEP. (2018). Single-use plastics: A roadmap for sustainability. United Nations Environment Programme, pp. 1–92. Retrieved from https://www.unenvironment.org/resources/report/single-use-plastics-roadmap-sustainability.

Varkey, P. S., Walker, T. R., & Saunders, S. J. (2021). Identifying barriers to reducing single-use plastic use in a coastal metropolitan city in Canada. *Ocean & Coastal Management*, *210*, 105663.

Walker, T. R. (2018). Drowning in debris: Solutions for a global pervasive marine pollution problem. *Marine Pollution Bulletin*, *126*, 338.

Walker, T. R. (2021). (Micro) plastics and the UN sustainable development goals. *Current Opinion in Green and Sustainable Chemistry*, *30*, 100497.

Walker, T. R. & Xanthos, D. (2018). A call for Canada to move toward zero plastic waste by reducing and recycling single-use plastics. *Resources, Conservation & Recycling*, *133*, 99–100.

Walker, T. R., McGuinty, E., Charlebois, S., & Music, J. (2021). Single-use plastic packaging in the Canadian food industry: Consumer behavior and perceptions. *Humanities and Social Sciences Communications*, 8(1), 1–11.

World Economic Forum, Ellen MacArthur Foundation, & McKinsey & Company. (2016). The new plastics economy — Rethinking the future of plastics. Ellen MacArthur Foundation, pp. 1–120. Retrieved from http://www.ellenmacarthurfoundation.org/publications.

Xanthos, D. & Walker, T. R. (2017). International policies to reduce plastic marine pollution from single-use plastics (plastic bags and microbeads): A review. *Marine Pollution Bulletin*, *118*(1–2), 17–26.

https://doi.org/10.1142/9789811259111_bmatter

Index

www.ingramcontent.com/pod-product-compliance
Ingram Content Group UK Ltd.
Pitfield, Milton Keynes, MK11 3LW, UK
UKHW022105170926
8624IPUK00024B/119

9 789811 259104